Managing Lakes and Reservoirs

Prepared by the

NORTH AMERICAN LAKE MANAGEMENT SOCIETY
Madison, Wisconsin

and

TERRENE INSTITUTE
Alexandria, Virginia

in cooperation with

U.S. ENVIRONMENTAL PROTECTION AGENCY
Office of Water, Assessment and Watershed Protection Division
Washington, D.C.

Third Edition
2001

Citation: Holdren, C., W. Jones, and J. Taggart. 2001. Managing Lakes and Reservoirs. N. Am. Lake Manage. Soc. and Terrene Inst., in coop. with Off. Water Assess. Watershed Prot. Div. U.S. Environ. Prot. Agency, Madison, WI.

ISBN 1-880686-15-5
EPA 841-B-01-006

Copies are available from

North American Lake Management Society
P.O. Box 5443
Madison, WI 53705
phone: (608) 233-2836
fax: (608) 233-3186
nalms@nalms.org
www.nalms.org

and the

Terrene Institute
4 Herbert Street
Alexandria, VA 22305
phone: (800) 726-4853
fax: (703) 548-6299
info@terrene.org
www.terrene.org

Preface

Barley straw was cow food when the first edition of this manual was published. And focus groups discussed camera lens, not how to control jet skiers. But that was 13 years ago — and *The Lake and Reservoir Restoration Guidance Manual* dealt primarily with restoration of our nation's lakes.

This third edition, *Managing Lakes and Reservoirs,* testifies to the success and the leadership of EPA's Clean Lakes Program. Since the publication of the first manual in 1988, hundreds of lakes have been restored, many of them as Clean Lakes projects and most of them with the assistance of the people who wrote this book.

Their expertise — based both on academic credentials and years of experience — has brought them to the consensus that management of lakes and their watersheds is ongoing — indeed, a forever process to achieve what people who use the lake want it to be. And those people, whether they live on the lake, drop their boats or lines in on weekends, occasionally pitch tents around it, or sell bait and soft drinks to those who do, must drive the management process.

So this manual is written for you, the lake user, by limnologists (look that up in the Glossary!), engineers, and sociologists who would empower you to care for your own lake.

Here is the work of nine professionals, reviewed by dozens more — collectively, that's about 500 years of experience. Their expertise ranges from working with homeowners to modeling lake problems, necessarily dictating chapters that differ in length and writing style. As editors, we've tried to make them both understandable and useful. Enjoy — and consider feeding the barley straw to your lake!

The Editors

Contents

CHAPTER 3
Planning: From Goals to Evaluation and Around Again 49

CHAPTER 4
Problem Identification 101

Acknowledgments

Managing Lakes and Reservoirs is the third edition of a manual originally titled *The Lake and Reservoir Restoration Guidance Manual* and produced to provide guidance to citizens on lake and reservoir management. The first two editions were published in 1988 and 1990.

Managing Lakes and Reservoirs was prepared under the guidance of Chris Holdren, Ph.D., project manager for the North American Lake Management Society (NALMS), with the assistance of Bill Jones, also with NALMS, and Judy Taggart of the Terrene Institute. Anne Weinberg has served as project manager for the U.S. Environmental Protection Agency. The authors by chapter are:

CHAPTER 1: Overview of Manual
G. Chris Holdren, Ph.D., CLM, *Bureau of Reclamation, Denver, Colorado*
Judith F. Taggart, *Terrene Institute, Alexandria, Virginia*

CHAPTER 2: Ecological Concepts
William W. Jones, *Indiana University, Bloomington*

CHAPTER 3: Planning: From Goals to Evaluation and Around Again
Lowell Klessig, Ph.D., *University of Wisconsin–Stevens Point*

CHAPTER 4: Problem Identification
Paul Garrison, *Wisconsin Department of Natural Resources, Madison*

CHAPTER 5: Predicting Lake Water Quality
Gertrud Nürnberg, Ph.D., and Bruce D. LaZerte, Ph.D.
Freshwater Research, Baysville, Ontario, Canada

CHAPTER 6: Watershed Management
Kent W. Thornton, Ph.D., *FTN Associates, Inc., Little Rock, Arkansas*
Clayton Creager, *Tetra Tech, Inc., Calistoga, California*

CHAPTER 7: Management Techniques Within the Lake or Reservoir
Kenneth Wagner, Ph.D., CLM, *ENSR, Inc., Wilbraham, Massachusetts*

CHAPTER 8: Developing and Implementing a Management Plan
Dick Osgood, *Ecosystem Strategies, Shorewood, Minnesota*

CHAPTER 9: Lake Protection and Maintenance
Ann Baughman, *Tip of the Mitt Watershed Council, Conway, Michigan*

In addition to the authors, several other individuals made valuable contributions to this document. Extensive comments on the outline for this revision were received from:

Frank X. Browne, Ph.D., P.E., F.X. Browne, Inc., Lansdale, Pennsylvania

Gertrud Nürnberg, Ph.D., Freshwater Research, Baysville, Ontario, Canada

Kent W. Thornton, Ph.D., FTN Associates, Inc., Little Rock, Arkansas

Anne Weinberg, U.S. Environmental Protection Agency, Washington, D.C.

Once the initial outline was developed, the following individuals provided written suggestions for improving the manual that were forwarded to chapter authors for consideration:

Chauncey Anderson, USGS-WRD, Portland, Oregon

Terry Anderson, Kentucky Division of Water, Frankfort

Roger Bachman, Ph.D., University of Florida, Gainesville

Bud Cann, New Jersey Department of Environmental Protection, Trenton

Timothy D. Feather, Ph.D., Planning and Management Consultants, Ltd., Carbondale, Illinois

Eileen Jokinen, University of Connecticut, Storrs

Shannon Lotthammer, Minnesota Pollution Control Agency, St. Paul

Gerd Marmulla, FAO/FIRI, Rome, Italy

Jenifer Parsons, Washington State Department of Ecology, Olympia

The original Chapter 2, written by Bruce Kimmel in *Lake and Reservoir Restoration Manual* (1988), served as the foundation for Chapter 2 in this manual. Other chapters also built on the work of the two previous editions, and acknowledge the contributions of those authors: Frank X. Browne, G. Dennis Cooke, William H. Funk, Douglas Knauer, Harvey Olem, Forrest E. Payne, William W. Walker, and Richard Wedepohl.

The following limnologists and lake management professionals reviewed Chapter 7 (In-lake Management Techniques section):

G.Dennis Cooke, Ph.D., Kent State University, Kent, Ohio

Tom Eberhardt, Sweetwater Technology, Brainerd, Minnesota

Richard Geney, General Environmental Systems, Summerfield, North Carolina

Robert Kortmann, Ph.D., Ecosystem Consulting Services, Coventry, Connecticut

Gerald Smith, Aquatic Control Technology, Sutton, Massachusetts

Eugene B. Welch, Ph.D., University of Washington, Seattle

Once a draft of the document was prepared, the entire document was evaluated through a formal peer review process to ensure that it would be acceptable to the target audience and that the information in the document was accurate. The document was reviewed by the following individuals, as well as several reviewers from U.S. EPA:

John Hains, Ph.D., U.S. Army Corps of Engineers Waterways Experiment Station, Calhoun Falls, South Carolina

Jeff Schloss, Cooperative Extension, University of New Hampshire, Durham

Al Sosiak, Alberta Environment, Calgary, Canada

Barbara Speziale, Ph.D., Clemson University, South Carolina

Steven Weber, Lake Las Vegas Resort, Henderson, Nevada

Donald D. Moores, Clearwater, Florida, and Lura Svestka, Terrene Institute, Alexandria, Virginia, developed the glossary for the document.

Once all suggested changes had been incorporated, the document went through a final review by the editors, chapter authors, and U.S. EPA. Nancy Page, Pinellas County Mosquito Control, Clearwater, Florida, also read the document to provide an additional check for clarity and grammatical errors.

Additional assistance throughout the process was provided by Philip Forsberg, North American Lake Management Society, Madison, Wisconsin, who assisted with project management and who helped solicit comments on the outline through the NALMS web site, and by Carlene Bahler, Terrene Institute, Alexandria, Virginia.

Lura Svestka of the Terrene Institute designed and typeset this manual and served as Production Manager for its publication.

All of these individuals are gratefully acknowledged for their efforts and contributions. The authors are particularly commended for their tireless work and their patience with the numerous reviews and revisions that led to the finished document.

CHAPTER 1

Overview

This manual deals with the very broad subject of protecting and managing lakes and reservoirs. It marks a complete revision of the first two editions of the *Lake and Reservoir Restoration Guidance Manual* (1988 and 1990). An enormous amount of information on lake management has been developed in the decade since their publication. So much information, in fact, that the burden on the authors of this third edition was not in finding good material to include, but in deciding where to stop.

Lake: a considerable inland body of standing water.

Reservoir: a lake created by artificially damming a stream or river.

The authors faced two additional substantive challenges:

1. To create a manual that deserves premier shelf space for its reference value, but remains compact enough to be eminently useful to the reader; and
2. To communicate that the term "management" is now more accurate than "restoration," largely because we don't always know the original conditions of natural lakes, and some reservoirs experience water quality problems from the moment they are impounded.

The North American Lake Management Society (NALMS) joined with the Terrene Institute and the U.S. Environmental Protection Agency to respond to these challenges and produce *Managing Lakes and Reservoirs.*

Who Should Use This Manual?

Although scientists and engineers will probably find this manual useful, it is written for informed citizens who are interested in protecting and managing lakes. Consequently:

- Technical jargon is kept to a minimum to help you grasp important points without stumbling over the words.
- Terms that are essential to lake management are defined in a sidebar the first time they appear, clearly explained in the text, and included in the glossary (Appendix A).
- The relatively simple word, lake, is used generically throughout this manual to include both natural and constructed lakes, which are called reservoirs. Distinctions between the two types of systems are discussed when they have important management implications.

- English units of measure are used, except for a few terms that are almost always reported in metric units (see Appendix B for information on the metric system).
- References to scientific literature and other sources of information are listed to provide more details.

The material presented in this manual fulfils a fourfold purpose:

1. To help you identify, describe, and define your lake problems;
2. To help you evaluate available lake and watershed management practices for addressing specific problems or for protecting water quality;
3. To describe the process of developing a site-specific lake or reservoir management plan; and
4. To illustrate how to put a lake management plan into practice and evaluate its effectiveness.

How the Manual is Organized

Each chapter is self-contained. You should be able to go directly to a chapter you're particularly interested in and find both adequate background information to understand the material discussed and guidance on how to apply the techniques described. While this does result in some repetition, it also provides you with more than one approach for managing your lake.

Information in this manual is arranged in much the same order you would follow in undertaking a lake management project. Topics fall into three general parts:

- Part 1. Understanding and Defining the Problem;
- Part 2. Lake and Watershed Management Techniques; and
- Part 3. Developing, Implementing, and Evaluating a Management Plan.

Part 1 — Understanding and Defining the Problem

▾ ***Chapter 2, Ecological Concepts,*** describes how inseparably lakes and watersheds are coupled and how lakes function as ecosystems. You must have some understanding of how the various components of a lake and watershed work and fit together before you can manage your lake. In other words, you don't have to be a mechanic to drive a car, but you do need to understand what makes the car go and what makes it stop. The eutrophication process, for example, can be accelerated or slowed down by various management techniques.

Chapter 2 describes key lake and watershed processes, eutrophication and other ecological concepts, and explains the importance of understanding ecological processes before undertaking a lake restoration project.

Eutrophication: the process of physical, chemical, and biological changes in a lake or reservoir associated with nutrient, organic matter, and silt enrichment. If the process is accelerated by human influence, it is called cultural eutrophication.

▾ ***Chapter 3, Planning: From Goals to Evaluation and Around Again,*** describes the importance of planning in the management process. The chapter deals with such key items as clarifying goals, developing alternatives, defining measurable objectives, and implementing and evaluating the management plan.

The planning process is a critical part of lake management that is often overlooked.

▼ ***Chapter 4, Problem Identification,*** discusses common lake problems, data collection and analysis techniques, typical water quality variables and monitoring techniques for assessing them, and how to use monitoring data to evaluate the effectiveness of lake and watershed management techniques.

▼ ***Chapter 5, Predicting Lake Water Quality,*** discusses modeling techniques for evaluating lake water quality and the effects of management techniques. While many users believe modeling takes place only after all data have been collected and a management plan implemented, this should not be the case. If modeling is to be used effectively, the data required by the model to evaluate the success of a management plan must be collected during the monitoring phase of the project. Chapter 5 focuses on modeling phosphorus concentrations in lakes, primarily because phosphorus is most often the nutrient involved in the eutrophication process, but also because phosphorus models are more highly developed than models for other water quality variables.

Part 2 — Management Techniques

▼ ***Chapter 6, Watershed Management,*** completely revised for this edition of the manual, recognizes that watersheds are critical in determining lake water quality. Point and nonpoint sources of pollutants are identified and discussed. In addition, best management practices for watershed management are described. Complete descriptions of various watershed management practices appear in Appendix 6-A.

Chapter 6 also discusses the Clean Water Act's Total Maximum Daily Load (TMDL) process that deals with pollutant loads to water bodies. Many states are using information developed for lake diagnostic studies to develop TMDLs.

▼ ***Chapter 7, Management Techniques Within the Lake or Reservoir,*** discusses in-lake management techniques for achieving a desired lake use. Information on advantages, disadvantages, mode of action, and approximate costs are given for the techniques presented. Tables help you find techniques particularly relevant to your problems, and numerous references guide you to additional sources of information.

Part 3 — Developing, Implementing, and Evaluating a Lake Management Plan

▼ ***Chapter 8, Developing and Implementing a Management Plan,*** describes how to develop a specific lake management plan and put it into practice. Chapter 8 differs from Chapter 3 in that Chapter 3 focuses on the steps necessary to develop consensus and set management goals, while Chapter 8 goes into more detail on the numerous practical details that are an important part of integrating lake and watershed information into an effective lake management plan.

▾ ***Chapter 9, Lake Protection and Maintenance,*** discusses how to protect lake quality, either as it exists or after the management plan is in place. You will learn how to form a lake association, conduct lake protection and stewardship activities, and develop a lake monitoring program. Chapter 9 emphasizes the fact that lake management is ongoing, not a short-term endeavor.

Appendices

Appendices supplement the material covered in Chapters 1 through 9, both at the end of the chapters themselves and at the conclusion of the manual. The glossary (Appendix A) contains definitions of many common lake management terms. Additional definitions can be found in the glossary maintained at the NALMS web site (www.nalms.org/glossary/glossary.htm). Appendix B shows how to convert English units to metric units, which are more common units of measure in lake management.

Focus of the Manual: Water Quality

This manual focuses on lake water quality. Particular attention is paid to the effects of excessive inputs of silt, nutrients, and organic matter — a process known as **eutrophication**.

The natural geologic process for moderately deep lakes is to gradually fill and become wetlands. The position of a lake along this geologic continuum from deep to shallow influences its natural water quality. Reservoirs are affected by similar problems. And because they're typically constructed along stream channels that naturally erode, reservoirs begin to fill with silt from the moment they are impounded and usually have much shorter life spans than natural lakes of similar size (see Chapter 2).

Many lakes would be eutrophic even without human activity.

So the **natural condition of a lake** — before home construction, before deforestation, before agriculture and other human activities — may not have been nearly as pristine as has been commonly believed. Many lakes would be eutrophic even without human activity.

- In the Southeast, for example, soil fertility, runoff patterns, and geology encourage a somewhat more eutrophic natural condition compared to northern lakes. Northerners expecting to see deep blue waters may be dismayed by the color of healthy southern lakes.

- Even comparing neighboring lakes may be misleading because they may differ in critical ways — depth, water source, erodibility of watershed soils, comparative watershed size, and local land use — whether they're across town from each other or across the state. For example, changes in lake quality from northern to southern Wisconsin or from eastern to western Minnesota reflect regional differences in soils and underlying geology.

Regional differences in climate, rainfall, topography (hills, valleys, plains), soils, geology, and land use all influence lake water quality. These factors have been studied and used to define areas with similar characteristics called ecoregions (Omernik, 1987). Each ecoregion has natural landscape features that can influence lake quality. Thus, to effectively manage your lake, you must identify the uses supported by these natural factors and develop a plan that will either restore the lake to this natural condition or protect its current condition.

Making Your Lake Work for You

This manual concentrates on how to determine what uses a reasonably managed lake can support. It is critical, therefore, to decide the way your lake will be used and to have those goals clearly in mind as you delineate existing lake and watershed problems.

Lake usage is a match between people's desires and the lake's capacity to satisfy them. Lake problems are defined in terms of the limits on desired uses — limitations that can be prevented or corrected with proper management. This is a critical definition for developing lake management programs: A lake problem is a limitation on the desired uses by a particular set of users. Before undertaking a management program, you must clearly define these desired uses, identify the limitations, and understand what causes them.

Lake usage is a match between people's desires and the lake's capacity to satisfy them.

A lake cannot be all things to all people. Desirable uses, even obtainable ones, can conflict. Lake organizations invariably would like to see their lake do everything. They want aesthetic pleasure, great fishing, clean water, sandy shorelines and bottoms, and a healthy wildlife population — all without pests, insects, or weeds. Unfortunately, almost no lake can meet all of these demands.

Individual lakes are suited to particular uses depending on physical characteristics of the lake basin and watershed and the quality of incoming water. Even when a lake can be used in several ways, it still may be necessary to manage for a specific use. Like cattlemen and sheepherders, motorboaters and trout fishermen don't necessarily get along.

Although it might be technically possible to drastically change a lake to meet the needs of a particular use, the cost will be high and the lake will usually require extensive maintenance; thus, the decision is usually unwise. You must understand a lake's capacity and attainable quality when developing a management plan. For example:

- Some lakes will never remain crystal clear, regardless of what you do, or
- If the drainage area is large relative to the lake surface and the soils in the watershed are highly erodible and nutrient-rich, the lake will quickly revert to a turbid state without constant management attention.

Defining — and Attaining — Desired Uses

While users obviously should decide how their lake should be used, they often lack sufficient knowledge to assess whether a use is practical. This manual will help you examine the feasibility of proposed lake management goals. **Chapter 3** focuses on the planning process, including setting management goals and determining their feasibility.

For many problems, you may need to consult an expert. If so, carefully examine the qualifications of groups or individuals being considered to determine who can best address the problem. State and federal agencies or universities may be able to furnish general lists of experts dealing with lake management issues, but references from groups with similar problems are usually more beneficial. In addition, NALMS maintains a list of Certified Lake Managers on its web site (www.nalms.org/member/nalmsclm.htm). The CLM program identifies individuals recognized by NALMS as having exceptional training and experience in lake management.

Lake and reservoir management is an active process. Informed citizens must become involved if you are to achieve desired and attainable lake uses for your lake. Getting people together and simply finding out what they want may require as much effort as figuring out how to do it. Since a given lake may serve many different groups of users, several methods might be required to involve them all.

Lake and reservoir management is an active process. Informed citizens must become involved if you are to achieve desired and attainable lake uses for your lake.

Lake homeowners and other local users can get involved with lake use decisions by belonging to one of several types of lake organizations. The legal and financial ability of these groups varies considerably from community to community and state to state (see Chapter 9 for additional discussion of legal authority and issues). The annual meeting of the local lake group is an obvious place to discuss and vote on priority uses for the lake. If the lake serves primarily local property owners and residents, such votes are likely to be respected by government agencies. Although large popular lakes are often managed by government agencies rather than by homeowners and lake associations, you can still provide information for managing these lakes by contacting the appropriate agencies.

Reaching a consensus on specific lake uses may be difficult, however, if more than one lake organization exists on the lake, especially if conflicting uses are already well established. Several procedures can be used to reach a consensus on desired lake uses and to identify various lake problems. These approaches, described in more detail in Chapter 3, include surveys, the nominal group process, and the Delphi process.

While consensus-building techniques can be very effective when properly used, most lake managers and informed citizens will need professional assistance. Lake associations typically include people of diverse occupations, however, so a member of the association may have the experience needed to use these methods.

Based on the fact that a lake problem is a limitation on the desired uses by a particular set of users, a definition of desired lake uses and the limitations on these uses represents the cornerstone of any lake management program.

Causes v. Symptoms: A Major Reason for this Manual

Lake users often tend to confuse the symptoms of problems with their causes. Although you will probably need professional help to identify causes of lake problems, you must understand lakes in general. The purpose of this manual is to help you define problems, understand underlying causes, evaluate techniques for addressing problems, develop an effective lake management plan, implement this plan, and evaluate its effectiveness.

In most cases, managing or restoring a lake eventually requires help from a professional lake manager, limnologist, or experienced professional. This manual will help you find and select qualified consultants.

This manual provides general guidance on lake and reservoir restoration and management techniques that have been proven on lakes throughout the United States and Europe. Some techniques might have to be modified for your particular lake in a specific region. This variability brings up a key point in lake management: whatever the starting conditions and the limitations on what can ultimately be achieved, the goal is always the same — managing the lake to minimize lake water quality problems.

Funding for Lake Management Projects

Two approaches can be used for financing lake management projects:

1. Determine the amount of funding available from various sources and then develop the best possible project that can be completed within the specified budget; or
2. Develop the management project and then raise money through any means available to fund it.

Neither approach is best for all circumstances. Which of these two approaches is used depends on numerous factors, including the availability of external funding, the extent of the problems identified, and the abilities and interests of the individuals and organizations involved in the management effort.

Although Chapters 8 and 9 suggest ways to fund lake management projects, this is not a major focus of the manual. There are several reasons for this omission:

- There is no single source of funding that can be used in all cases.
- Different groups often have widely varying ideas of how much funding is needed to implement a management program.
- Funding opportunities often change and any specific information provided here would quickly become out of date.

It is often possible to obtain state or federal funding to implement a lake management project, particularly for lakes with public access. Contact your state environmental agency to find out if any programs exist in your area and, if so, what requirements (local match, public access, state priority lists, etc.) must be met.

Local fundraising efforts can also be very effective. Many lake associations have been able to raise hundreds of thousands of dollars for lake management projects without any state or federal assistance. The amount of money raised through local efforts is limited only by the ingenuity of those involved.

Sources of Additional Information

Previous editions of this manual included extensive lists of federal agencies and state sources of information that could be used for assistance with lake management projects. Such lists quickly become outdated through personnel changes and even changes in the names of the agencies involved. The U.S. Environmental Protection Agency, local offices of the U.S. Department of Agriculture Natural Resources Conservation Service, and state environmental agencies remain good starting points for information. Some good technical (Cooke, et al. 1992) and popular (McComas, 1993; Phillips et al. 2000) books on lake management also are excellent sources. Additional sources of information are described in Chapter 9.

The growth of the internet has replaced many traditional means of finding information. For individuals without personal computers, most local libraries now offer computers and assistance with internet access. Search engines, such as Yahoo or AltaVista, can be used as a starting point for obtaining information on specific topics, while home pages for many organizations involved with lakes offer links to other information sources. Examples include the home page of the EPA Office of Wetlands, Oceans, and Watersheds (www.epa.gov/owow), which has links to many sources of information; the resources section on the NALMS home page (www.nalm.org/resource/resource.htm); and the links page on the Terrene home page (www.terrene.org).

Lakes as Resources

Lakes are important natural resources. As sources of recreation, they support fishing, boating, swimming, and aesthetic enjoyment. Fishing and swimming are among the fastest growing and most popular forms of outdoor recreation in the United States and Canada. Lakes' commercial value in food supply, tourism, and transportation is worth many billions of dollars each year.

Lakes also provide life-sustaining functions such as flood protection, generation of electricity, and sources of drinking water.

And, as places of beauty, they offer solitude and relaxation: for example, over 60 percent of Wisconsin lake property owners asked what they valued in lakes rated aesthetics as especially important.

Yes, you value your lake. We understand that, and offer you this manual as a tool that will empower you to protect and manage your precious resource.

References

Cooke, G.D., E. Welch, S. Peterson, and P. Newroth. 1993. Restoration and Management of Lakes and Reservoirs. 2nd ed. Lewis Publishers, Boca Raton, FL.

McComas, S. 1993. Lake Smarts: The First Lake Maintenance Handbook. Terrene Institute, Alexandria, VA.

Moore, L. and K. Thornton, eds. 1988. Lake and Reservoir Restoration Guidance Manual. EPA 440/5-88-002. Prep. by N. Am. Lake Manage. Soc. for Office of Res. and Devel., Envir. Res. Lab., Corvallis, OR, and Office of Water, Criteria Stand. Div., Nonpoint Sources Branch, U.S. Environ. Prot. Agency, Washington, DC.

Olem, H. and G. Flock, eds. 1990. Lake and Reservoir Restoration Guidance Manual, 2nd ed. EPA 440/4-90-006. Prep. by N. Am. Lake Manage. Soc. for U.S. Environ. Prot. Agency, Washington, DC.

Omernik, J.M. 1987. Aquatic ecoregions of the conterminous United States. Ann. Ass. Am. Geogr.

Phillips, N., M. Kelly, and J. Taggart. 2000. The Lake Pocket Book. Terrene Institute, Alexandria, VA.

CHAPTER 2

Ecological Concepts

Lake and Reservoir Ecosystems

The condition of a lake at any one particular time is the result of a complex interaction of many different physical, chemical, and biological factors. Rainfall cycles, watershed characteristics, lake basin shape and depth, the lake water itself, and bottom sediments all contribute to this condition. These physical and chemical factors, in turn, support a community of biological organisms that is unique to lakes (Fig. 2-1). In fact, the plants and animals, along with the physical and chemical components of their immediate environment, define the lake ecosystem.

We can gain an understanding of aquatic ecosystems through limnology: the study of the physical, chemical, geological, and biological factors that affect aquatic productivity and water quality in freshwater ecosystems — lakes, reservoirs, rivers, and streams.

Understanding these fundamental ecosystem processes is critical if we are to effectively manage a lake. Lakes are highly interactive systems. It is impossible to alter one characteristic — for example, the amount of rooted plants or algae — without affecting some other part of the system, such as fish production.

The components of lake ecosystems are constantly changing. Light increases and decreases, plants grow and die, nutrients are used and released, eggs hatch, insects emerge, the water circulates — in regular or irregular, long or short cycles. Recognizing the natural changes in your lake not only helps you better understand lake ecology but also gives you reasonable expectations for lake management outcomes.

For example, a lake association on an otherwise high-quality lake might overreact to a late summer algal bloom by treating the algae chemically, causing a massive algal die-off that can consume vital oxygen and disrupt food chains. Had they understood how the lake works, they would have known that many lakes experience a late summer increase in algae but the algae will die shortly, as days grow shorter and water temperatures drop below the optima for algae — so the condition will clear up on its own. The chemical treatment in this case was a waste of effort and money, added unnecessary chemicals to the lake, and treated a symptom (algae growth) rather than the cause (nutrients).

This chapter is not intended to be a complete text on limnology. Rather, it will give you the background information to understand lake processes, interpret lake data, and understand the causes of lake degradation — to prepare you to make sound lake management decisions.

Ecosystem: A system of interrelated organisms and their physical-chemical environment. In this manual, the ecosystem is usually defined to include the lake and its watershed.

Biota: All living organisms (e.g., plants, animals, fungi, bacteria) occurring in a specified area.

Ecology: Scientific study of relationships between organisms and their environment. Also defined as the study of the structure and function of nature.

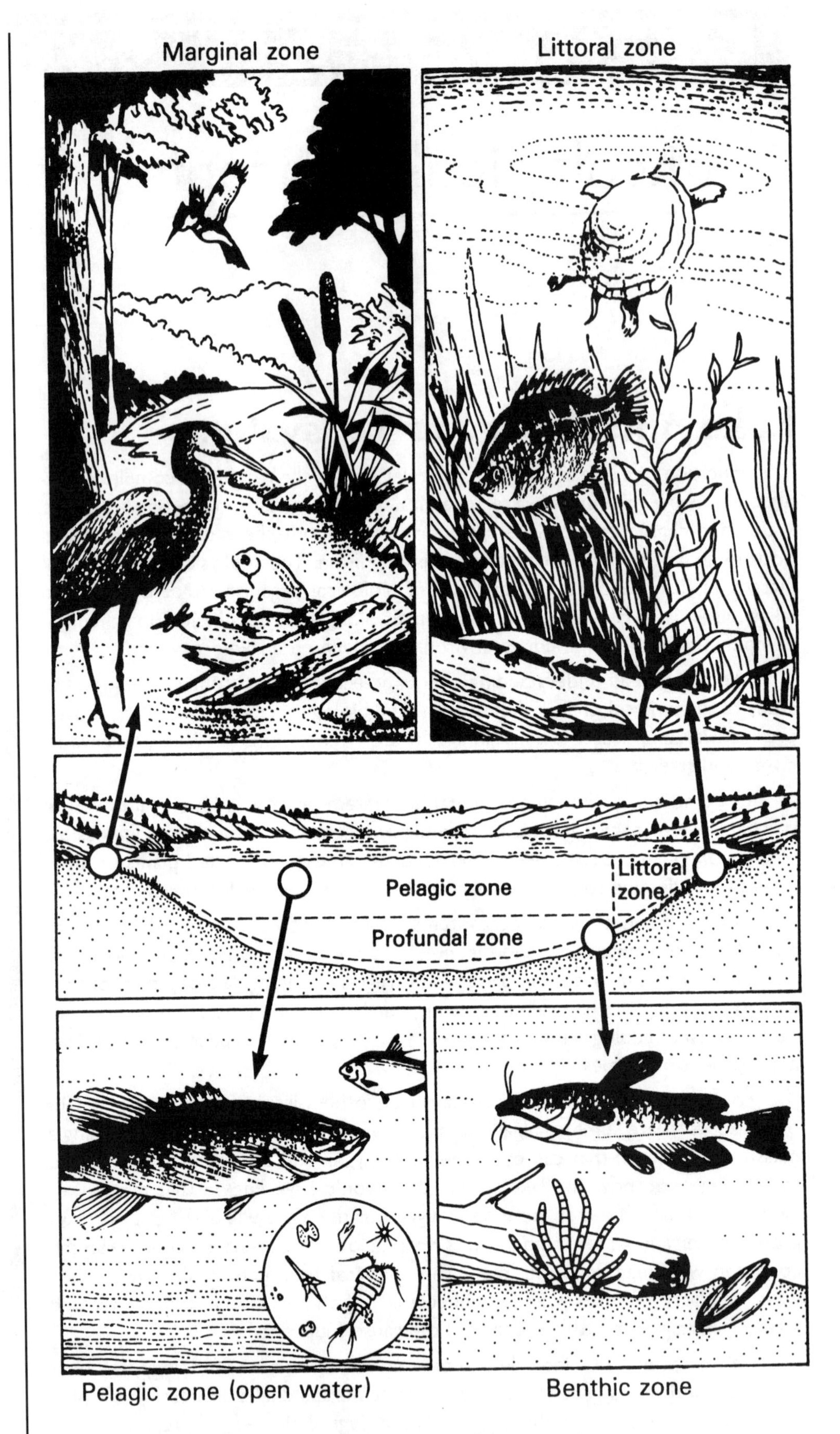

Figure 2-1.—The location and nature of typical lake communities, habitats, and organisms. In addition to the lake's watershed, all of these components are part of the lake ecosystem. Source: Moore and Thornton, 1988.

Lake Formation and Distribution

The physical characteristics of lakes — such as shape, area, depth, and watershed size — are largely established when the lake forms. In his classic work, *A Treatise on Limnology* (1957), G.E. Hutchinson described 76 different ways in which lakes form. These can be condensed into a few basic types of lakes.

Limnology is the scientific study of the physical, chemical, geological, and biological factors that affect aquatic productivity and water quality in freshwater ecosystems—lakes, reservoirs, rivers, and streams.

▼ ***Glacial Lakes.*** Most of the natural lakes in the Upper Great Lakes Region were created during the last period of glaciation some 14,000 years ago. As the massive ice sheets advanced and then retreated back to the north, they left several different types of lakes:

- Depressions scoured by the glacier filled with meltwater: ice scour lakes (the Great Lakes are a good example);
- Rock and soil debris (terminal and lateral moraines) deposited by the receding glacier dammed up valleys creating lakes (for example, New York's Finger Lakes); and
- Large blocks of ice left behind formed and filled in depressions, creating ice-block or kettle lakes (Fig. 2-2).

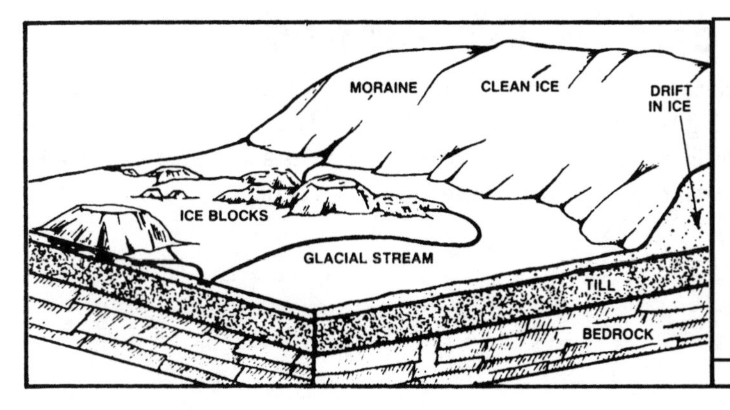

DURING GLACIATION
About 3,000 years ago the last glaciers began to retreat from the North American continent. Many of the small lakes in the upper midwest and north central states as well as Canada were formed by huge ice blocks buried in the loose rock and soil and deposited by the glaciers. When the buried ice blocks melted, they left holes in the glacial till which filled with water from the melting glaciers.

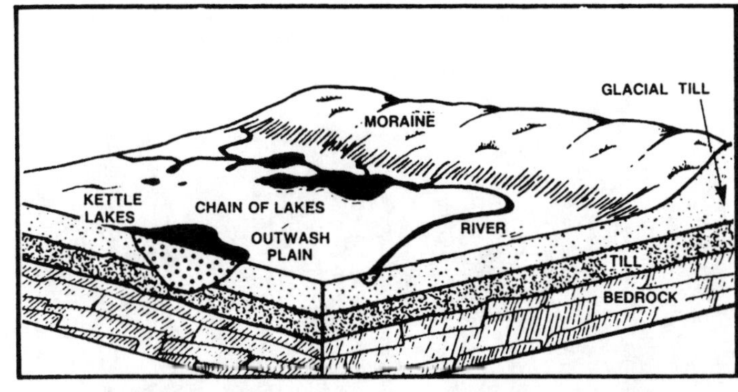

AFTER GLACIATION
Chains of lakes formed along some streams that drained the melting glaciers. Other lakes were created between the moraines and the retreating ice mass from the melting water.

MORAINE
A moraine is a ridge of low rolling hills made up of unsorted rocks and soil deposited when the glacial ice mass melted.

Figure 2-2.—The effects of glaciation in shaping lake basins. Source: Moore and Thornton, 1988.

▼ ***Tectonic Basins.*** Some of the deepest and oldest lakes in the world resulted from movements deep within the earth's crust that created faults and depressions in which "tectonic basin" lakes formed. Lake Baikal in eastern Siberia, the deepest lake in the world at 1,620 meters (5,300 feet), was created in this way. Lake Tahoe in California is another tectonic basin lake.

▼ ***Uplifted Seabeds.*** Moderate uplifting of ancient sea beds created shallow lakes in the West (Great Salt Lake) and in Florida (Lake Okeechobee).

Caldera lake: a large, usually deep basin that forms in the center of an inactive volcano.

- ***Volcanic.*** Volcanic activity creates lakes when water fills an inactive caldera (large crater) (Crater Lake, Oregon) or when lava flows dam up valleys (lakes within Lassen National Park, California).

- ***Reservoirs.*** Much of the central and southern United States has been relatively untouched by recent glaciation, landslides, and volcanoes. In these areas people create lakes, more correctly called reservoirs or impoundments. Federal agencies have impounded major rivers and operate the resulting large reservoirs for multiple purposes: water supply, flood control, recreation, wildlife, and hydroelectric power generation.

- ***Other Types.*** Other lakes have been formed by earthquakes, landslides, and dissolution of soluble rock formations (solution lakes). Solution lakes are most common in the karst regions of Indiana, Kentucky, and Florida.

Just how many lakes are there in the United States anyway? Most reliable estimates have set the figure at 100,000 lakes larger than 100 surface acres — excluding Alaska, which has several million alone (Horwitz, 1980).

The Lake and Its Watershed

A lake is not an isolated body of water, but part of a larger ecosystem that includes the surrounding land that drains into the lake. The land from which rain and surface water drain toward a central collector such as a stream, river, or lake is called a watershed, catchment, or drainage basin (Fig. 2-3).

Watershed: A drainage area or basin in which land and water areas drain or flow toward a central collector such as a stream, river, or lake at a lower elevation.

Figure 2-3.—A lake's watershed includes all wetlands, streams, and upland areas from which water flows into the lake (dark arrows). Source: Horwitz (1980) *Our Nation's Lakes, pg. 11.*

Watershed Inputs

It is often said that a lake is a reflection of its watershed. This is most certainly true. Lakes receive water, dissolved materials carried in water, and particulates, such as soil, from their watersheds; particulates and gases from the atmosphere; and energy from the sun and wind. Therefore, the condition of a lake at any one time is determined not only by what is already in the lake, but also by these outside factors — further confirming lakes as complex ecosystems.

▼ ***Water.*** In a continuous cycle, rain or snow falls on the land; fills rivers, streams, and lakes; evaporates and returns to the atmosphere to eventually fall on the land again (see Fig. 2-4). The volume of water entering lakes can have important consequences for lake processes and management.

Direct precipitation, surface water runoff, and groundwater flow all contribute to keeping lakes filled with water. Like a bathtub, when the entering water exceeds the capacity of the lake, it overflows. In most lakes, the overflow leaves via a surface outlet stream. The relationship between the volume of a lake and the volume of water entering the lake over time determines the lake's hydraulic (water) residence time (see Fig. 2-5).

If the water entering the lake from all sources during the year equals the volume of the lake, the hydraulic residence time is one year. Lake Shafer, a shallow Indiana reservoir with a very large watershed, has a hydraulic residence time of only three days. Lake Tahoe, California, a deep lake with a relatively small watershed, has a hydraulic residence time of approximately 700 years!

▼ ***Particulates.*** Soil, leaves, twigs, and other organic debris are particulates (also known as TSS: total suspended solids) carried into lakes by water running off the watershed. Human activities that disturb the natural, vegetated land cover promote these losses (Table 2-1).

Particulates may surge following major storms as high velocity water erodes stream beds and banks. Other particulates fall into lakes from the atmosphere — wind-blown soil from newly-tilled agricultural fields being a major source.

Particulates add to the turbidity of lake water, thus decreasing water transparency and the light algae need to grow — conversely, nutrients attached to particulates can promote excessive algal growth.

Fish are also affected by particulates. Some find it difficult to locate prey in muddy waters, reducing their growth rates and overall health. Fine particulates irritate the gills of some fish species. And particulate deposits often cover fish spawning sites and smother their eggs. Particulate deposits also damage habitat for aquatic insects, which are a major food source for some fish.

Lakes trap sediments very efficiently. Flowing stream water keeps particulates suspended, but when it empties into a lake, the flow diminishes and the particulates eventually settle onto the lake bottom. This accumulation decreases lake volume, creates shallow water areas where rooted plants can grow, and may interfere with boats.

Motorboats may actually resuspend newly-deposited fine particulates, further decreasing transparency. For example, a 75-hp boat motor can resuspend fine clay sediments to a depth of over 10 feet (Yousef et al. 1978).

Organic matter: Molecules manufactured by plants and animals containing linked carbon atoms and elements such as hydrogen, oxygen, nitrogen, sulfur, and phosphorus.

The Hydrologic Cycle

Because precipitation and surface water runoff directly influence the nature of lake ecosystems, a good way to begin to learn about lakes is to understand the hydrologic (water) cycle. The circulation of water from atmosphere to Earth and back to the atmosphere is a process that is powered by the sun. About three-fourths of the precipitation that falls on land is returned to the atmosphere as vapor through evaporation and transpiration from terrestrial plants and emergent and floating aquatic plants. The remaining precipitation either is stored in ice caps, or drains directly off the land into surface water systems (such as streams, rivers, lakes, or oceans) from which it eventually evaporates, or infiltrates the soil and underlying rock layers and enters the groundwater system. Groundwater enters lakes and streams through underwater seeps, springs, or surface channels and then evaporates into the atmosphere.

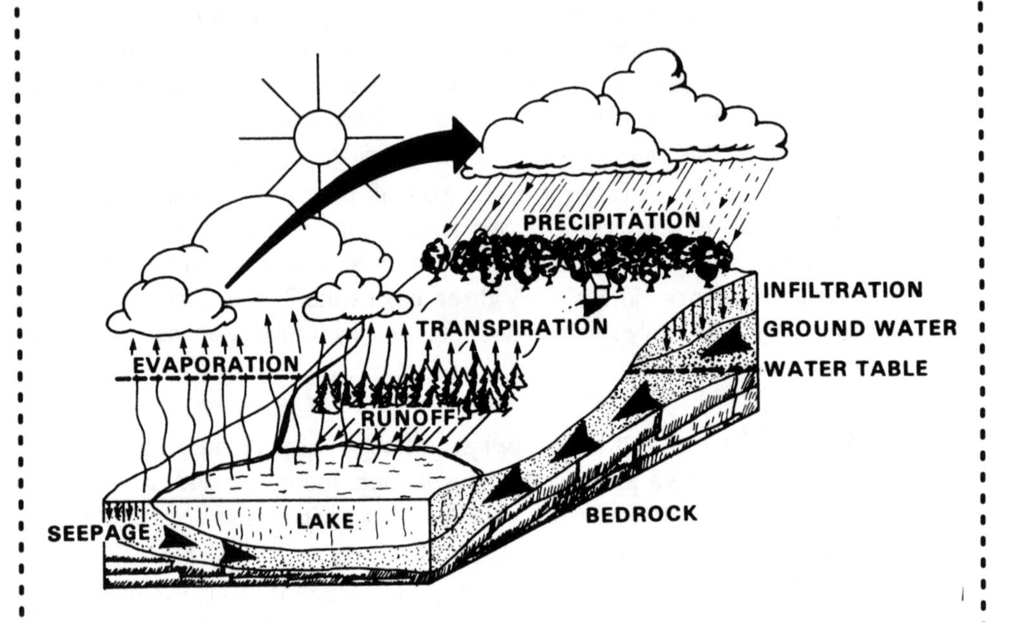

Figure 2-4.—The hydrologic cycle. Source: Moore and Thornton, 1988.

Lakes and reservoirs have a water "balance," as described in this simple equation: water input = water output +/– the amount of water stored in the lake. Inputs are direct precipitation, groundwater, and surface stream inflow, while outputs are surface discharge (outflow), evaporation, losses to groundwater, and water withdrawn for domestic, agricultural, and industrial purposes. If inputs are greater than outputs, lake levels rise as water is stored. Conversely, when outputs are greater — for example, during a summer drought —lake levels fall as losses exceed gains.

Some lakes, called *seepage lakes*, form where the groundwater flow system intersects with the land surface. Seepage lakes are maintained primarily by groundwater inflow, and their water levels fluctuate with seasonal variations in the local water table. *Drainage lakes*, on the other hand, are fed primarily by inflowing rivers and streams; therefore, their water levels vary with the surface water runoff from their watersheds. In both cases, the balance between hydrologic inputs and outputs influences the nutrient supply to the lake, the lake's water residence time, and, consequently, the lake's productivity and water quality. ♦

Hydraulic Residence Time

The average time required to completely renew a lake's water volume is called the hydraulic residence time. For instance, it might take 5 minutes to completely fill a bathtub with the tap fully open and the bottom drain closed. The hydraulic residence time of the tub, then, is 5 minutes. With the tap and drain only half open, the hydraulic residence time would be 10 minutes.

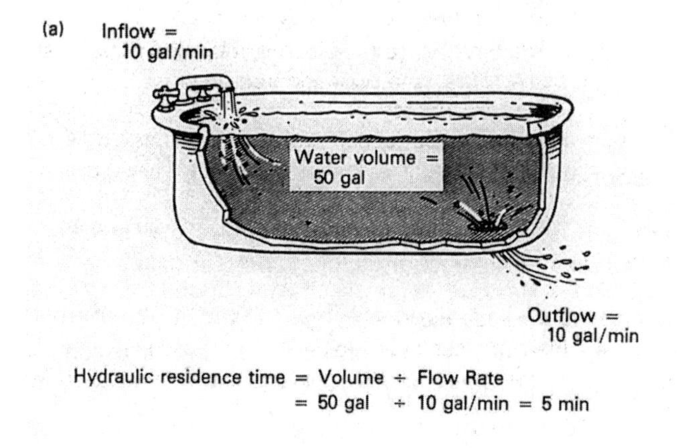

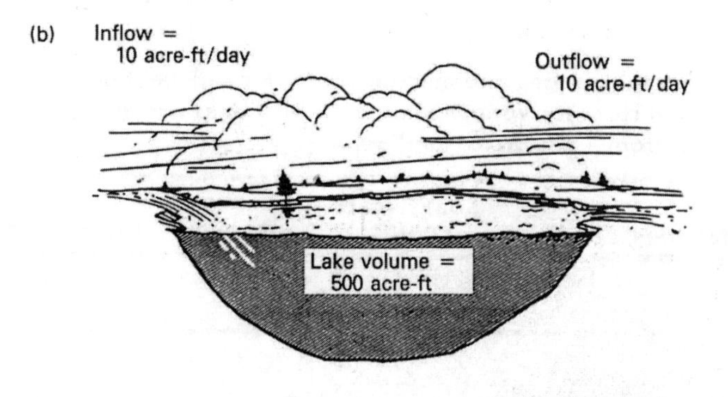

Figure 2-5.—Hydraulic residence time is an important factor to consider in restoration programs. The simple formula given in the figure assumes that inflow is equal to outflow. Source: Moore and Thornton, 1988.

If the lake basin volume is relatively small and the flow of water is relatively high, the hydraulic residence time can be so short (10 days or less) that algal cells produced in the water column are washed out faster than they can grow and accumulate.

An intermediate water residence time allows both an abundant supply of plant nutrients and adequate time for algae to assimilate them, to grow, and then accumulate.

Longer water residence times from 100 days to several years provide plenty of time for algal biomass to accumulate if sufficient nutrients are present. The production of algae may ultimately be limited by the supply of nutrients. If the nutrient supply is high, algal biomass will be very large. The combined effects of nutrient income (or "nutrient loading") and hydraulic residence time on the production of algae is the basis of methods for predicting changes in the lake's condition following variations in one or both of these processes (such as the diversion of wastewater flows). These concepts are discussed in more detail in Chapter 5. ♦

Algae: generally microscopic plants found suspended in lakes (phytoplankton) or attached to solid structures (periphyton).

Acre-foot: volume measure equal to 1 foot of water covering 1 acre of area; 1 ac-ft = 382,850 gallons.

▼ ***Dissolved Material.*** Surface runoff and groundwater discharges carry a variety of dissolved materials to the lake, among them:

- Minerals dissolved from bedrock deposits, such as calcium carbonate (an important component of a lake's buffering capacity);
- Metals derived from bedrock (e.g., iron and aluminum) or human activities (e.g., zinc and lead) that are important to many chemical and biological functions within lakes;
- Pesticides, herbicides, toxic pollutants, chemicals in wastewater discharges, and industrial waste materials;
- Phosphorus and nitrogen, the primary plant nutrients that encourage growth of algae and rooted plants; and
- Oxygen, needed by fish and other aquatic organisms, and important in many chemical reactions in lakes.

Buffering capacity: the ability of lake water to resist changes in pH or acidity. The buffering capacity comes from alkaline materials (for example, carbonates, bicarbonates, sulfates) that neutralize acids.

Phosphorus and nitrogen are two of the more important dissolved materials in most runoff. While phosphorus and nitrogen occur naturally in leaf litter, soil, and bedrock, they are readily lost in runoff when people disturb the land, primarily by growing row crops or building structures (Table 2-1).

Table 2-1.—Example ranges and median values for sediment and nutrient export and input rates for various land uses. All values are in kg/ha-yr (kilograms lost from 1 hectare of land per year).

LAND USE	SUSPENDED SOLIDS	TOTAL PHOSPHORUS	TOTAL NITROGEN
RURAL EXPORT			
Cropland	20 — 5,100	0.2 — 4.6	4.3 — 31
Improved pasture	30 — 80	0.1 — 0.5	3.2 — 14
Forest/Woodland	1 — 820	0.02 — 0.67	2.0 — 20
Feedlot	—	10 — 620	100 — 1,600
URBAN EXPORT			
Residential	620 — 2,300	0.4 — 1.3	5.0 — 7.3
Commercial	50 — 830	0.1 — 0.9	1.9 –11
Industrial	450 — 1,700	0.9 — 4.1	1.9 — 14
Developing urban	27,500	23	63
ATMOSPHERIC INPUT			
Forest	—	0.26	6.5
Agricultural	—	0.28	13.1
Urban industrial	—	1.01	21.4

Sources: Novotny and Olem, 1994; Reckhow et al. 1980; Sonzogni et al. 1980.

Airsheds

A lake's airshed is often overlooked as an important contributor to its water quality. The airshed represents the area of the atmosphere from which airborne materials are transported and deposited on a lake and its watershed. Generally larger than its watershed, an airshed's size depends on the direction and speed of the wind that carries gases and particulates.

Among the airborne contaminants affecting lakes is phosphorus. Winds pick up and carry large amounts of dust and soil from agricultural fields and urban areas. Phosphorus attached to these particles can be transported many miles and eventually deposited in lakes as dry particulates or mixed with rain. In rural northeastern Connecticut the annual atmospheric deposition of phosphorus on the land was measured at 126 mg/m^2 (Kortmann, 1991). This alone can contribute a significant amount of phosphorus to shallow lakes and ponds.

Airsheds are Important Too!

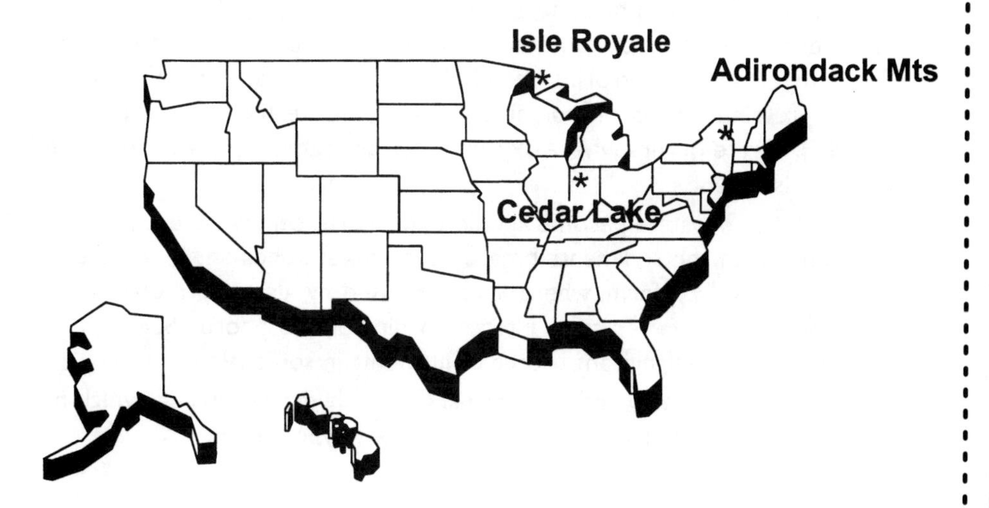

Many other lake pollutants are carried in the air. Steel mills 20 miles north of Cedar Lake, Indiana were the suspected source of elevated zinc and lead concentrations in its surficial sediments (Echelberger and Jones, 1984). Southerly winds carry these and other airborne pollutants downwind to Cedar Lake. Atmospheric transport of mercury from combustion sources has been implicated in mercury fish consumption advisories that now include 41 states (U.S. EPA, 2000).

The airshed of New York's Adirondack mountain lakes extends all the way down to the lower Ohio River valley. Emissions from coal-fired electrical generating stations in Indiana, Kentucky, and Ohio produce sulfur dioxide gas that is converted to sulfuric acid in the atmosphere and falls as acid rain hundreds of miles away in the Adirondack Mountains.

Findings at Siskiwit Lake on Isle Royale in Lake Superior illustrate that the entire earth is the airshed for some organic pollutants. Isle Royale is a national park and has no sources of a group of toxic chemicals known as polychlorinated dioxins and furans, which are produced by combustion of domestic and industrial waste. However, these contaminants were measured in surficial sediments in Siskiwit Lake (Czuczwa et al. 1984). These and other chemicals are now known to be distributed worldwide by global atmospheric transport (Simonich and Hites, 1995).

The bottom line is that we must consider airshed effects in managing our lakes.

Photosynthesis: process in which plants use sunlight, nutrients, water, and carbon dioxide to produce energy necessary for growth. Oxygen is a byproduct of photosynthesis.

Inorganic: chemical compounds that do not include hydrocarbons.

Oxygen saturation: when dissolved oxygen in water is in equilibrium with oxygen in the atmosphere it is considered saturated. The amount of oxygen that fresh water can contain varies with temperature. Cold water can contain more oxygen at saturation than can warm water. During algal blooms, photosynthesis rates can be high enough that dissolved oxygen can become super-saturated (greater than 100% of the saturation value).

Surface and subsurface drainage from fertile (nutrient-rich) watersheds results in biologically productive lakes, and drainage from infertile (nutrient-poor) watersheds results in biologically unproductive lakes. The relative fertility of watersheds and, thus, of lakes varies locally and regionally.

Oxygen enters lakes from the atmosphere and as a byproduct of plant photosynthesis. The oxygen content of water remains in approximate equilibrium with the oxygen content in the atmosphere: excess oxygen produced by photosynthesis diffuses into the atmosphere while oxygen consumed by respiration is replaced by oxygen diffused from the atmosphere back into the water.

Diffusion is a relatively slow process and it sometimes cannot keep up with oxygen consumption and production reactions within the lake. For example, bacterial decomposition of organic wastes in lake sediments may consume oxygen at a much greater rate than it can diffuse or mix back in, resulting in low oxygen levels.

When the loss of oxygen from the water exceeds its input, the oxygen content of the lake water declines. If the dissolved oxygen becomes severely depleted, the lake can become anoxic, causing odors, fishkills, and increases in undesirable forms of some nutrients. For example, dissolved oxygen is very important in controlling chemical reactions in lake water. Of the two most prevalent forms of inorganic nitrogen, nitrate (NO_3^-) and ammonia (NH_4^+), nitrate occurs where oxygen is plentiful; ammonia, when dissolved oxygen is absent.

If bottom waters become anoxic, phosphorus otherwise bound in iron compounds in the sediments breaks its chemical bonds and re-enters the water in dissolved form, where it can be used by algae for more growth. This process is referred to as internal loading of phosphorus. Such internal loading may be a significant source of nutrients in some lakes.

Other chemically reduced forms, such as hydrogen sulfide, which has a rotten egg odor, may also form in an anoxic hypolimnion.

Watershed Characteristics

Characteristics of watersheds — particularly size, topography, local geology, soil fertility and erodibility, and land cover — influence the amount and nature of materials entering lakes.

▼ ***Size.*** Lakes with large watersheds can be more affected by people's activities than lakes with small watersheds and the same intensity and nature of human activity (Fig. 2-6). This is true simply because more land is draining into the downstream lake. When such large watersheds are disturbed, they deliver more water, sediments, nutrients, and other pollutants to streams and lakes.

On the other hand, lakes with large watersheds may respond more quickly to watershed improvements. For example, improved runoff quality resulting from land management in large watersheds will more rapidly flush out poor-quality lake water than similar improvements in small watersheds. A large watershed is generally considered as one with more than 7 to 10 acres of watershed for each acre of lake. For example, using this criterion, a watershed of 1,000 acres would be considered large for a 100-acre lake.

Decomposition: the transformation of organic molecules (e.g., sugar) to inorganic molecules (e.g., carbon dioxide and water) through biological and non-biological processes.

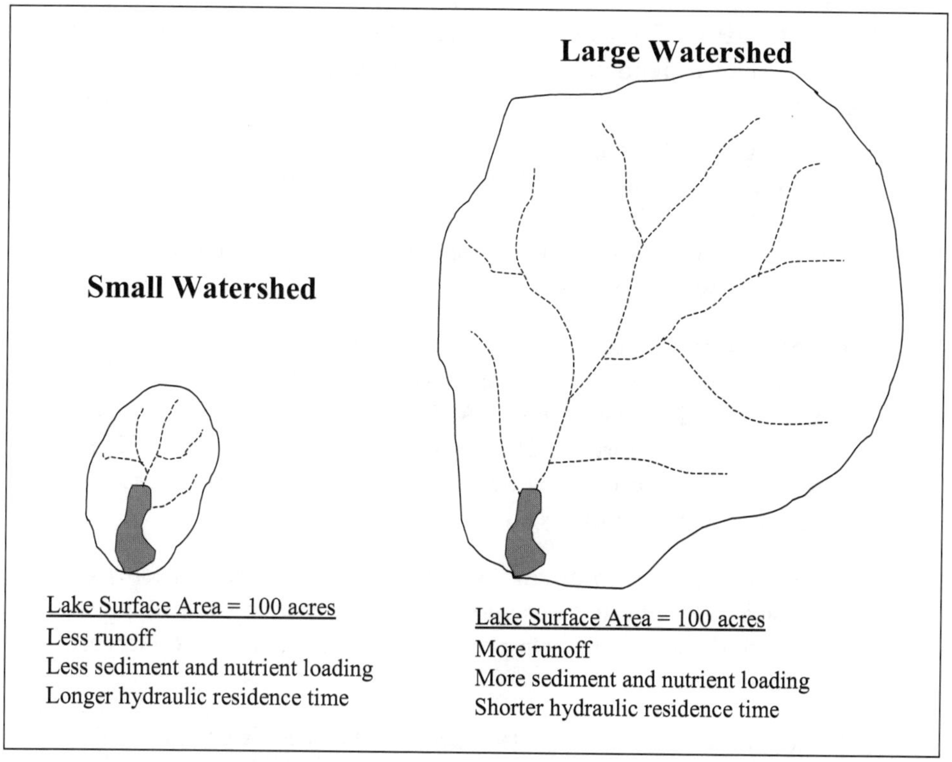

Large Watershed

Small Watershed

Lake Surface Area = 100 acres
Less runoff
Less sediment and nutrient loading
Longer hydraulic residence time

Lake Surface Area = 100 acres
More runoff
More sediment and nutrient loading
Shorter hydraulic residence time

Figure 2-6.—Watershed size can affect many important lake characteristics. Lakes receive water, nutrients, sediments, and other materials from their watersheds. Larger watersheds, relative to lake size, deliver proportionately more of these materials to the lake.

▼ ***Land slope.*** Steep slopes increase the velocity of surface water runoff, thus increasing the likelihood that soils and other particulates will detach, be caught up by the runoff, and be transported down the watershed and into the water.

▼ ***Geology.*** Regional geology controls the type of minerals and nutrients that may enter a lake. For example, a deep alpine lake located in a granitic watershed in the Colorado Rockies is almost certain to have pristine, crystal clear, high quality water but very low biological productivity and poor fishing. On the other hand, a lake basin situated in rich alluvial soils will likely have poor water quality because of its high turbidity, high concentrations of nutrients and organic matter, frequent algal blooms, and abundant aquatic macrophytes.

▼ ***Soil type.*** The erosion and runoff potential of soils is related to their cohesiveness and permeability. Loose-structured soils of intermediate particle size like silts are most susceptible to erosion because they are less cohesive than clays (smallest particles) but smaller than sands (Novotny and Olem, 1994). Soils of larger particle size like sands are much more permeable so water soaks into them more quickly rather than running off. Runoff is greater on clay soils that are more tightly packed and contain smaller pores.

▼ ***Land cover.*** The amount and type of materials carried by runoff from a watershed depend primarily upon the cover on the land (Fig. 2-7). For example:

- **Bare ground** has no protection from the erosive force of rain drops. Rain drops hitting bare soil strike the ground with such force that soil is ejected into the air, leaving a tiny "impact crater." As more and more rain falls, these craters connect to form channels that carry away the water and the loose soil.

- **Impermeable surfaces** (roads, parking lots, roofs) do not allow water to percolate into the soil. Instead, water flows off the surface and enters drainage channels. A greater proportion of impermeable surfaces yields more runoff. For example, in an urban area, 90 percent of the rainfall would likely become runoff while in a forest only 10 percent might run off (Fig. 2-7).

- **Dense vegetation cover** (high grasses, shrubs, trees) reduces the erosive force of raindrops and provides resistance to flowing water. Both of these actions help protect soil from erosion and increase infiltration.

- **Human activities** that place additional materials on the land (fertilizers, pesticides, septic systems, solid waste dumps, road salt, etc.) greatly increase the amounts and types of materials reaching lakes and streams.

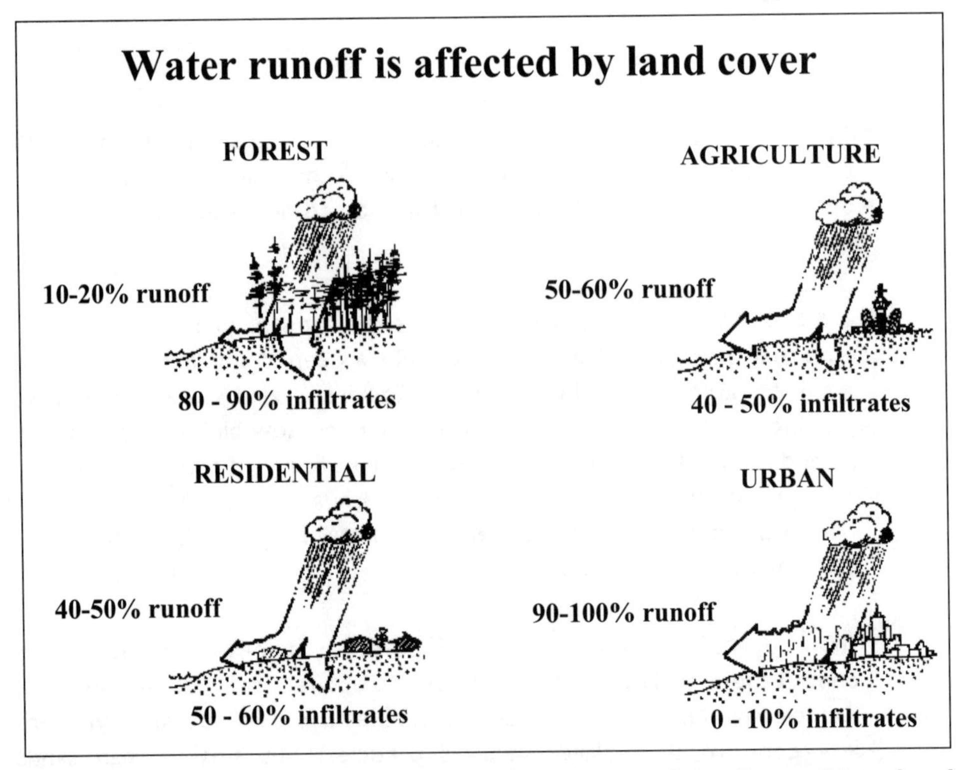

Figure 2-7.—General relationships between land cover and runoff: runoff is reduced on land with vegetative cover and increased on land with impermeable surfaces such as roads, parking lots, and roofs. Adapted from Marsh and Borton, 1975.

Lake Morphometry

A lake's physical dimensions (morphometry), which influence lake water quality and productivity levels in a number of ways, are largely established when a lake forms. A bathymetric map (Fig. 2-8) depicts the depth contours of a lake and allows you to determine its surface area, maximum depth, mean depth, and shoreline length. Bathymetric maps clearly show drop-offs and bars — features important to anglers.

Some of the more important morphometric parameters to consider for lake management include:

- ***Surface area (A_o);*** units = acres (ac) or hectares (ha): important for managing lake uses.

- ***Maximum depth (Z_{max});*** units = feet (ft) or meters (m): the location where water is deepest — influences stratification and the proportion of water in which algae can grow. For example, most water in shallow lakes may have enough light for algae to grow, but in deeper lakes, much of the deep water does not have enough light for algae (Fig. 2-9).

- ***Mean depth ($\bar{z}$)*** units = feet (ft) or meters (m): average depth of the lake calculated by dividing the volume by the surface area ($\bar{z} = V/A_o$).

- ***Volume (V);*** units = cubic meters (m^3) or acre-feet: total amount of water in lake — influences dilution of inflowing nutrients, affects hydraulic residence time.

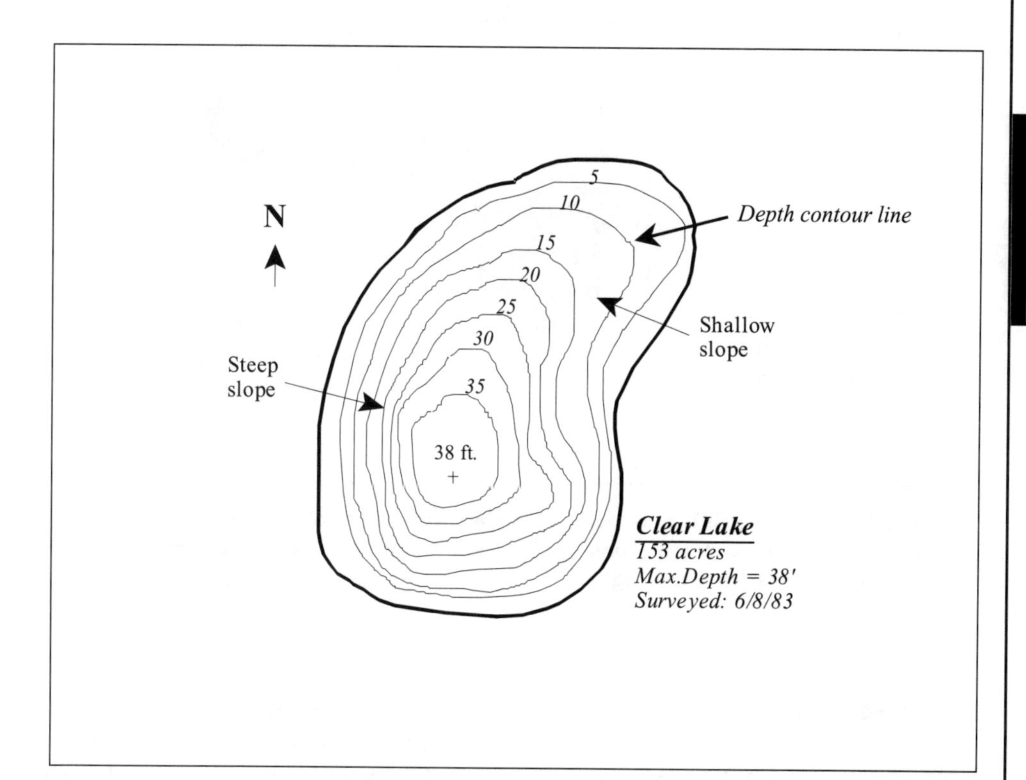

Figure 2-8.—Bathymetric map for Clear Lake. Large shallow sloped areas could support an extensive littoral zone. Steeper sloped areas indicate drop-offs important for fishing. The maximum depth is indicated by the cross.

Littoral zone: that portion of the lake lying within the photic zone. Rooted plants can grow here and sediments support a rich biological community.

Photic zone: that area of a lake having sufficient light to support photosynthesis.

Aphotic zone: that area of the lake too dark to support photosynthesis.

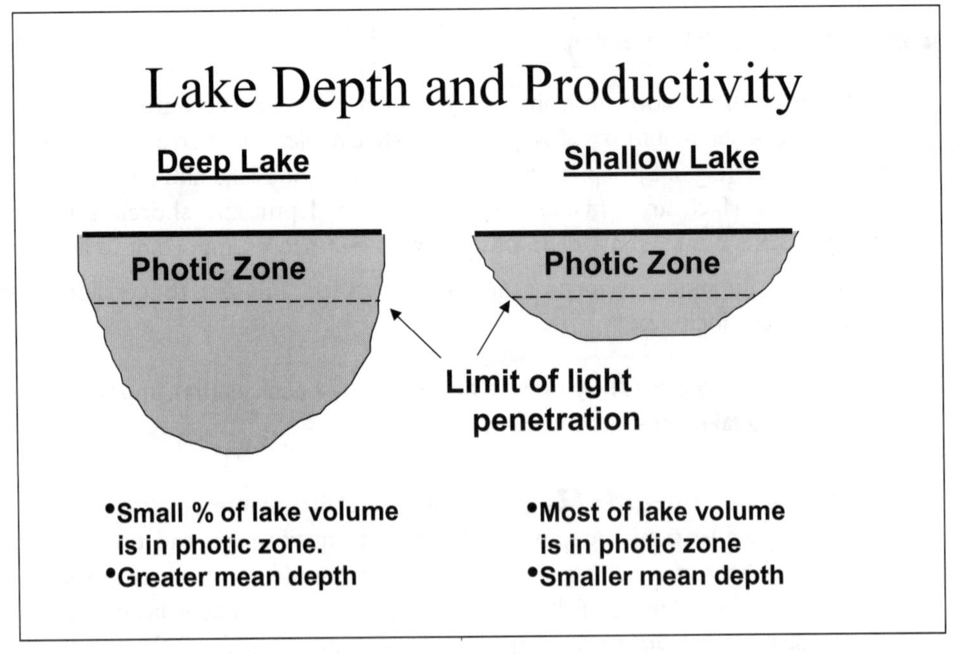

Figure 2-9.—Because less of their overall volume is in the photic zone, deep lakes are less able to support plant growth, while shallow lakes with a proportionally larger lake volume within the photic zone can support plant growth.

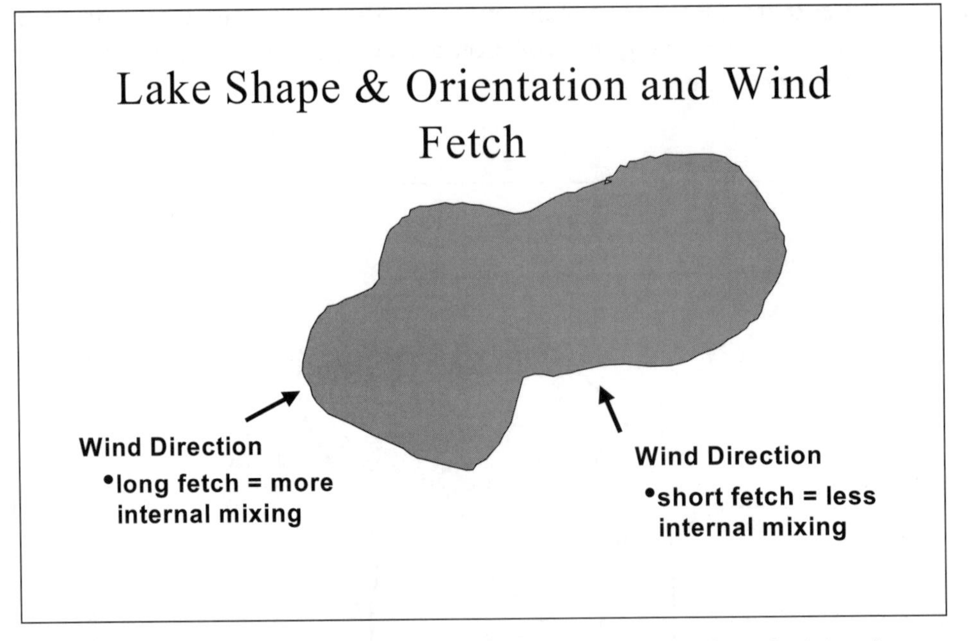

Figure 2-10.—Wind fetch is one factor that affects the amount of wind mixing that occurs in surface waters. A long, unimpeded length of lake along the primary wind direction allows waves to grow higher.

▼ ***Fetch*** (ft. or m.): longest uninterrupted distance over which the wind blows across the lake — varies with wind direction, and affects wave height and vertical mixing of water (Fig. 2-10). A lake with a very long wind fetch may be more turbid than a lake of the same size where the fetch is shorter.

▼ ***Shoreline Length (L);*** units = meters (m) or feet (ft): the total length of the shoreline encompassing the lake (lake circumference). This distance represents the total amount of lake front available for activities such as building homes and for lake edge effects (shoreline erosion, littoral habitat, etc.).

▼ ***Shoreline Development (D_L);*** unitless: measure of "roundness" of lake, is the ratio of the length of the shoreline to the circumference of a circle whose area is equal to that of the lake. A perfect circle has a shoreline development value of 1.0. As the shape of the shoreline becomes more irregular, the value increases. Irregularly-shaped reservoirs with numerous embayments (coves) can have shoreline development values exceeding 3. Lakes with very irregular shorelines have more nearshore shallows for rooted plant growth and more shoreline for houses and shoreline erosion — all of which may increase productivity in the lake (Fig. 2-11). Shoreline development can be determined by the following equation:

$$D_L = \frac{L}{2\sqrt{\pi A_o}}$$

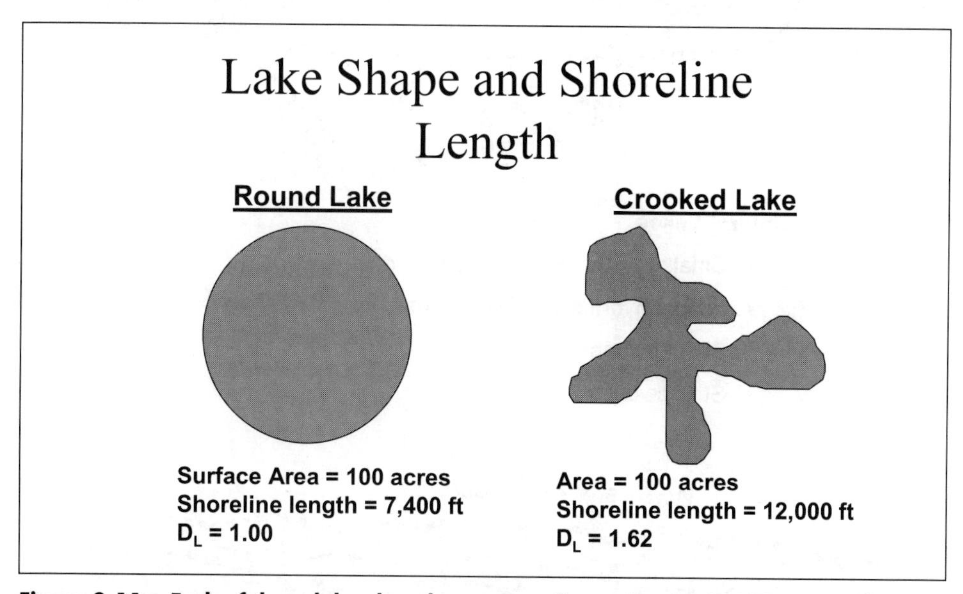

Figure 2-11.—Each of these lakes has the same surface area size — 100 acres. The irregular-shaped Crooked Lake has a much larger shoreline length for littoral plant growth, home sites, and shoreline erosion.

Lakes v. Reservoirs

In contrast to glacial lakes that may be many thousands of years old, most reservoirs in the western hemisphere have been constructed within the past 100 years, although some are much older. For example, the Romans built Proserpina Dam in Spain in the second century and the lake is still being used for recreation and as a water supply. Reservoirs have been built for many purposes, often for multiple uses, including: water supply, flood control, power generation, recreation, and navigation.

While the same basic physical, chemical, and biological processes occur in reservoirs and natural lakes, the age, morphology, location in the drainage basin, and hydrological characteristics make them unique ecosystems (Cooke and Kennedy, 1989).

- Because reservoirs are often flooded river valleys, they often are long and narrow rather than circular or ovid like glacial lakes, and they tend to have irregular shorelines (Figure 2-12).
- While natural lakes often receive water from several small streams and groundwater, a single large tributary usually supplies reservoirs.
- Water leaves drainage lakes via an unregulated surface discharge but reservoir releases are most often discharged through submersed, controlled gates (Cooke and Kennedy, 1989).
- Many reservoirs are sited such that they have large watersheds to meet water supply or flood storage needs. Therefore, reservoirs often have watersheds many times larger than those of natural lakes. These larger watersheds deliver more water, sediments, and nutrients to reservoirs.

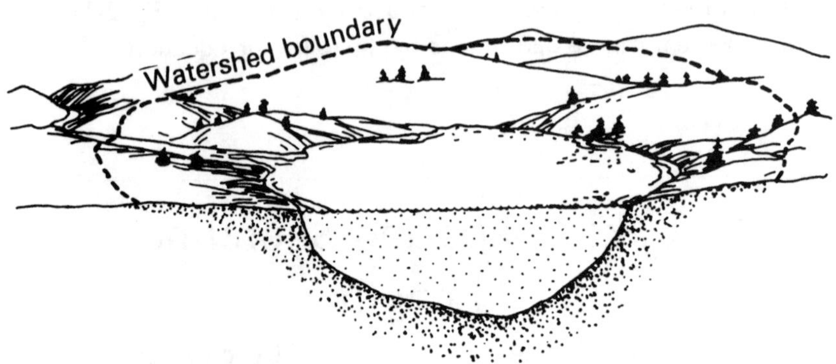

Natural Lakes

- Smaller watershed area: lake surface area ratio
- Longer hydraulic residence time
- Simpler shape
- Surface outlet

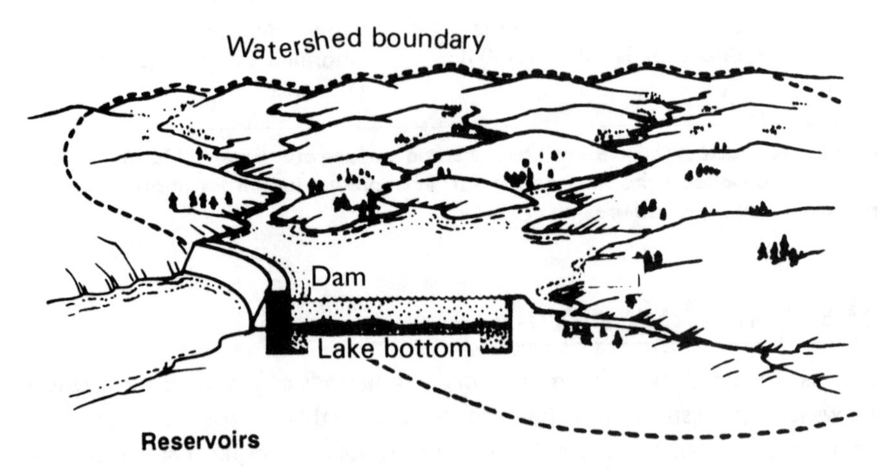

Reservoirs

- Larger watershed area: lake surface area ratio
- Shorter hydraulic residence time
- More complex shape, higher D_L value
- May have surface and/or subsurface outlet(s).

Figure 2-12.—General comparison between natural lakes and reservoirs. Source: Moore and Thornton, 1988.

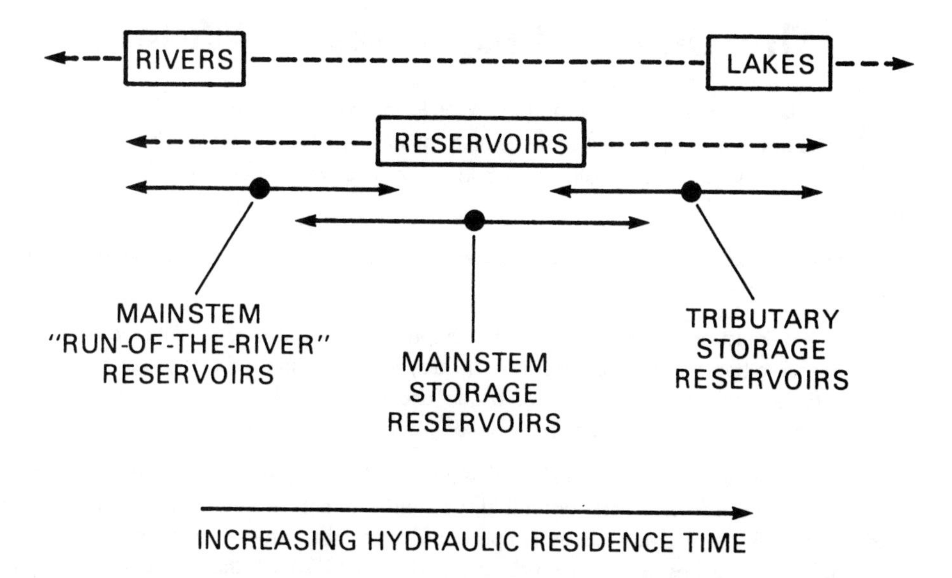

Figure 2-13.—Reservoirs occupy an intermediate position between rivers and natural lakes along a continuum of aquatic ecosystems ranging from rivers to natural lakes. Water residence time and the degree of riverine influence are primary factors determining the relative positions of different types of reservoirs (mainstem-run-of-the-river, mainstem storage, and tributary storage impoundments) along the river-lake continuum. Modified from Kimmel and Groeger (1984).

Actually, reservoirs and natural lakes are probably more alike than different. Their physical, chemical, and biological conditions overlap greatly, as illustrated in Figure 2-13. With regard to the environmental factors that control water quality and biological productivity, reservoirs occupy an intermediate position between natural lakes and rivers (Kimmel and Groeger, 1984).

Lake Processes

Lake Stratification and Mixing

The thermal properties of lakes and the annual circulation events they cause probably have a greater influence on lake biology and chemistry than any other factor. Lake water absorbs heat energy directly from sunlight and additional heat from the air. The mixing action of wind helps distribute this heat throughout a lake's surface waters.

Stratification: process in which several horizontal water layers of different density form in some lakes.

As surface water warms up in the spring, it becomes lighter than the cooler, more dense water at the lake bottom that does not receive this heat. This is just one of the unique properties of water — its density changes with temperature (see special box on Unique Properties of Water [next page]).

As the surface water continues to warm, the density difference between the surface and bottom waters becomes too great for the wind energy to mix, and the lake becomes stratified. Thermal stratification defines the condition in many temperate lakes where warm surface waters overlie cold bottom waters.

- The well-mixed and uniformly warm surface waters are called the epilimnion, while
- The uniformly cold, unmixed bottom waters are called the hypolimnion (Fig. 2-14).

The Unique Properties of Water

Water is a unique substance, and to understand how lakes behave, it is useful to understand water's physical and chemical properties. The molecular structure of water and the way in which water molecules associate with each other dictate these properties:

1. Water is an excellent solvent; many gases, minerals, and organic compounds dissolve readily in it.

2. Water is a liquid at natural environmental temperatures and pressures. Although this property seems rather common and obvious, in fact, it is quite important. If water behaved at ordinary temperatures and pressures as do chemically similar inorganic compounds, it would be present only as a vapor, and lakes would not exist.

3. The temperature-density relationship of water is also unique. Most liquids become increasingly dense (more mass, or weight, per unit volume) as they cool. Water also rapidly becomes more dense as its temperature drops, but only to a certain point (Fig. 2-14). Water reaches its maximum density at 39.2°F (3.94°C), then it decreases slightly in density until it reaches 32°F (0°C), the freezing point. At this point, ice forms and its density decreases sharply. Ice, therefore, is much lighter than liquid water and forms at the surface of lakes rather than at the lake bottom.

   A second important consequence of the temperature-density relationship of water is the thermal stratification of lakes. Energy is required to mix fluids of differing densities, and the amount of energy necessary is related to the difference in density. In the case of the water column mixing in lakes, this energy is provided primarily by wind. Therefore, the changes in water density that accompany rapidly decreasing water temperatures in the metalimnion during summer stratification are of great importance. The metalimnetic density gradient provides a strong and effective barrier to water column mixing.

** The layer of greatest temperature change, the metalimnion, presents a barrier to mixing. The thermocline is not a layer, but a plane through the point of maximum temperature change. The epilimnion and hypolimnion are relatively uniform in temperature. As the graph illustrates, ice is much less dense (lighter) than water. Warm water is less dense than cold water, but more dense than ice. Density changes most rapidly at warm temperatures.*

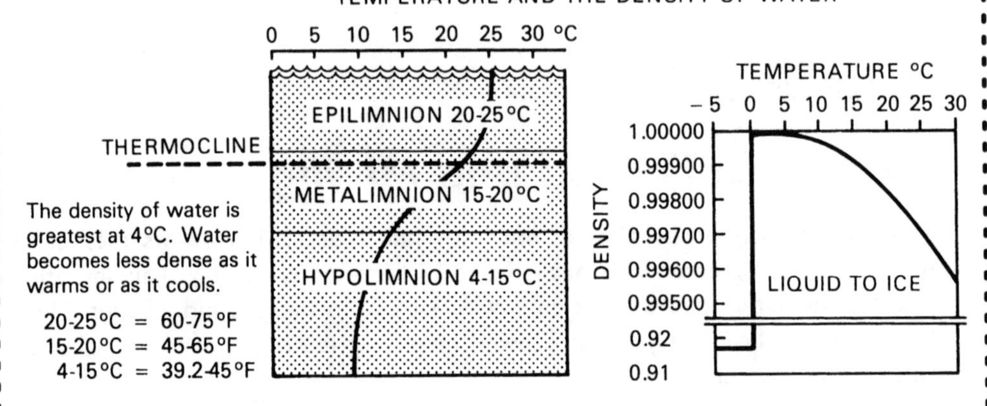

Figure 2-14.—The temperature-density relationship of water enables deep lakes to stratify during summer. (* See explanation in side column.)

4. Water also has an unusually high specific heat. Specific heat is the amount of energy required to change the temperature of 1 g of water by 1°C. Water also has a high latent heat of fusion, which is the energy required to melt 1 g of ice at 0°C. These properties make lakes slow to thaw and warm in the spring and slow to cool and freeze in the fall, thus providing exceptionally stable thermal environments for aquatic organisms.

   Additionally, because water gains and loses heat slowly, the presence of large lakes can exert a significant influence on local and regional climate. A good example is the Great Lakes, which have a dramatic effect on both the air temperature and on the precipitation in the states and provinces surrounding them.

- The two layers are separated by a zone of rapidly changing temperature and density called the metalimnion — that layer of water where the temperature changes by more than one degree Celsius for each meter of water depth. In effect, the metalimnion acts as a lid separating the epilimnion from the hypolimnion during stratification.
- Another term, the thermocline, defines that particular depth within the metalimnion where the rate of change in temperature is greatest.

As the epilimnion cools in the late summer and fall, the temperature difference between layers decreases, and mixing becomes easier. With the cooling of the surface, the mixing layer gradually extends downward until the entire water column is again mixed and homogeneous (Fig. 2-15D). This destratification process is referred to as the fall overturn.

Under ice cover during winter, the lake may again stratify but in reverse order. The cooler, less dense water (< 4^0C) overlies the dense water at 4^0C, the temperature of maximum water density (Fig 2-15A). When the ice melts and the surface waters begin to warm up, the density differences with depth are minimal and the lake again circulates during spring turnover (Fig 2-15B).

The extent of summertime thermal stratification in lakes depends on several factors, including lake depth, wind fetch, wind exposure, and spring temperatures. Most temperate lakes 5 to 7 meters deep stratify. However, even relatively deep lakes may not stratify if they have large wind fetches or are unprotected by trees or other wind barriers. On the other hand, a relatively shallow lake may stratify if it is surrounded by bluffs or has a small wind fetch. A long, cool spring can allow a deep epilimnion to develop, because slow warming allows for a longer period of deep circulation. The thickness of the epilimnion may be very shallow when hot weather causes early stratification after only a short circulation period.

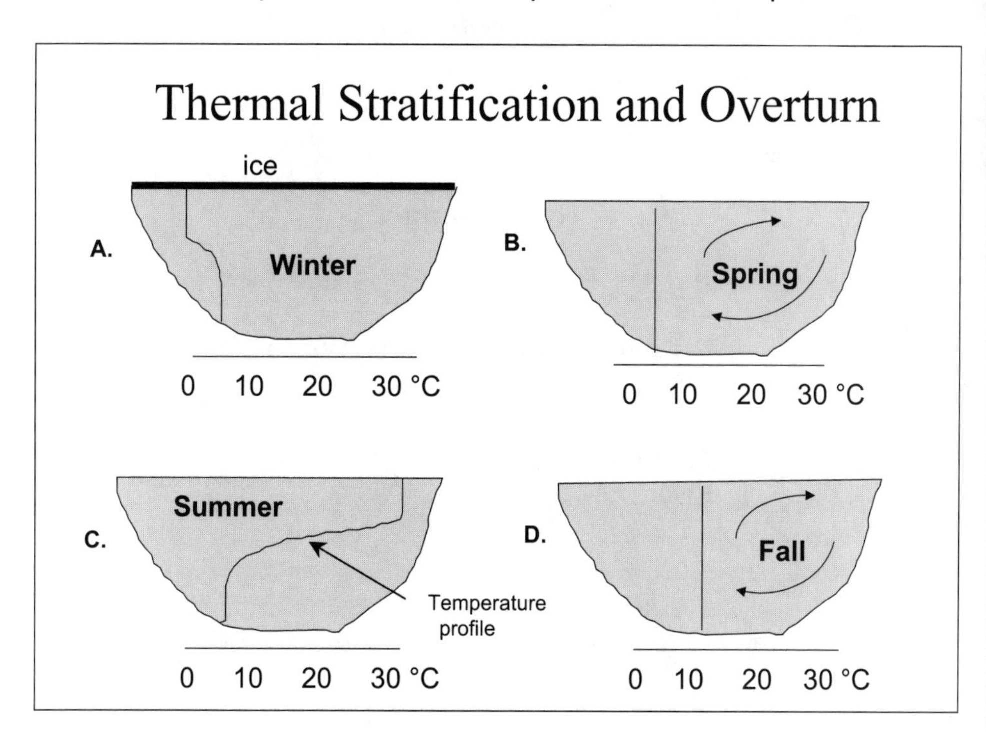

Figure 2-15.—The changing of the seasons in north temperate areas and the subsequent changes in lake temperature provide for alternating periods of thermal stratification and circulation.

Dimictic: undergoing two periods of mixing per year. For example, north temperate lakes often stratify during the summer and winter, and have periods of circulation in the fall and spring.

Monomictic: undergoing only one period of mixing per year. For example, lakes in the southern U.S. may never get cold enough to stratify during the winter. Instead, one long period of circulation lasts from fall to spring.

Polymictic: shallow lakes and reservoirs that circulate frequently and stratify only when winds are calm. Deeper lakes subject to strong, uninterrupted winds may also not stratify.

Anoxia: a condition of no oxygen in the water. Often occurs near the bottom of fertile, stratified lakes in the summer and under ice in late winter.

▼ ***Reservoirs and Stratification.*** Reservoirs may exhibit varying degrees of thermal stratification. These differences are related to geographic location, operation, and morphometry (Cooke and Kennedy, 1989). Reservoirs located in southern states gain significant heat due to the warmer climate and longer duration of the summer period. Temperature profiles for southern reservoirs often lack a pronounced thermocline, and temperatures in bottom waters may be similar to those in the surface waters. Others may lack a distinct hypolimnion but will have both an epilimnion and metalimnion. Shallow reservoirs exposed to the wind may circulate much of the year and might stratify for brief periods during calm, hot weather.

Spatial patterns in thermal structure are often observed in reservoirs. The upper basin of reservoirs is often shallow and well mixed because of tributary inputs and wind action. Deep water near the dam may be thermally stratified throughout the summer, creating a reservoir with two distinctive habitats.

When the inflowing tributary water is warmer, and thus lighter, the waters will flow over the reservoir's surface and ultimately mix with the surface waters. Colder and denser inflowing water will eventually settle into the reservoir's bottom, with extensive mixing possible. Inflowing water with a temperature and density intermediate to the reservoir's epilimnion and hypolimnion can flow as a layer within the metalimnion (Cooke and Kennedy, 1989). When several tributaries of different densities enter a reservoir, several distinct density layers may be apparent (Fig. 2-16).

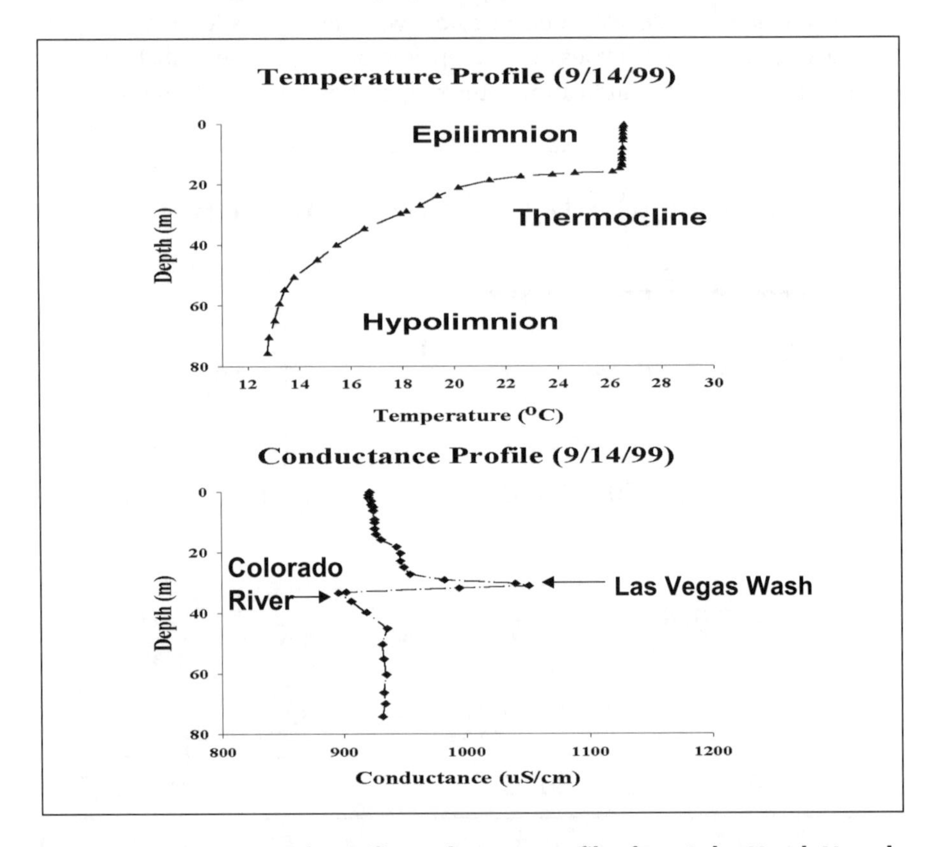

Figure 2-16.—Temperature and specific conductance profiles from Lake Mead, Nevada. The effects of surface inflows are easily detected as the high-conductance but warmer discharge from the Las Vegas Wash overlies the lower-conductance but cooler discharge from the Colorado River. Density layers such as these are evident in lakes and reservoirs having multiple inlets of different temperatures (LaBounty, 2000).

▼ ***Shallow Lakes.*** Shallow lakes, for example, those with a mean depth of less than 10 feet, may have many periods of circulation. Lakes and reservoirs that circulate frequently and rarely stratify except during conditions of calm wind are called **polymictic.** Deeper lakes having large fetches may also be polymictic because strong winds blowing uninterrupted over the water surface have enough energy to completely mix the lake. For example, Florida's Lake Okeechobee, the largest lake (427,000 acres) in the southeastern United States, is shallow (mean depth = 8.9 feet), has a large wind fetch (nearly 37 miles), and is polymictic (Hanlon, 1999).

▼ ***Some Potential Consequences of Stratification.*** When a lake initially stratifies, the hypolimnion is rich in dissolved oxygen from early spring mixing of the water column and from oxygen production by plants. However, because of the metalimnetic barrier, the hypolimnion is isolated from gas exchanges with the atmosphere during the summer and is often too dark for algae to photosynthetically produce oxygen. So a hypolimnion can become anoxic during summer or winter thermal stratification as decomposing organic matter consumes its reserve of dissolved oxygen.

Hypolimnetic anoxia has very important consequences for lake productivity and fishery management and thus is a major target for lake restoration. For example, most fish need relatively high concentrations of dissolved oxygen and cannot survive in an oxygen-deficient hypolimnion; but the midsummer epilimnion may be too warm for them. In addition, under anoxic conditions, nutrients such as nitrogen and phosphorus are released from the bottom sediments to the water column, where they ultimately promote more algae production, organic matter decomposition, and thus, more severe hypolimnetic oxygen depletion.

Is summertime thermal stratification good or bad for a lake? While stratification of a productive lake may lead to an anoxic hypolimnion, phosphorus release from sediments, and loss of fish habitat, the metalimnion generally confines these problems to the hypolimnion during the summer growing season.

Once the lake turns over in the fall, mixing these nutrients into the surface waters rarely causes algal blooms, since, at that time, light and temperature have more effect on algal growth than nutrients. Thus, stratification helps keep the summertime surface waters less productive than they otherwise might be.

The bottom line: remember that thermal stratification does not necessarily cause lake problems. Lake problems are caused by excess phosphorus, nitrogen, and sediments entering the lake from the watershed.

Photosynthesis and Respiration

Light energy from the sun fuels the lake ecosystem. As light strikes the lake surface, a large proportion is reflected back and never enters the water (Fig. 2-17). Of the remaining light some is scattered by suspended particles, some is absorbed by dissolved color and organic material in the water, and finally, some is used by algae and macrophytes (rooted plants) in photosynthesis to convert the light energy to chemical energy.

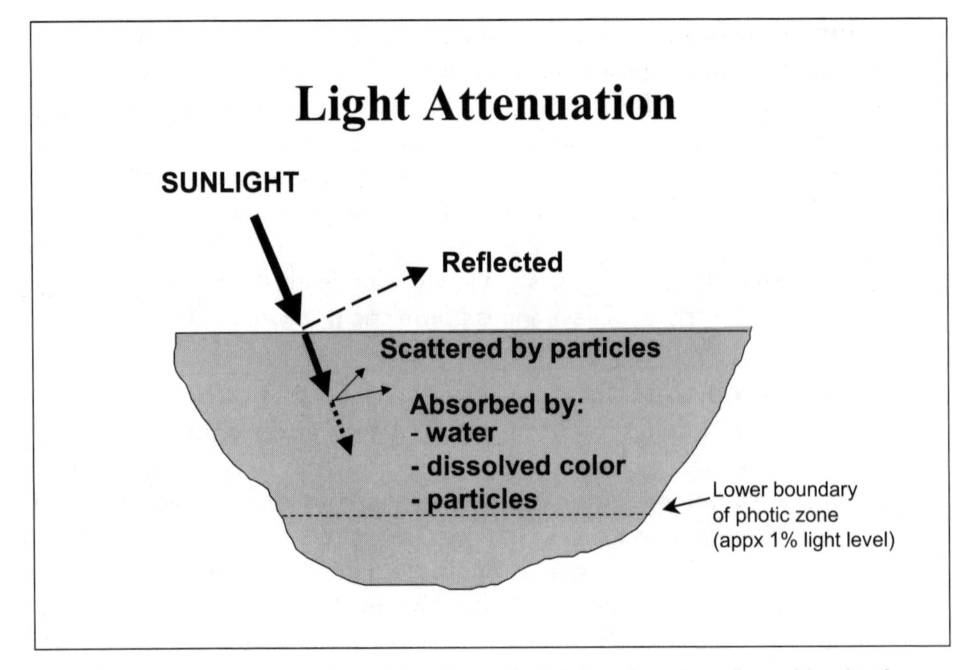

Figure 2-17.—Much of the sunlight striking the lake surface is reflected back. The remaining light is scattered and absorbed — with only a small portion available for photosynthesis.

In photosynthesis, aquatic plants and algae use the energy from sunlight, carbon dioxide, and water to produce sugar, water, and molecular oxygen (Fig. 2-18). The sun's energy is stored in the sugar as chemical energy.

Plant cells then convert the sugar, along with certain inorganic elements such as phosphorus, nitrogen, and sulfur, into organic compounds such as proteins, carbohydrates, and fats.

The rate of photosynthetic uptake of carbon to form sugar is called primary productivity. The amount of plant material produced and remaining in the system is called primary production — analogous to a crop in a farmer's field. While in-lake photosynthesis normally produces the organic matter for the lake's food

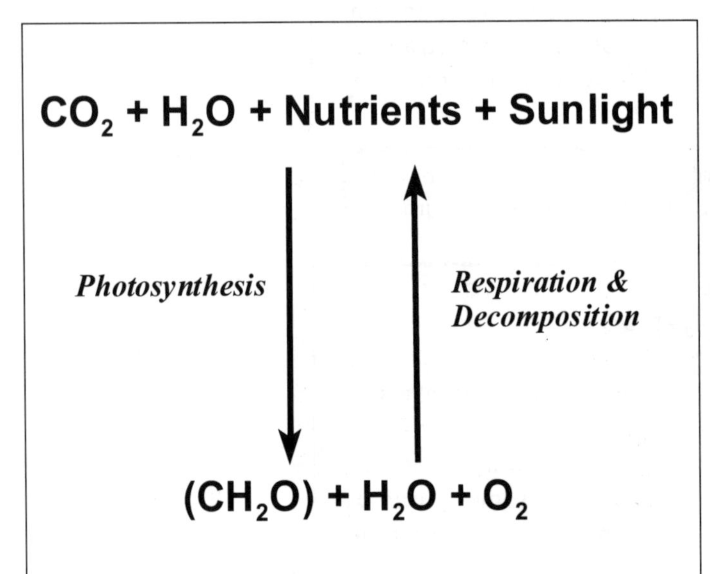

Figure 2-18.—The equilibrium relationship between photosynthesis and respiration-decomposition processes. The photosynthetic conversion of light energy, carbon dioxide (CO_2), water (H_2O), and nutrients into organic matter produces oxygen (O_2), carbon, nitrogen, sulfur, and phosphorus in organic compounds of high potential energy. Respiration-decomposition processes tend to restore the equilibrium by consuming oxygen and decomposing organic materials. Adapted from Moore and Thornton, 1988.

web, most lakes also receive significant inputs of energy in the forms of dissolved and particulate organic matter from their watersheds.

Molecular oxygen, which is also produced by photosynthesis, is the primary source of dissolved oxygen in the water and of oxygen in the atmosphere. Oxygen is required to completely break down organic molecules and release their chemical energy. This process is called oxidation.

Plants and animals release this energy through a process called respiration. Its end products — energy, carbon dioxide, and water — are produced by the breakdown of organic molecules in the presence of oxygen (Fig. 2-18).

Respiration: process by which oxygen is conveyed to tissues and cells, and carbon dioxide and water are given off.

Because light is required, the primary (photosynthetic) production of organic matter by aquatic plants is restricted to the photic zone (see Fig. 2-9). The thickness of the photic zone depends upon the transparency of the lake water and corresponds to the depth to which at least one percent of the surface light intensity penetrates. Below this, in the aphotic zone, the light is too weak to support a significant amount of photosynthesis.

The thickness of the photic zone may vary seasonally; for example, it may be relatively thick in spring but can thin as the growth of plankton or presence of suspended solids limits transparency and light penetration.

Aquatic Plant Productivity

▼ ***Algae.*** Algae are photosynthetic organisms that form the base of the aquatic food chain and are grazed upon by zooplankton and herbivorous fish. Microscopic algae that have little or no resistance to water currents and live suspended in the water are called *phytoplankton.* Although some phytoplankton species have flagella to help them move, even the flagellated forms cannot overcome water currents. In other words, they are *planktonic*.

Zooplankton: Microscopic animals that float freely in lake water, graze on detritus particles, bacteria, other zooplankton, and algae, and may be consumed by fish.

Other lake or stream algae may be found primarily attached to substrates in the water such as piers, rocks, and rooted plants. These attached algae are known as *periphyton*.

The abundance (production) of photosynthetic algae is controlled primarily by water temperature, light, nutrients, hydraulic residence time, and consumption by herbivorous predators.

Algae are classified according to color: green algae, blue-green algae, golden algae, and so on. The colors refer to the most visible pigment in the cells — chlorophyll in green algae; chlorophyll, carotene, and phycobilins in blue-green algae; and abundant carotene in golden algae. The green and golden algae contain these pigments in specialized structures called *chloroplasts*.

Planktonic: passively floating or drifting. If mobile, unable to move against water currents.

Chlorophyll: the green pigment in many types of algae. The primary photosynthetic pigment.

When viewed under a microscope, these distinct colored structures (chloroplasts) are very visible. The blue-green algae, on the other hand, do not have specialized structures and their pigment is evenly spread throughout each cell.

During intense algal blooms of a single species, you can identify the particular group of algae by the color of the lake water. Limnologists measure the amount of chlorophyll pigments in lake water to determine the abundance of phytoplankton.

Blue-greens are the most primitive algae; they date back as far as 2.8 billion years (St. Amand, 1995). They were the first photosynthetic organisms on Earth. Like bacteria, blue-greens have no membrane-bound organelles within cells. Because they are more like bacteria, they are more correctly referred to as Cyanobacteria.

Despite this primitive heritage, blue-greens possess several amazing competitive advantages over other algal groups (Table 2-2):

- Some blue-greens can regulate their buoyancy by retaining gases, thus allowing them to maintain their position in the photic zone.
- Some blue-greens have specialized structures called heterocysts that they use to absorb atmospheric nitrogen (N_2) from the water — a process called nitrogen fixation, also used by some bacteria in the roots of legumes on land.
- Blue-green algae are often colonial, forming masses too large for zooplankton predators to handle.
- Others have a mucilaginous outer covering that is unpalatable to zooplankton.
- Some blue-greens produce neurotoxins or liver toxins when stressed (Kotak et al. 1994). They inhibit predation by zooplankton, but also have been implicated in numerous livestock and domestic animal deaths.

Table 2-2.—Algae are classified according to color. General characteristics of the more common divisions are given.

ALGAE GROUP	CHARACTERISTICS
Diatoms (*Bacillariophyta*)	Have cell walls made from silica, chloroplasts are yellow-green, often have early season dominance, cell walls may persist for thousands of years in sediments where they are useful for historic water quality studies (paleolimnology).
Green algae (*Chlorophyta*)	Chloroplasts are bright green, population growth often follows spring diatoms, important food for zooplankton, occasionally cause nuisance blooms especially when nitrogen concentrations in the lake are high.
Golden algae (*Chrysophyta*)	Optimize their growth during the relatively cool waters of late winter-early spring, taste and odor problems are often associated with golden algae, may supplement photosynthetic production by capturing food particles in the water.
Blue-green algae (*Cyanobacteria*)	Similar biologically to bacteria, no chloroplasts — color is distributed evenly throughout cells, often form filamentous or spherical colonies, are considered nuisance organisms, often dominate summer plankton in productive lakes, most can regulate buoyancy, many are nitrogen-fixers, some produce toxins.
Dinoflagellates (*Dinophyta*)	Free-swimming organisms having two flagella, have forward-spiraling swimming motion, blooms are associated with organic pollution, may cause taste and odor problems in water, marine dinoflagellates are responsible for "red tides."

Paleolimnology: the study of lake sediments and the relics preserved in them.

When there is enough light for photosynthesis, the availability of nutrients often controls phytoplankton productivity. The difference between how much of a particular element a photosynthetic organism needs and how much is available exerts the most significant influence on lake productivity. Table 2-3 compares the supply of essential nutrients relative to their demand. Phosphorus and nitrogen are the most used elements, and, therefore, they are the most likely to affect lake productivity.

Table 2-3.—The listed elements are required for plant growth. Plant demand is represented by the percentage of these essential elements in the living tissue of freshwater plants. Supply is represented by the proportions of these elements in world mean river water. The imbalance between demand and supply is an important factor in limiting plant growth (Valentyne, 1974).

ELEMENT	SYMBOL	DEMAND BY PLANTS (%)	SUPPLY IN WATER (%)	DEMAND SUPPLY RATIO[1]
Oxygen	O	80.5	89.0	1
Hydrogen	H	9.7	11.0	1
Carbon	C	6.5	.0012	5,000
Silicon	Si	1.3	.00065	2,000
NITROGEN	**N**	**.7**	**.000023**	**30,000**
Calcium	Ca	.4	.0015	< 1,000
Potassium	K	.3	.00023	1,300
PHOSPHORUS	**P**	**.08**	**.000001**	**80,000**
Magnesium	Mg	.07	.0004	< 1,000
Sulfur	S	.06	.06	.0004
Sodium	Na	.04	.0006	< 1,000
Iron	Fe	.02	.00007	< 1,000

[1] Percent of element in plant tissue ÷ percent in available water. The higher the ratio, the scarcer the nutrient. Phosphorus, in particular, is likely to limit plant growth in a lake. If more phosphorus is supplied, however, plant growth is likely to accelerate unless and until limited by some other factor.

Phosphorus, in particular, can severely affect a lake's biological productivity. Modern society produces rich sources of this element; wastewaters, fertilizers, agricultural drainage, detergents, and municipal sewage all contain high concentrations of phosphorus. If they enter the lake, they can stimulate algal productivity — and then you may have nuisance algal blooms, noxious tastes and odors, oxygen depletion in the water column, and undesirable fishkills during winter and summer.

Many lake management efforts target phosphorus when they're trying to reduce algal production and improve lake water quality. You can prevent phosphorus from entering a lake by controlling its sources; for example:

- Using chemical flocculation in advanced wastewater treatment plants;
- Using proper agricultural and land management practices;
- Improving septic systems; and
- Applying fertilizer carefully (see Chapter 6).

Over the past 20 years, in our attempt to curb eutrophication we have increasingly tried to minimize phosphorus inputs to lakes.

- Methods for precipitating or inactivating phosphorus within the lake are discussed in Chapter 7.
- A method for determining the amount of phosphorus loading from the watershed is discussed in Chapter 4.
- Formulas for predicting lake phosphorus concentrations are given in Chapter 5.

For infertile lakes with poor fishing, improving the fishery may be the highest priority. So the managers of those lakes may choose to add phosphorus- and nitrogen-containing fertilizers to enhance fish production — another lake management tool!

▼ ***Zooplankton.*** Freshwater lakes contain a richly diverse array of microscopic and macroscopic animals in the water column that float, drift, or swim weakly (i.e., they are at the mercy of currents). These animals are collectively known as zooplankton. The most significant groups of freshwater zooplankton are the cladocerans, copepods, protozoa, and rotifers. Zooplankton populations face the same challenges as all other animals — the need for food and oxygen, excretion of wastes, and reproduction. Although they are tiny, the relative abundance and diversity of these organisms dramatically influence energy flow, nutrient cycling, and community dynamics within aquatic ecosystems.

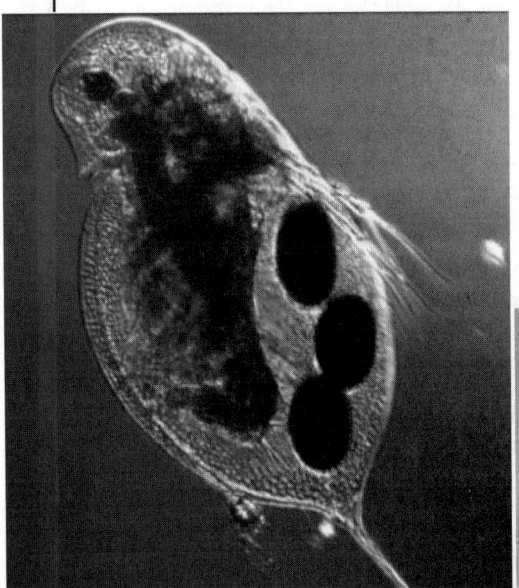

(Left)
Daphnia, a large (about 2 to 3 mm long) filter-feeding zooplankton. Note the three embryos in the brood pouch.

(Bottom)
Bosmina, a small (about 0.5 mm long) filter-feeding zooplankton.

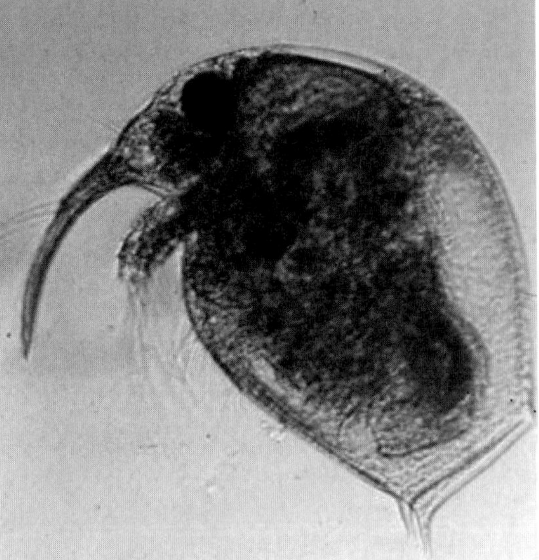

Planktivores: fish and invertebrates that feed on plankton.

Zooplankton are the primary consumers of algae. Some are grazers, or more specifically herbivores, since they consume plants. Many of the zooplankton are filter feeders. They filter large quantities of water and collect edible algae contained therein. Others are raptorial — they seize specific algal particles or other zooplankton. At the community level, zooplankton are of critical importance. They are the primary mechanism for transmitting the energy of the primary producers (phytoplankton) to organisms at higher trophic levels in the food chain. The young of many gamefish, minnows, and other fish such as shad feed on zooplankton. These fish are known collectively as planktivores.

▼ ***Macrophytes.*** Macrophytes are vascular plants — that is, they have conducting cells to transport nutrients and liquid through their stems. Aquatic macrophyte growth is controlled by many factors, including temperature, light, sediment texture, sediment slope, and hydrostatic pressure.

Most rooted macrophytes obtain their nutrients from the bottom sediments rather than the water and are restricted by light to the shallow littoral water. The littoral zone (Fig. 2-1) is defined as that area of the lake at and above the 1 percent light level. In other words, the littoral zone defines the area where there is enough light at the sediments to support rooted plant growth (Fig. 2-19).

Another aquatic macrophyte called duckweed is not rooted in the sediments but floats at the water surface. Duckweed floats with the wind or water currents and is not dependent on depth, sediment type, or water clarity (Borman et al. 1997). It is often associated with eutrophic waters and can multiply to large populations when adequate nutrients are present.

Too often, people refer to all rooted aquatic plants as weeds and their goal is to eradicate them all from the lake. This thinking is short-sighted and

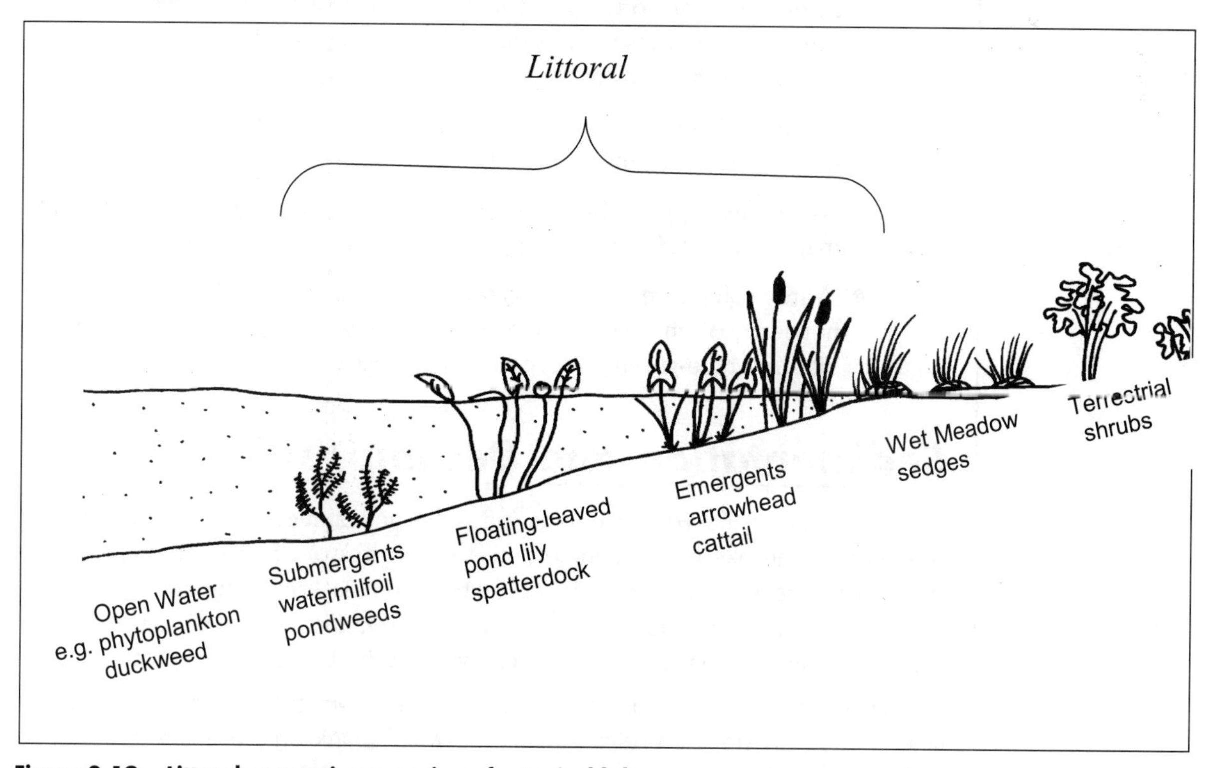

Figure 2-19.—Littoral vegetation zonation of a typical lake.

disregards the importance of macrophytes to the lake ecosystem. The rooted macrophytes with all their positive attributes make the littoral zone the most productive and important aquatic habitat in freshwater lakes.

- Aquatic plant communities protect the shoreline from erosion by dampening the force of waves and stabilizing soils.
- Rooted macrophytes also provide fish habitat and spawning sites, waterfowl cover and food, and habitat for macroinvertebrates (i.e., insects, snails, etc.) (Nichols and Vennie, 1991).
- Many species of macrophytes, such as the white water lily and pickerelweed, are aesthetically pleasing because they have beautiful flowers or interesting forms.

But even some of the most beautiful aquatic plants can be undesirable exotics (non-natives). These include species such as Eurasian watermilfoil, curly-leaf pondweed, water hyacinth, hydrilla, and purple loosestrife. These plants can invade lakes and rapidly crowd out desirable native plants. Lake managers should be able to identify these species and should develop a plan to manage them if they are found in a lake.

Therefore, producing stable, diverse aquatic plant communities containing high percentages of desirable species should be a primary lake management goal from an economic and environmental perspective.

Phytoplankton Community Succession

As the growing season proceeds, a succession of algal communities typically occurs in a lake (Fig. 2-20).

- Phytoplankton biomass tends to be low in the early spring because of cool water temperatures and low light levels. Diatoms and sometimes golden algae are often the dominant algal group at this time as they can withstand these conditions and silica is more available following spring (and fall) turnover.
- In late spring to early summer, algal biomass increases, and green algae may replace diatoms.
- As summer approaches and the lake water warms up, blue-green algae may dominate, forming dense growths in productive lakes.
- Water column mixing during fall overturn increases the supply of nutrients, often resulting in a late season, but short-lived, bloom of diatoms, blue-green algae, or dinoflagellates.

Sedimentation and Decomposition

Sedimentation occurs when particles (silt, algae, animal feces, and dead organisms) sink through the lake water column onto the lake bottom. Sedimentation is a very important process that affects phytoplankton biomass levels, phytoplankton community succession, and transfers of organic matter, nutrients, and particle-associated contaminants from the lake's upper layers to the bottom sediments.

Sedimentation of particulate organic matter from the water column to the lake bottom provides a critical linkage between planktonic primary production and the growth of bottom-dwelling organisms (such as aquatic insect larvae,

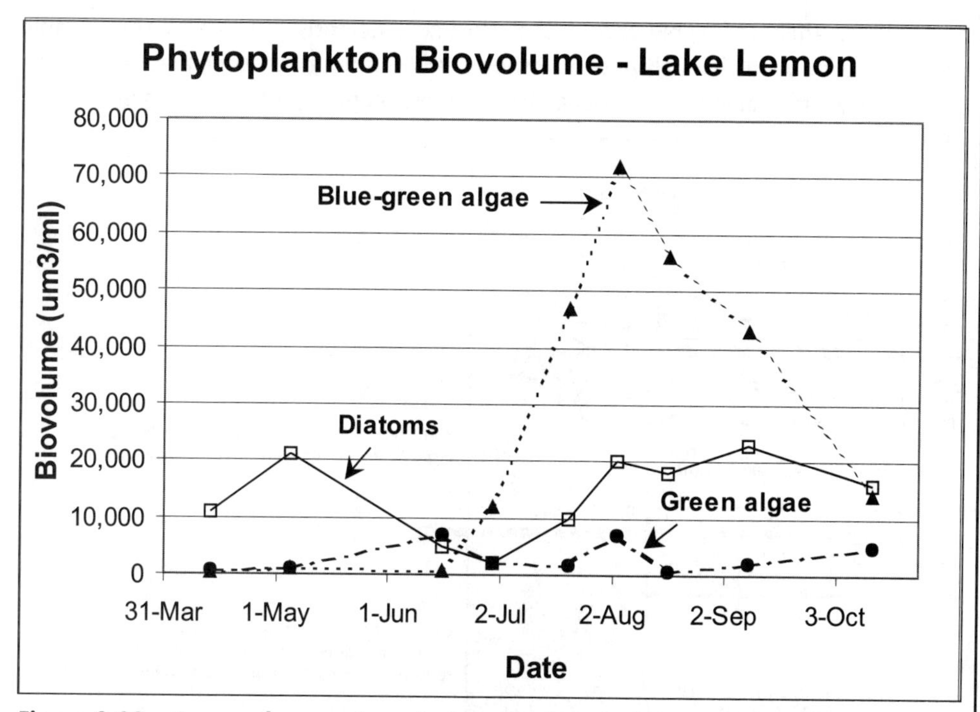

Figure 2-20.—Seasonal succession of phytoplankton in Lake Lemon, Indiana, a productive 1,600-acre reservoir (from Zorgorski and Jones, 1986).

clams, and crayfish) that eat this detrital organic matter and, in turn, are eaten by larger predatory organisms, such as fish and turtles. The sedimentation of living and dead plankton down through the water column is so pervasive that it is referred to as plankton rain.

Oxygen-consuming decomposition processes degrade settling plankton, zooplankton feces, and other organic detritus particles in the water column and bottom sediments. The respiration of all organisms (including bacteria, fungi, and other microbes) drives organic matter decomposition, a collective term for the net conversion of organic material back to inorganic compounds (see Fig. 2-18).

Plankton rain: the almost constant settling of plankton, live and dead, through the water to the bottom sediments.

In the hypolimnion of productive lakes, sedimentation of organic matter from the surface waters is extensive. And because there's so much algae and other suspended particles, light penetrates the water column to the hypolimnion only minimally (or not at all) so photosynthesis cannot occur. Under these conditions, the oxygen consumed in the hypolimnion and bottom sediments during decomposition (respiration) of this organic matter greatly exceeds the oxygen produced. Also, as described earlier, the hypolimnion is isolated from the atmosphere by a water density barrier known as the metalimnion. The result, in productive thermally stratified lakes, is a depletion — even complete absence — of dissolved oxygen in the hypolimnion. Shallow productive lakes with a prolonged snow and ice cover can suffer the same effects.

Chemical oxygen demand (COD): Nonbiological uptake of molecular oxygen by organic and inorganic compounds in water.

Marked chemical and physical changes are associated with oxygen depletion. They include:

- Increased release of nutrients (phosphorus and ammonia), certain metals (iron and manganese), and compounds (hydrogen sulfide and methane) from the anoxic bottom sediments;
- Destruction of oxygenated habitats for aquatic animals; and
- Incomplete decomposition of sedimented organic matter (Fig. 2-21).

The inability of highly productive lakes to efficiently process all the organic material produced through photosynthesis led limnologist Arthur Hasler to refer to these overfed and obese lakes as being "physiologically senile" (Hasler, 1969).

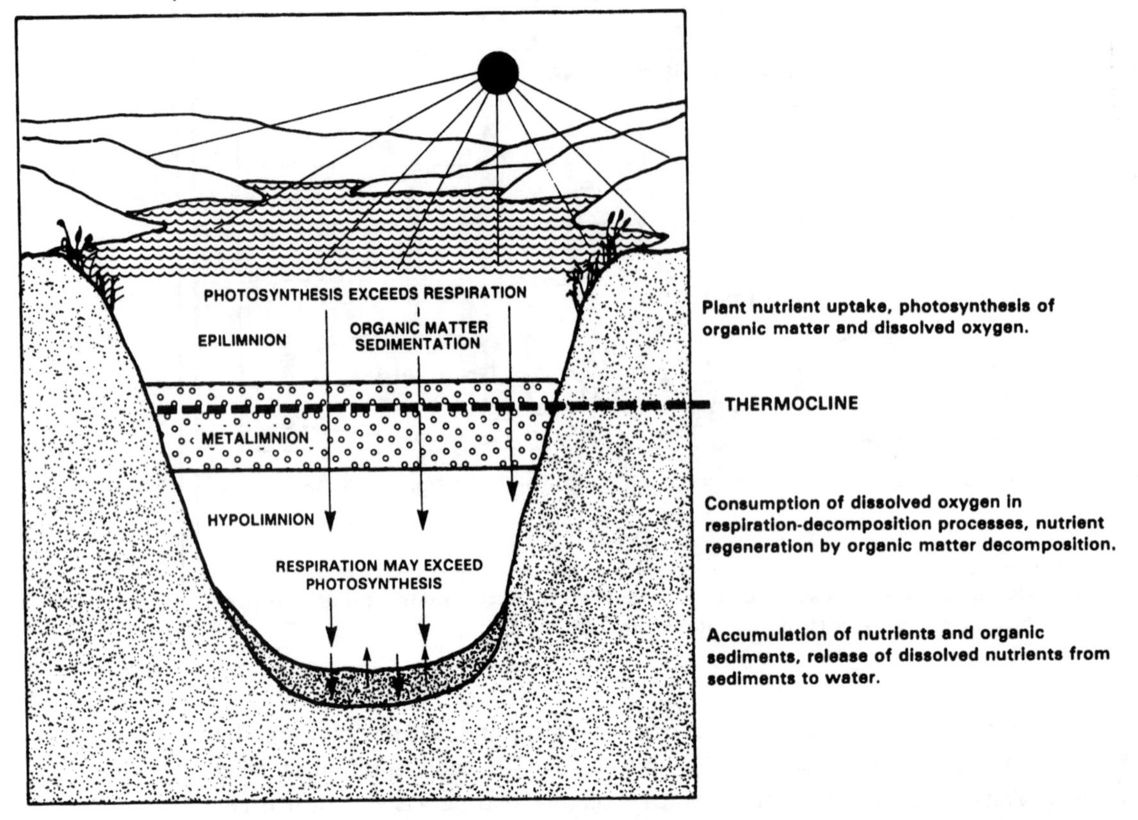

Figure 2-21.—Influence of photosynthesis and respiration-decomposition processes and organic matter sedimentation on the distribution of nutrients, organic matter, and dissolved oxygen in a stratified lake. Source: Moore and Thornton, 1988.

Food Web Structure, Energy Flow, and Nutrient Cycling

In-lake plant production usually forms the organic matter base of the lake's food web. Although some water bodies (especially rapidly flushed reservoirs) receive important supplements of organic matter from river and stream inflow, most lakes require a reliable level of algal and macrophyte production to maintain productive food webs (Adams et al. 1983).

Some of the organic matter produced photosynthetically by the lake's primary producers (algae and macrophytes) is consumed by herbivores (grazers) that range from tiny zooplankton to snails to grazing minnows.

Herbivores, such as the zooplankton, are fed on by planktivores (including predatory zooplankton, aquatic insects, and planktivorous fish) that, in turn, provide a food source for the higher-level consumers such as piscivorous (fish-eating) fish (bass, walleye, pike) and fish-eating birds (kingfishers, herons, ospreys, eagles).

This general progression of feeding levels (also called trophic levels) from primary producers, to herbivores, to planktivores, to increasingly larger preda-

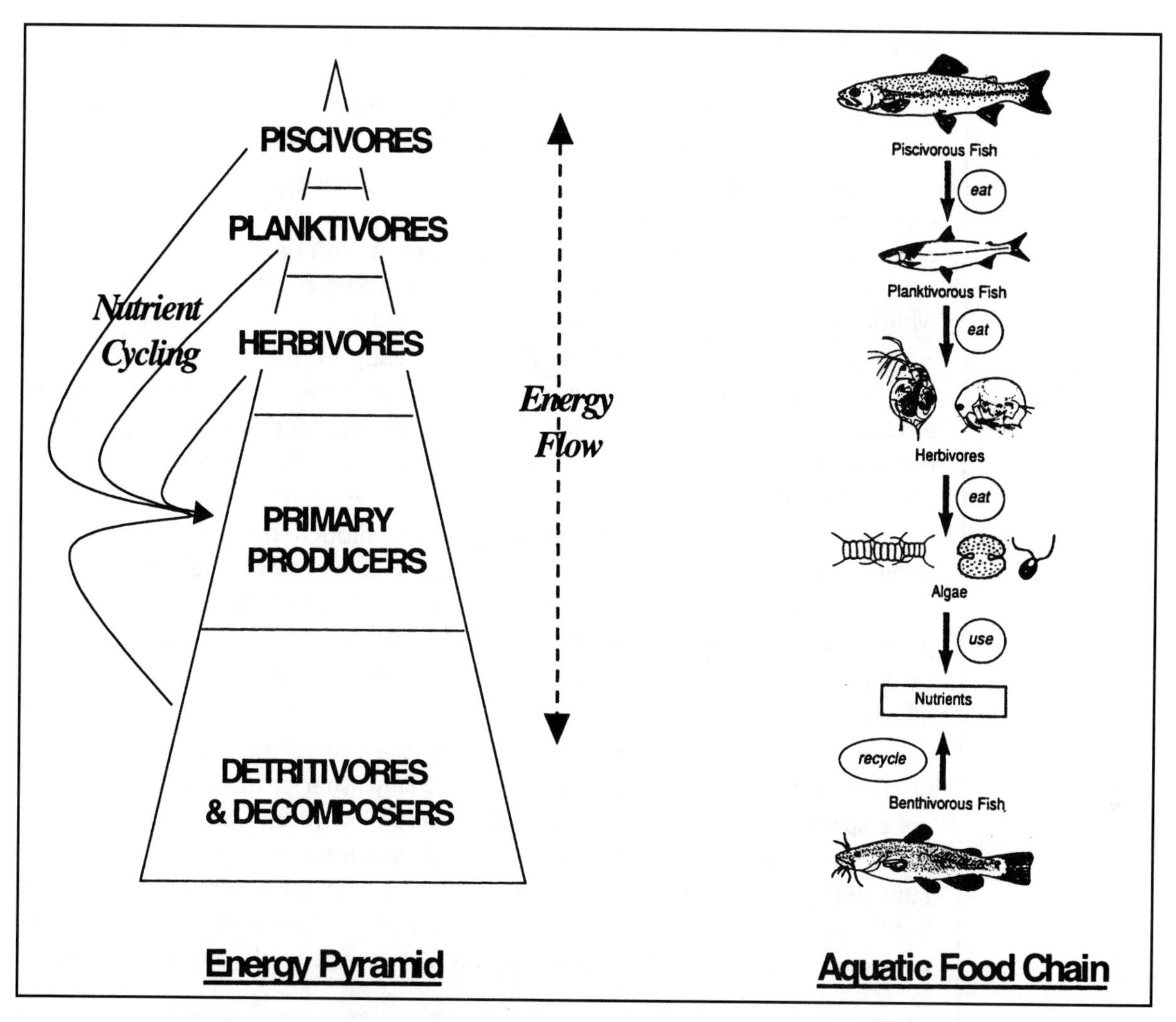

Figure 2-22.—The food-chain concept refers to the progression of feeding (or trophic) levels from primary producers, to herbivores, to higher predators. As shown, this process involves both the transfer of energy among lake organisms and the recycling of nutrients. Because the available energy decreases at each trophic level, a large food base of primary producers, herbivores, and planktivores is required to support a few large game fish.

tors, constitutes a food chain (Fig. 2-22). The actual complex of feeding interactions among all the lake's organisms is called a food web.

As shown in Fig. 2-22, the food chain concept also involves the flow of energy among the lake organisms and the recycling of nutrients.

The energy flow begins with light energy from the sun, which is converted by green plant photosynthesis into the chemical energy represented by the organic matter produced by plants. Each subsequent consumer (herbivore, planktivore, piscivore) transfers only a fraction (usually about 10 to 20 percent) of the energy received on up the chain to the next trophic level (Kozlovsky, 1968; Gulland, 1970).

Because of this inefficient energy transfer, a few large game fish depend on a large supply of smaller fish, which depend on a very large supply of smaller herbivores, which depend on a successively much larger base of photosynthetic production by phytoplankton and other aquatic plants.

Finally, by constantly producing wastes and eventually dying, all of these organisms nourish detritivores (detritus-eating organisms), bacteria, and fungi, which derive their energy by decomposing organic matter. Organic matter decomposition recycles nutrients required to produce plants. This nutrient regeneration occurs at virtually every level of the food web, and only a small fraction of the organic matter produced ultimately accumulates as permanent bottom sediment.

In an actual food web within a lake, many different types of organisms live at each trophic level, directly affecting their "neighbors" and indirectly influencing other populations with which they interact. Species with significant influence in the food web are called keystone species; they cannot be removed from the system without causing a dramatic change in the community.

For example, piscivorous fish (bass, pike, etc.) are frequently keystone species in lakes. They eat planktivorous fish that, in turn, eat zooplankton. If few piscivores exist, either because of overfishing or degraded habitat, planktivorous fish will increase, eliminating the large herbivorous zooplankton that eat algae. Thus, algal populations increase and transparency decreases.

Figure 2-23 shows a more complex view of biotic interactions and nutrient cycling in a lake or reservoir ecosystem. Biota in one trophic level affect those in other trophic levels in many ways — directly through predation or changes in the physical or chemical characteristics of the water, or indirectly through nutrient cycling pathways. Understanding the complexity of these pathways is essential for effective lake management.

In addition to their "keystone" role at the top of the aquatic food chain, piscivorous fish are also valued as sport fish. Many fisheries managers try to maximize sport fish biomass to make anglers happy. However, optimal warm water sport fish yields require higher nutrient levels and more overall lake productivity (and thus poorer transparency) than may be desirable for other lake uses (Fig. 2-24). This situation becomes more problematic when both warmwater and coldwater fisheries are desired as increasing nutrient levels can have significant

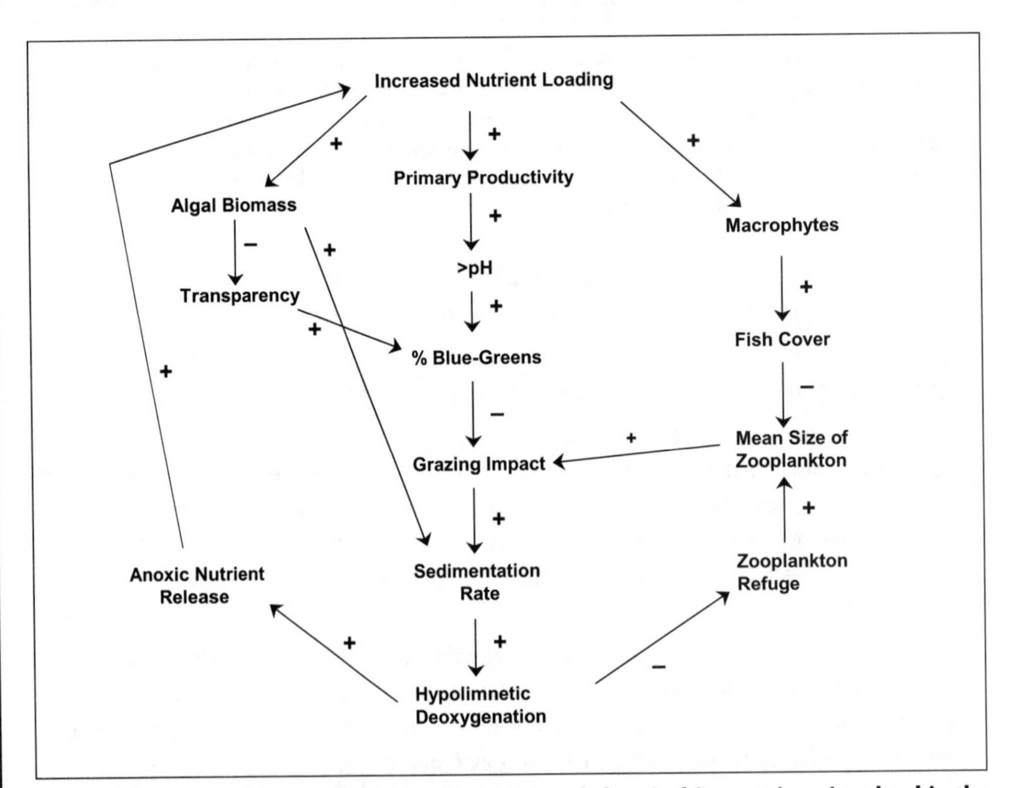

Figure 2-23.—A conceptual model of the biotic and chemical interactions involved in the food web and their influence on eutrophication. A plus (+) indicates a positive influence; a minus (−) indicates a negative influence.

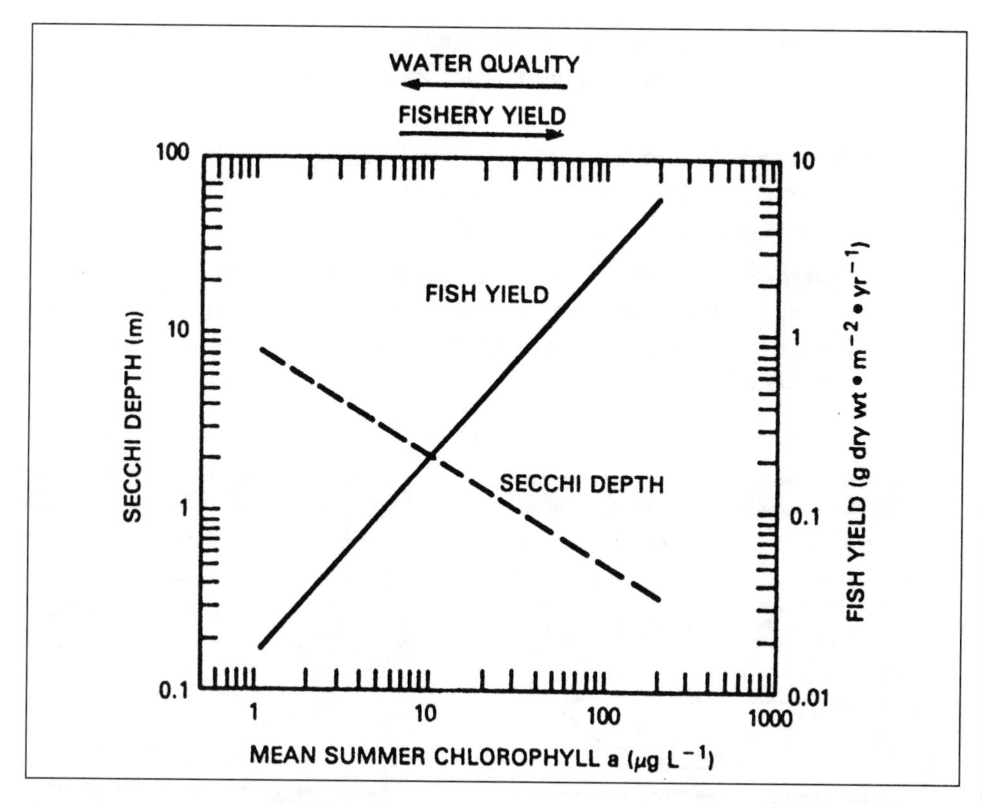

Figure 2-24.—Relationship between lake characteristics (e.g., Secchi depth, chlorophyll) and management objectives (e.g., water quality, fishery yield). Arrows denote decreased clarity and increased fish yield. Modified from Wagner and Oglesby (1984).

impacts on the coldwater fisheries (see Chapter 5 and Fig. 5-9 for additional discussion).

For example, a lake can become eutrophic at total phosphorus concentrations of 40 µg/L, but Ney (1996) found that sport fish biomass in reservoirs probably does not peak at less than 100 µg/L, The upshot of this is that your lake cannot necessarily provide both clear water quality and abundant fish populations at the same time. Consequently, you must clearly establish your lake management priorities to avoid conflicts among lake uses.

Eutrophication

Lakes are temporary (in a geologic sense) features of the landscape. The Great Lakes, for example, have existed in their current shapes for only about 12,000 years. Over tens to many thousands of years, lake basins change in size and depth as a result of climate, movements in the earth's crust, shoreline erosion, and the accumulation of sediment. Eutrophication is the term used to describe this process.

Eutrophication is defined as the excessive addition of inorganic nutrients, organic matter, and silt to lakes, thereby increasing biological productivity (Cooke et al. 1993). The key part of this definition is the biological response of the lake to these additions; high phosphorus concentrations alone do not determine eutrophication.

The classical lake succession sequence (Fig. 2-25) is usually depicted as a unidirectional progression through the following series of phases (trophic states) having these general characteristics:

- **Oligotrophy:** lack of plant nutrients keeps productivity low; lake contains oxygen at all depths; clear water; deeper lakes can support trout.
- **Mesotrophy:** moderate plant productivity; hypolimnion may lack oxygen in summer; moderately clear water; warmwater fisheries only — bass and perch may dominate.
- **Eutrophy:** contains excess nutrients; blue-green algae dominate during summer; algae scums are probable at times; hypolimnion lacks oxygen in summer; poor transparency; rooted macrophyte problems may be evident.

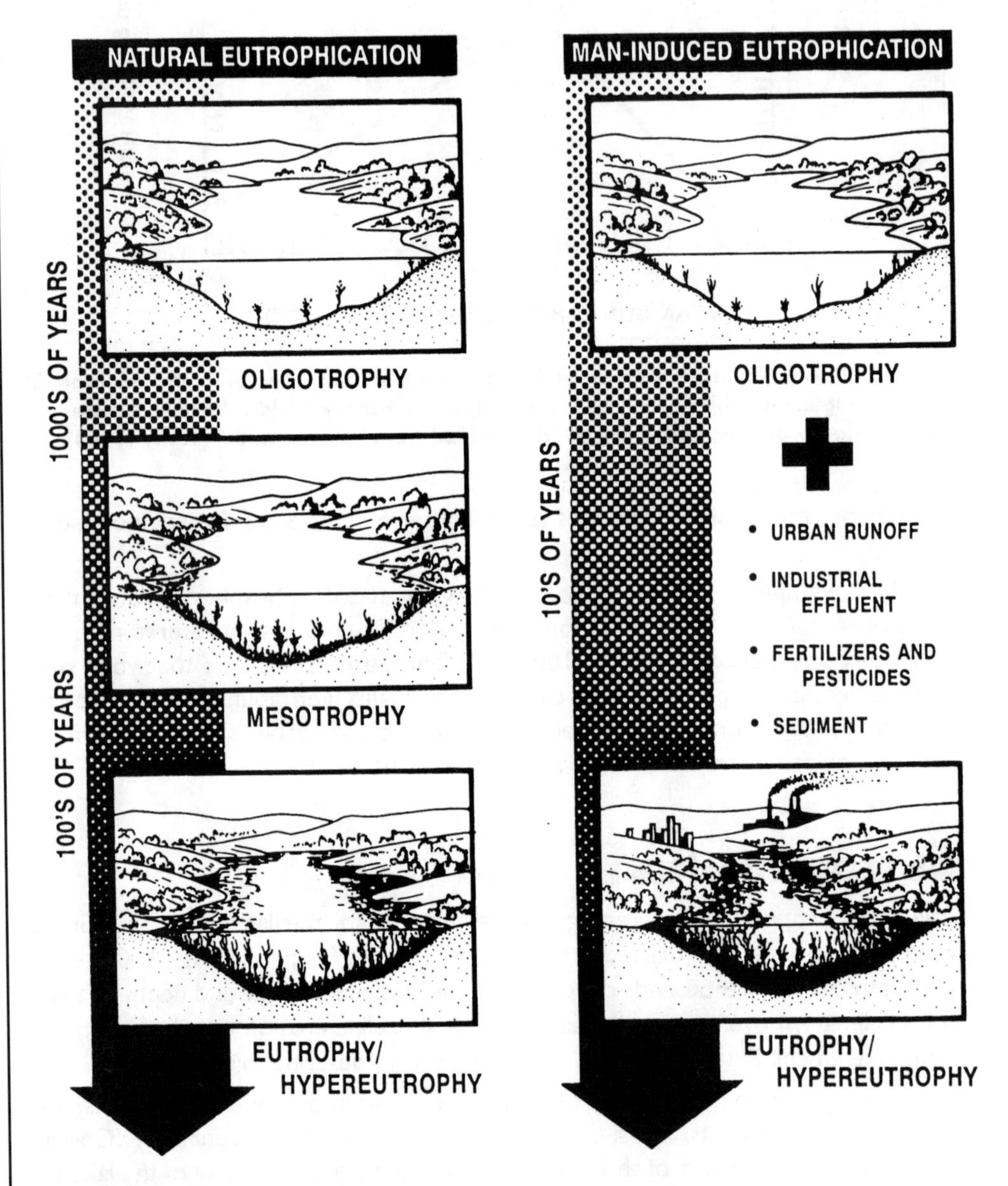

Figure 2-25.—

(left column) **The progression of natural lake eutrophication through nutrient-poor (oligotrophy) to nutrient-rich (eutrophy) stages. The diagram depicts the natural process of gradual nutrient enrichment and basin filling over a long period of time (e.g., thousands of years).**

(right column) **Human-induced or cultural eutrophication in which this process is greatly accelerated by increased inputs of nutrients and sediments into a lake, as a result of watershed disturbance by humans.**

- **Hypereutrophy:** algal scums dominate in summer; few macrophytes; no oxygen in hypolimnion; fishkills possible in summer and under winter ice.

These lake trophic states correspond to gradual increases in lake productivity from oligotrophy to eutrophy.

Evidence obtained from sediment cores, however, indicates that changes in lake trophic status are not necessarily gradual or unidirectional. If their watersheds remain relatively undisturbed, lakes can retain the same trophic status for many thousands of years. For example, the Italian lake, Lago di Monterosi, remained oligotrophic for 22,000 years until the Romans built a road through its watershed in 171 B.C. This disturbance boosted productivity in the lake to eutrophic proportions (Frey, 1980). Productivity subsequently declined as the watershed restabilized and delivered less nutrients to the lake (Fig. 2-26).

Lessons to learn from this and other examples like it are that eutrophication is not necessarily a natural process, and when it does happen, it is often reversible. In contrast, rapid changes in lake nutrient status and productivity do result from human disturbances to the watershed, rather than gradual enrichment and filling of the lake basin through natural means.

People cause cultural eutrophication when their use of the land dramatically increases nutrient, soil, or organic matter loads to the lake. They can drastically shorten a lake's lifespan by clearing forests, building roads, cultivating fields, developing residential communities, and discharging wastewater — all these activities increase the nutrients and soil that eventually move into the lake (see Chapter 6's sections on nonpoint and cultural sources).

Some lakes, however, are naturally eutrophic. In fact, if they lie in naturally fertile watersheds, they have little chance of being anything other than eutrophic. Unless some other factor such as higher turbidity or an increase in the hydraulic flushing rate intervenes, these lakes will naturally have very high rates of primary production.

Natural and constructed lakes age by the same processes — nutrient enrichment and basin filling — but at very different rates. As a rule, reservoirs become eutrophic more rapidly than natural lakes, because most reservoirs receive higher sediment and nutrient loads than do most natural lakes. They may even be eutrophic when initially filled. Reservoirs, especially those with hypolimnetic outlets, trap sediments more efficiently than they retain nutrients; therefore, they age primarily by the filling of their basins with riverborne silts and clays.

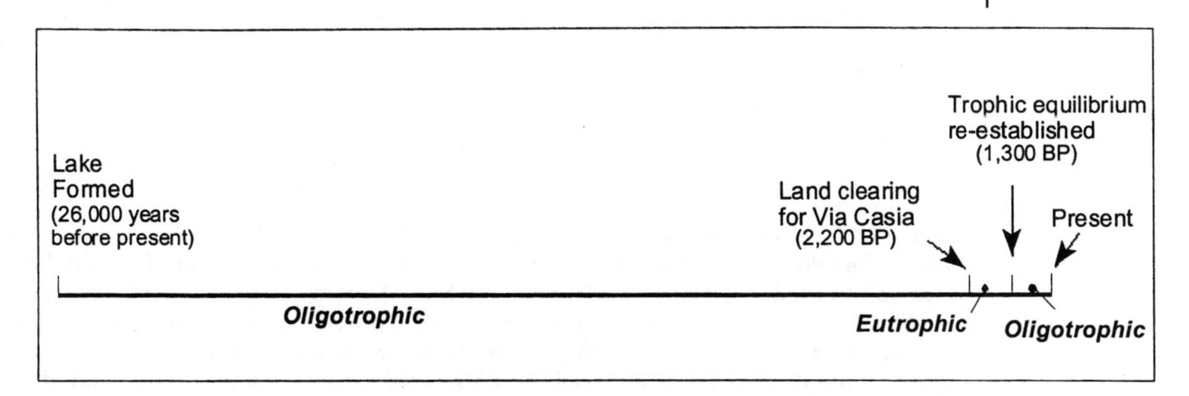

Figure 2-26.—Trophic state timeline for Lago di Monterosi, Italy. This lake remained oligotrophic for over 20,000 years. Following a brief period of eutrophy, caused by human land use, the lake returned to oligotrophy naturally.

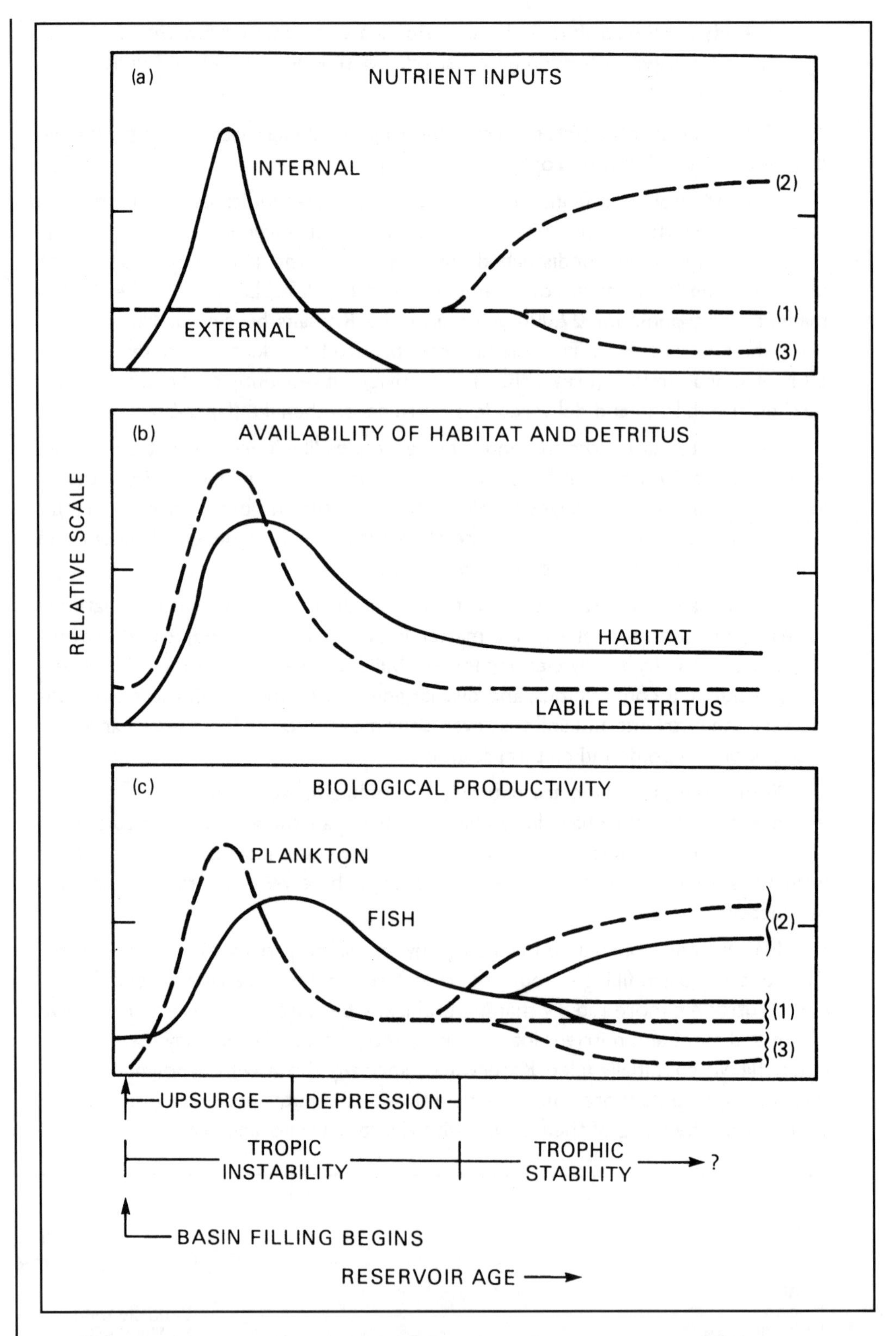

Figure 2-27.—Factors influencing biological productivity or ±trophic progression≤in a reservoir in the initial years after impoundment: (a) internal nutrient loading from the flooded reservoir basin and external nutrient loading from the watershed, (b) availability of habitat (flooded vegetation) and unstable terrestrial detritus supporting macroinvertebrates and fish, and (c) plankton and fish production. The initial period of trophic instability (i.e., upsurge and depression) is followed by a less productive, but more stable, period in the maturing reservoir (1). However, disturbances or land-use changes in the watershed can result in increases (2) or decreases (3) in external nutrient loading and, consequently, in reservoir productivity. Modified from Kimmel and Groeger (1986).

However, unlike natural lakes, reservoirs often do not go through the classical trophic progression from oligotrophy to eutrophy. In fact, newly filled impoundments usually go through a relatively short period of trophic instability in which a highly productive period (the "trophic upsurge") is followed by a decline in lake productivity (the "trophic depression"), eventually reaching a less productive but more stable trophic state (Fig. 2-27).

The trophic upsurge results largely from nutrient inputs from both external sources (the watershed) and internal sources (leaching of nutrients from the flooded soils of the reservoir basin and decomposition of terrestrial vegetation and litter).

The trophic depression is, in fact, the system's first step toward its natural productivity level dictated by the level of external nutrient inputs. Flooding of soils, vegetation, and litter as the new reservoir fills contributes to both abundant food and expanding habitat. As the reservoir matures, both food and habitat resources decline, fish production decreases, and the fish community stabilizes.

This trophic upsurge and depression — the "boom and bust" period — inevitably raises simultaneous concerns about poor water quality and false hopes for a higher level of fishery yield than can be sustained over the long term. Ultimately, in reservoirs and in natural lakes, the nature of the watershed (and what people do there) will determine the water quality, biological productivity, and trophic status of the system.

Finally, we must remember that while eutrophication is an important ecological process in lakes, it doesn't cause all lake ills. High and low water levels, acid precipitation, mercury, pesticides, and other toxins are among the many other problems that may affect lakes (see Chapter 4). Keep these factors in perspective as you assess lake conditions and develop management plans.

Ecology's Place in Lake Protection, Restoration, and Management

We must establish the ecological basis for lake management if we are to succeed in managing and maintaining sustainable lake ecosystems. As Osgood (1996) said so well, "ecologically-based lake management is an active process that identifies and engages the lake community in meaningful, measurable, and sustainable actions that seek to attain a desired condition which, at a minimum, stabilizes the lake ecosystem."

The understanding gained from this chapter will help you evaluate the potential benefits and limitations on lake protection and restoration approaches and techniques described in the rest of this manual.

Most of what we know about lake and reservoir management has been learned in the last 30 years through experience gained from many studies conducted throughout the world. Experience gained from previous management efforts leads to the following conclusions:

1. There is no panacea for lake management or restoration problems; different situations require different approaches and solutions.

2. A complex set of physical, chemical, and biological factors influences lake ecosystems and affects their responsiveness to restoration and management efforts.

3. Because of the tight coupling between lakes and their watersheds, good conservation practices in the watershed are essential for improving and protecting lake water quality.

4. Internal nutrient loading and recycling may continue to hamper lake recovery even after watershed sources are controlled. The relative importance of external versus internal nutrient sources must be evaluated by a diagnostic lake study.

5. The physical, chemical, and biological components of lake ecosystems are intricately linked. Actions targeting one component will likely affect another. For example, lake renovation to enhance water quality by limiting nutrients and thereby reducing algal production will also decrease fish production. You must decide what you want to accomplish and set priorities.

6. To be sustainable, lake renovation and management objectives must consider the uses that the natural condition of the lake (and its watershed) can support most readily.

In summary, the character of a lake or reservoir is determined by a complex set of physical, chemical, and biological factors that vary with lake origin, its regional setting, and the nature of the watershed. Important factors include hydrology, climate, watershed geology, watershed to lake ratio, soil fertility, hydraulic residence time, lake basin shape, external and internal nutrient loading rates, presence or absence of thermal stratification, lake habitats, and lake biota.

In some situations, a natural combination of these factors may dictate that a lake will be highly productive (eutrophic) and efforts to transform it to an unproductive, clear-water (oligotrophic) state would be ill-advised. However, if a lake has become eutrophic or has developed other water quality problems as a result of, for example, increased nutrient loading from the watershed, then this condition can be reversed and the lake improved by an appropriate combination of watershed and in-lake management. But the best thing to do is to take steps to protect your lake and its watershed before problems develop.

The following chapters will discuss and compare a variety of lake and watershed management techniques. While reading through this information, remember that the effectiveness of any lake management method or combination of methods will depend entirely on the ecological soundness of its application.

References

Adams, S.M., B.L. Kimmel, and G.R. Ploskey. 1983. Organic matter sources for reservoir fish production: A trophic-dynamics analysis. Can. J. Fish Aquat. Sci. 40:1480-95.

Borman, S., R. Korth, and J. Temte. 1997. Through the Looking Glass — A Field Guide to Aquatic Plants. Reindl Printing, Inc., Merrill, WI.

Cooke, G.D. and R. H Kennedy. 1989. Water Quality Management for Reservoirs and Tailwaters. Report 1: *in* Reservoir Water Quality Management Techniques. Tech. Rep. E-89-1. U.S. Army Corps Eng., Washington, DC.

Cooke, G.D., E.B. Welch, S.A. Peterson, and P.R. Newroth. 1993. Restoration and Management of Lakes and Reservoirs. 2nd ed. Lewis Publishers, Boca Raton, FL.

Czuczwa, J.M., B.D. McVeety, and R.A. Hites. 1984. Polychlorinated dibenzo-p-dioxins and dibenzofurans in sediments from Siskiwit Lake, Isle Royale. Science 226:568-9.

Echelberger, W.F., Jr. and W.W. Jones. 1984. Cedar Lake Restoration Feasibility Study. ESAC-84-01. School Publ. Environ. Affairs, Indiana Univ., Bloomington.

Frey, D.G. 1980. The heritage of our lakes. The Key Reporter 45(4):2-4.

Gulland, J.A. 1970. Food chain studies and some problems in world fisheries. Pages 296-315 *in* J.H. Steele, ed. Marine Food Chains. Univ. Calif. Press, Los Angeles.

Hanlon, C.G. 1999. Relationships between total phosphorus concentrations, sampling frequency, and wind velocity in a shallow, polymictic lake. Lake Reservoir Manage. 15(1):39-46

Hasler, A.D. 1969. Cultural eutrophication is reversible. BioScience 19(5):425-31.

Horwitz, E.L. 1980. Our Nation's Lakes. EPA 440/5-80-009. U.S. Environ. Prot. Agency, Washington, DC.

Hutchinson, G.E. 1957. A Treatise on Limnology. John Wiley and Sons, New York.

Kimmel, B.L. and A.W. Groeger. 1984. Factors controlling primary production in lakes and reservoirs: a perspective. Pages 277-281 *in* Lake and Reservoir Management. N. Am. Lake Manage. Soc. and U.S. Environ. Prot. Agency. EPA 440/5/84-001. Washington, DC.

———. 1986. Limnological and ecological changes associated with reservoir aging. Pages 103-109 *in* G.E. Hall and M.J. Van Den Avyle, eds. Reservoir Fisheries Management: Strategies for the 80s. Reservoir Comm., Am. Fish. Soc., Bethesda, MD.

Kotak, B.G., E.E. Prepas, and S.E. Hrudey. 1994. Blue-green algal toxins in drinking water supplies — Research in Alberta. Lake Line 14(1):37-40.

Kortmann, R.W. 1991. Phosphorus loading — look to the sky. Lake Line 11(4):19-20.

Kozlovsky, D.G. 1968. A critical evaluation of the trophic level concept. I. Ecological efficiencies. Ecology 49: 48-60.

LaBounty, J.F. 2000. Pers. Commun. Southern Nevada Water Authority, Las Vegas.

Marsh, W.M. and T.E. Borton. 1975. Inland Lake Watershed Analysis — A Planning and Management Approach. Inland Lake Manage. Unit, Mich. Dep. Nat. Resour., Lansing.

Moore, L. and K. Thornton, eds. 1988. Lake and Reservoir Restoration Guidance Manual. EPA 440/5-88-002. Prep. by N. Am. Lake Manage. Soc. for Office of Res. and Devel., Envir. Res. Lab., Corvallis, OR, and Office of Water, Criteria Stand. Div., Nonpoint Sources Branch, U.S. Environ. Prot. Agency, Washington, DC.

Ney, J.J. 1996. Oligotrophication and its discontents: effects of reduced nutrient loading on reservoir fisheries. Am. Fish. Soc. Symp. 16:285-95.

Nichols, S.A. and J.G. Vennie. 1991. Attributes of Wisconsin Lake Plants. Inf. Circ.73. Wis. Geolog. Nat. Hist. Surv., Madison.

Novotny, V. and H. Olem. 1994. Water Quality — Prevention, Identification, and Management of Diffuse Pollution. Van Nostrand Reinhold, New York.

Osgood, D. 1996. The ecological basis for lake and reservoir management. Lake Line 16(2):18-19,30-32.

Reckhow, K.H., M.N. Beaulac, and J.T. Simpson. 1980. Modeling Phosphorus Loading and Lake Response Under Uncertainty: A Manual and Compilation of Export

Coefficients. EPA 440/5-80-011. Off. Water Reg. Stand. U.S. Environ. Prot. Agency, Washington, DC.

Simonich, S.L. and R.A. Hites. 1995. Global distribution of persistent organochlorine compounds. Science 269:1851-54.

Sonzogni, W.C. et al. 1980. Pollution from land runoff. Environ. Sci. Technol. 14(2):148-53.

St. Amand, A. 1995. Algae — nature's artwork. Lake Line, 15(3): 10-11, 24-26.

U.S. Environmental Protection Agency. 2000. Update: National Listing of Fish and Wildlife Advisories. EPA-823-F-00-016. Washington, DC.

Valentyne, J.R. 1974. The Algal Bowl: Lakes and Man. Misc. Special Publ. 22. Dep. Environ. Fish. Res. Board Canada, Ottawa.

Wagner, K.J. and R.T. Oglesby. 1984. Incompatibility of common lake management objectives. Pages 97-100 *in* Lake Reservoir Management. EPA 440/5-84-001. U.S. Environ. Prot. Agency, Washington, DC.

Yousef, Y. et al. 1978. Mixing effects due to boating activities in shallow lakes. Draft Rep. OWRT, U.S. Dep. Int. Technol. Rep. ESEI 78-10, Washington, DC.

Zorgorski, J.S. and W.W. Jones. 1986. Lake Lemon Diagnostic/Feasibility Study. ESAC-86-02. School Publ. Environ. Affairs, Indiana Univ., Bloomington.

CHAPTER 3

Planning: From Goals to Evaluation and Around Again

Planning is the opposite of fate. If we don't accept *c'est la vie* (that's life — that's how things happen), then we believe in planning. We believe in systematically charting a course for the future to minimize conflict and maximize the attainment of our goals.

Without a plan, year-to-year decisions are often inconsistent and management is invariably inefficient. To effectively manage your lake, you must have a systematic plan. And, you must also manage your watershed. Examples can be found in *Watershed Protection: A Project Focus* (U.S. EPA, 1995).

Planning is conducted formally or informally in all realms of life. While the specifics vary, the planning process generally conforms to the model shown in Figure 3-1.

With this common sense approach, most citizens will readily accept the process if they are involved from the beginning. In contrast, presenting them with a plan prepared by professionals violates the basic premise of citizen participation and seldom results in community ownership of the plan.

Citizens must be involved in the whole process because that's when various — even contradictory approaches — can be aired and eventually reconciled into

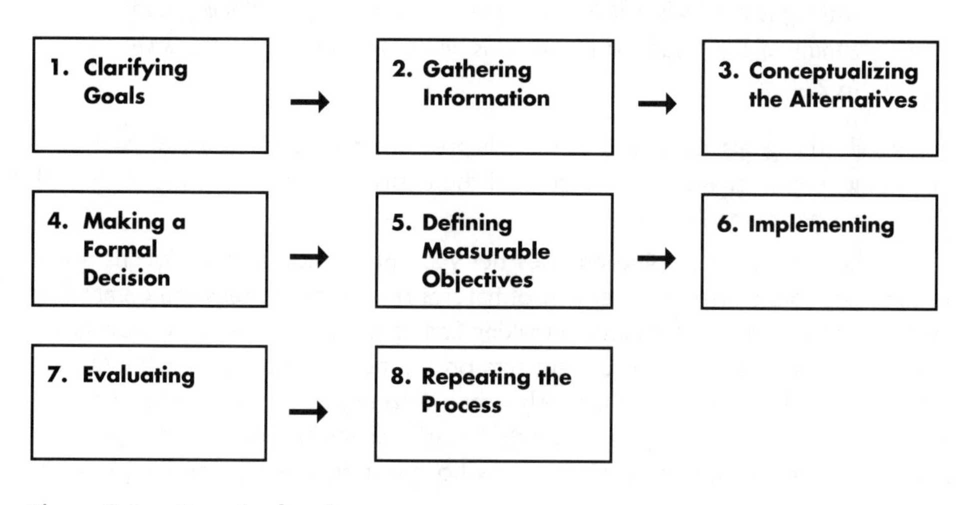

Figure 3-1.— Generic planning process.

a plan that can work for the entire community. Otherwise, some citizens will not only disagree with elements of the plan, but more importantly, will feel alienated from the whole process of decision-making.

STEP 1: Clarifying Goals

Why We Care About a Lake

A goal is a desired state of affairs that is sufficiently broad and multifaceted to insure unanimity.

Whose Goals?

The answer to this question depends on our definition of community. Some definitions are very narrow: one subdivision or just riparian owners. Broader definitions would include the people living around the lake or even beyond.

Aldo Leopold suggested in 1949 that we go beyond the local area, beyond our borders, beyond our time, and beyond our species to include future generations, the biotic community of plants and animals, and even the inorganic elements (air, water, and soil) upon which all life depends. His Land Ethic embraced the concept of Homo sapiens as part of a larger community rather than as masters of the universe.

> *We abuse land because we regard it as a commodity belonging to us. When we see land as a community to which we belong we may begin to use it with love and respect (p. viii).*
>
> *All ethics so far evolved rest on a single premise: that the individual is a member of a community of interdependent parts. His instincts prompt him to compete for a place in the community, but his ethics prompt him also to cooperate (perhaps in order that there may be a place to compete for) (p. 203).*
>
> *The Land Ethic simply enlarges the boundaries of the community to include soils, water, plants, and animals or collectively: the land (p. 206).*
>
> *A thing is right when it tends to preserve the integrity, stability, and beauty of the biotic community. It is wrong when it tends otherwise (p. 224).*

While the goals discussed in this chapter are the social goals of lakeshore communities, concern for the integrity of the entire ecosystem must underpin all efforts to meet human needs.

While the goals discussed in this chapter are the social goals of lakeshore communities, concern for the integrity of the entire ecosystem must underpin all efforts to meet human needs.

"Whose goals" also differentiates between professionals and citizens. After decades of professional management of natural resources, citizens are demanding a voice in setting the agenda and in making final decisions. Some professionals are comfortable with the notion of "experts on tap, not on top"; others find it difficult to be "public servants" to a largely urban citizenry rather than resource managers. Their training to apply the single "best" technical solution to a problem often makes them intolerant of citizens who question their "scientific prescriptions."

But an increasingly sophisticated citizenry, with access to Internet information and to other citizens dealing with similar situations, will no longer blindly follow the recommendations of well-meaning, competent professionals. Increasingly, the term "partnership" is being used to describe a new shared responsibility. Citizens now determine the goals for their lake (within its legal and limnological limits), decide which techniques to use, and help make it happen. They gather information, raise funds, monitor water quality, and volunteer their physical labor on lake projects.

Primary Social Needs

In 1974 the Technical Committee of the Water Resource Centers of the Western States focused on a very broad question: "How does water contribute to society?" They decided to first determine society's primary social needs and then relate water to each need.

Twenty years later, research at the University of Wisconsin-Stevens Point expanded the Committee's original list to the following conditions that must be present in a balanced social structure, as shown in Table 3-1.

Table 3-1.—Conditions that must be present in a balanced social structure.

Condition	Examples
Aesthetic Opportunity	Natural beauty/landscapes Attractive buildings/cities Soothing interior spaces
Collective Security	International peace National defense Protection from internal disorder
Cultural Opportunity	Music Art Heritage
Economic Opportunity	Good wages/job security Return on investment Efficient production
Educational Opportunity	Formal schools Continuing education Wisdom of elders
Emotional Security	Family bonds Friendships Sense of community
Environmental Security	Clean water Clean air Biodiversity
Individual Freedom and Variety	Private property rights Consumer choices Unrestricted speech/media
Individual Security	Health care Police protection/prisons Fire protection
Recreational Opportunity	Outdoor sports Indoor leisure activities Relaxation time
Spiritual Dimension	Humility before Supreme Being Awe of nature Set of moral principles

Lakes contribute to most of these needs. People buy lakefront property — and tourists visit the lake — because of its natural beauty (aesthetics). The money they spend supports many rural communities. As the population ages, retirees living on social security, investments, and pensions further enhance the economic development of lake-rich areas.

Lakes provide an excellent opportunity for environmental education. Students not only learn about geology, chemistry, and biology, but they can also get involved in the political and legal aspects of lake management.

"Going to the cottage" has many overtones — many of them emotional. The special times at the lake with family and friends become part of a rich emotional heritage that enriches generation after generation. Families and friends need special places to build their emotional ties — a cabin on the lake a long way from the cares of everyday life is one of the very best.

Environmental security means clean lakes, a biodiverse ecosystem of plant and animal life. These special parts of the environment are especially vulnerable to pollution because they lie at the bottom of watersheds — and so many people use them (two-thirds of Wisconsin adults each year).

The high value Americans put on personal freedom and individual rights often leads to conflicts over how we use our land and water. We cherish the freedom lakes offer, but misusing that freedom can destroy many other lake benefits. Personal watercraft (jet skis) expand the freedom of a few but diminish that of many others. Individual submarines may cause a similar controversy in coastal waters and clear lakes.

While lake aesthetics typically affects more people in the community (Klessig, 1973; Klessig et al, 1983; Shifferd, 1997), lakes also make a huge recreational contribution. Long home to traditional fishing, swimming, and boating, lake water sports now also encompass sailing, scuba diving, and wind surfing.

And some people cherish lakes as a spiritual setting — a place to reflect — a place to look into the soul of the earth and find meaning in this world and the next.

When a community is clarifying its goals, it must determine what goals the lake can meet. For example, a lake may be useless for fishing or boating but be an aesthetic gem. All lake planning should explicitly — and by consensus — state which goals are being addressed.

STEP 2: Gathering Information

What We Know About the Ecological and Sociological Systems

Step 2 gathers information needed by professionals and citizens to take the succeeding steps. Data-gathering assesses social needs, identifies problems, and inventories physical resources. If little information is available regarding the lake and its users, this step will require a significant investment of time and money.

Needs Assessment: Gathering Information on the Local Community

Part of diagnosing any problem is to ask the patient (in this case the lake user) "where it hurts." Unless the lake manager and the ultimate decision-making body know what people expect from the lake and which expectations are not being met, it will be impossible to develop and implement a plan that meets human needs.

Unless the lake manager and the ultimate decision-making body know what people expect from the lake and which expectations are not being met, it will be impossible to develop and implement a plan that meets human needs.

Many pieces of social data could be relevant. You can use numerous public involvement techniques to obtain such information:

▼ ***Census Data:*** These data, available every 10 years, provide a demographic baseline, a formal snapshot of the local population (age structure, family composition, length of residency, urbanization of the area, migration patterns, income and education levels). Population projections and other trends can be graphed (Roser and Serow, 1984).

▼ ***Target Populations:*** A "needs assessment" systematically determines what people in the community feel they need. It begins with broad discussions of community needs and ends with a prioritized list.

Again, ask "whose needs." Should lake planning focus on businesses on the lake, on people who live around the lake, on everybody in the community — or on the people who use public access for boating and fishing, or enjoy the public beaches, parks, and drives?

▼ ***Nominal Group Process:*** The nominal group process (Delbecq, et al. 1975) is an excellent way to begin the needs assessment. Unlike the standard group meeting procedure, it does a good job of identifying all concerns.

A typical group meeting makes a decision through the following sequence: a motion, discussion, and a vote. This standard procedure frustrates many people either because they feel intimidated about speaking up before the group, or because a few dominant personalities monopolize the discussion. And the motion is made before the discussion, which often necessitates numerous amendments.

In addition to its effectiveness in eliciting concerns and ideas, the nominal group process can also be used to prioritize uses, problems, and projects.

The process has many variations. In its simplest form, each participant is first asked to write down a list of issues. The moderator then asks each person to volunteer one issue, proceeding around the group until all issues are transferred from individual written lists to sheets of paper hung in view of the group. During this time, no one is allowed to discuss or debate the appropriateness of anyone's suggestion.

In addition to its effectiveness in eliciting concerns and ideas, the nominal group process can also be used to prioritize uses, problems, and projects.

After all issues are listed, the group might debate whether certain issues should be combined. The discussion is led by the persons who suggested the issues and is designed to help others understand them more fully. The moderator must be forceful in keeping the discussion focused on understanding each issue and eliminating duplication if the "authors" of those issues agree. The discussion is never allowed to become a debate on the merits of the issue.

Following the discussion, the moderator allows each person to select a small number of issues to "save" by placing a mark or sticker next to those issues. (The physical act of getting up and placing marks provides a nice, refreshing break in the process.) Only the issues with votes are "saved." At this stage each participant may be asked to give reasons why they feel strongly about their top two or three items. After everyone has shared those feelings, each participant is asked to indicate which two or three of the marked items might be less important than the others. Again, no debate is allowed, but if time allows, each participant may indicate why these items are less important.

Participants then rank the saved issues by assigning 10 points with a maximum of 4 points to any one item. The group then focuses on those items (2–5) receiving the most votes (usually, there is a clear break between a top set and the rest). Be sure, however, to preserve all issues in the minutes of the meetings. Thus, every person can see his/her contribution even if it didn't make the top set.

The nominal group method is designed to allow equal participation by all members of the group; it neutralizes dominant personalities. If a group exceeds 15 people, it is advisable to split the group into smaller subgroups and proceed until each subgroup has identified its "saved" pool. The "saved" pools are then combined and the entire group ranks the issues in the combined pool by the same 10–4 procedure.

The nominal group method is designed to allow equal participation by all members of the group; it neutralizes dominant personalities.

In larger lake communities, not everyone may be able to participate directly, so you may want to form a task force or advisory committee to represent the community before a city council or county board. The nominal group process may still be a useful procedure for the task force or advisory committee itself to use.

In addition to identifying issues, participants leave the process with a much higher sense of ownership than they do after participating in a standard meeting. After the nominal group experience, citizens identify with the top concerns because they've actively helped select them.

▼ ***Social Surveys of Users – Behavior, Attitudes, and Preferences:*** The most systematic way to obtain social data is to develop a well-designed survey and have a professional survey agency conduct interviews or send out mail questionnaires. These results are not biased by group discussion at the time of the survey, peer pressure, or the feelings of the person conducting the interviews.

Surveys are most useful in getting data about actual behavior, preferences, or concrete choices. A survey can determine how many boats a person owns and how many days s/he uses their cottage each year and whether the annual meeting of the lake district should be held on Saturday afternoon or Sunday morning.

Surveys are most useful in getting data about actual behavior, preferences, or concrete choices.

Surveys are less useful for policy options, which are complex and hard to describe in a single question; too often, answers are based on factors not considered by the survey. Surveys should not substitute for group discussion, voting in person at community meetings, and democratic decision-making.

The lake organization officers, with professional help from a local university or Extension office, could do their own survey by sending a questionnaire to residents, property owners, and possibly even public users of the lake. The results would identify concerns and tell the board what activities are most common and what activities are desired but not being fulfilled.

When you survey lake residents, you should contact the entire population; part of the purpose of such a survey is to give people a sense of participation in the planning process. You can sample other lake users by using contacts at the lake, tracking down owners of vehicles parked at access points (through their license plates), or random sampling people in the broader community.

Response rate is much more critical than the size of the sample in getting accurate information. A good pretest will not only clarify confusing questions but may identify factors that would reduce response rate if not corrected. Don Dillman in *Mail and Telephone Surveys: The Total Design Method* (1978) provides excellent guidance on social surveys.

Two follow-ups, including a new questionnaire each time, will typically be required for a mail survey. Telephone surveys must include callbacks until the targeted respondent completes the interview or refuses to do so. Don't just interview "any adult in the household."

Response rates in general have declined but with a special local focus like a lake, you should be able to get a 60 to 75 percent response rate. If it falls under 50 percent, the data may be flawed since the non-respondents are likely to be quite different than the respondents. You can find out how different by making a special sample of the non-respondents — perhaps by phoning non-respondents to a mail survey, or personally contacting non-respondents to a telephone survey.

▼ ***Focus Group:*** This technique has become very common, especially in market research. The focus group typically assembles a small group (5–10) of people for either one or a series of meetings to discuss complex new products or social issues. While the focus group cannot claim to be representative, it facilitates in-depth understanding and provides the convenor with thoughtful insights from a group of citizens. In market research and larger population contexts, participants are often paid for their time. In a typical lake setting, food, refreshments, and a sincere "thank you" should be sufficient reward.

Like the Nominal Group Process, the Focus Group Technique is useful at the beginning of the process, and again during the implementation phase. As shown in Table 3-2, it cannot substitute for the quantitative data that a social survey provides.

Table 3-2.—Strengths and weakness of four social survey research techniques.

	MAILED QUESTIONNAIRE	TELEPHONE INTERVIEW	IN-PERSON INTERVIEW	FOCUS GROUP INTERVIEW
Generalize to larger population	Yes	Yes	Yes	No
Minimum time for completion	Moderate	Fast	Slow	Fast
Cost	Low	Moderate	High	Low
Cost impact of size of geographic area covered	Low	Low to Moderate	High	High
Interviewer effects	Low	Moderate	High	High
Ability to deal with complex issues	Moderate	Low to Moderate	High	High
Response rate	Moderate	Moderate to High	Moderate to High	NA

Source: Barbara Burnell, Wisconsin Survey Research Laboratory, University of Wisconsin-Extension (1998).

▼ ***Delphi Process:*** The Delphi technique (Delbecq, et al. 1975) is based on the premise that any one expert (or citizen) has incomplete knowledge and is inherently biased. Therefore, a panel of experts is expected to produce a more complete range of issues or solutions and more balanced recommendations than a single expert.

This procedure is useful in setting research priorities, summarizing current knowledge, and making policy recommendations for public bodies. For instance, you could use it to design a management plan for a new reservoir.

The first stage of the process is to solicit the full range of issues, ideas, and concerns associated with the topic. The experts (at either a meeting or through correspondence) simply provide a "laundry list" of all items that might be appropriate.

In the second stage, the same experts rank the list developed in Phase I by some criterion of importance. You can use additional phases to obtain greater specificity and consensus regarding the highest ranked items. Some concerns will require many phases before agreement is reached.

The results of the Final Phase are communicated to the organization that initiated the effort.

This procedure is too complicated and expensive for most lakeshore communities; but it does emphasize that lake organizations should get a second opinion on major recommendations they receive from a consultant or public agency.

▼ ***Other Public Involvement Methodologies:*** Prior to 1970 citizens had limited input in natural resource policies. Some agencies held public hearings but they were primarily an opportunity for the agency to tell the public what they planned to do or were already doing.

Since 1970 the major federal natural resource agencies have been mandated to provide for public participation. As a result, a whole set of public participation processes has evolved. Examples include the Consensus Model, Key Community Contact Interviews, Futures Retreats, Citizen Advisory Committees, Simulation Games, and Guided Tours. Kathy Carman and Ken Keith (1994) describe 69 *Community Consultation Techniques* in their guide for planners and facilitators in Australia. University Extension offices and private organizational consultants can provide guidance in using many of these techniques.

Citizens are sometimes more concerned about how decisions are being made than the decisions themselves.

Citizens are sometimes more concerned about how decisions are being made than the decisions themselves. Schmuck and Runkel (1994) have developed a 10-item instrument (see Table 3-3) to gauge how an organization's board of directors, committee members, employees, or citizen members feel about the process of decision-making.

Problem Identification

The needs assessment is just that; it gives you a list of concerns that need attention. Some concerns may relate to the social system; others to the ecological system. Most lakeshore communities will have a mix of both — and both will need attention. Problems in both systems reduce the quality of life for citizens by diminishing the lake as an aesthetic, economic, educational, emotional, recreational, and spiritual resource.

The needs assessment is just that; it gives you a list of concerns that need attention.

A lake problem is a limitation on use. Historically, the most common problems have been associated with the lake itself: too many weeds, too much algae, murky water, odor, or poor fishing. During the past two decades, conflicts over its use have increased dramatically.

As the size of motors increased, so did the conflicts between motorized users and non-motorized users (canoeists, swimmers, anglers, scuba divers, and wind surfers). And then, personal watercraft (jet skis) appeared. Personal watercraft users can damage shallow lake areas and harass waterfowl, in addition to polluting the air and water. But their overwhelming impact is on other people: user conflicts are social system problems.

Another social system problem for numerous lakes is the lack of an effective organization to manage and protect the lake. Many plans developed with excellent professional guidance are never (or poorly) implemented because the lake management organization was not properly structured, had too little power, lacked public support, or its leaders were not capable of carrying out the responsibilities. To ad-

Table 3-3.—Decision-making: Scorecard on the process.

THE WAY THINGS ARE Always					Never	DECISION-MAKING ATTRIBUTES	THE WAY THINGS SHOULD BE Always					Never
1	2	3	4	5	6	A. Decisions are made through teamwork.	1	2	3	4	5	6
1	2	3	4	5	6	B. Facts from those who know are used to make decisions.	1	2	3	4	5	6
1	2	3	4	5	6	C. You take a part in making decisions that affect you.	1	2	3	4	5	6
1	2	3	4	5	6	D. You or your peers help make decisions.	1	2	3	4	5	6
1	2	3	4	5	6	E. When decisions are made, they are based on information that you think is right and fair.	1	2	3	4	5	6
1	2	3	4	5	6	F. Decisions are made by those who know most about the problem.	1	2	3	4	5	6
1	2	3	4	5	6	G. The people who make decisions that affect you are aware of the things you face.	1	2	3	4	5	6
1	2	3	4	5	6	H. Decisions are made in such a way that you do not mind carrying them out.	1	2	3	4	5	6
1	2	3	4	5	6	I. Leaders work with their peers and people below them to make the decision.	1	2	3	4	5	6
1	2	3	4	5	6	J. Things are organized so that you or your peers can help make decisions.	1	2	3	4	5	6

Source: Schmuck, R.A. and P.J. Runkel. 1994. The Handbook of Organizational Development in Schools and Colleges, p. 285-86.

dress this problem in Wisconsin, the Wisconsin Lakes Partnership has begun a Lake Leaders Institute to enhance the leadership capacity of local lake leaders.

The process of Problem Identification is dealt with in detail in Chapter 4 — especially problems with the physical resource.

Inventory

The third aspect of Step 2 in the planning process is to inventory the lake and the watershed and to clarify the relationship between them. The inventory should also include state and local regulations that apply to both. Boating regulations in particular are becoming more common and more critical.

Start the inventory with a thorough search for records of previous studies on the lake. Such data are useful as a historical baseline and may reduce the need for certain new inventories. For instance, documentation of previous land use is essential for understanding the impact of recent land-use changes on water quality.

In addition to searching lake association and government files, you may want to interview several senior citizens and retired professionals whose long-term memory can lead you to "fugitive literature" (non-scientific, often anecdotal or journalistic) on the lake, and whose insights can enrich discussions.

The type of data needed will vary with the needs and the problems. Typically, an inventory includes information on hydrology (e.g., hydraulic resonance time), water quality (e.g., dissolved oxygen, temperature, nutrient concentration, vegetation, transparency, hardness, and sedimentation), land use and land-use regulations, fish and wildlife populations, sensitive ecological areas, and shoreland visual quality. The remaining chapters of this book will detail how to collect such data and how to interpret the results and understand the interrelationships.

STEP 3: Conceptualizing the Alternatives

What We Could Do

In Step 1 the professional's only role is to help citizens and community leaders think broadly about all the potential social goals that relate to their lake and encourage them to clarify those goals and keep them in balance.

Step 2 involves citizen monitoring, but professionals bear the primary responsibility for most of the data gathering.

In Step 3 professionals play a lead role supplemented by the ideas of citizens. And in this step no idea should be disregarded.

Typically, interpreting the information from Step 2 will result in a series of possible actions to protect or rehabilitate the lake. They will vary in cost. Some will be sure bets with very predictable results. Some techniques may have undesirable side effects or not produce enough change or be very expensive or too risky. For example, the same weed problem could be attacked with dredging, winter drawdown (in northern climates), herbicides, harvesting, grass carp (where legal), erosion control ordinances, stream fencing, winter manure storage (in northern climates), sedimentation traps, or benign neglect. The selection of a technique in Step 4 should be based on the results of all the previous steps plus affordability.

Be careful to conceptualize the full set of alternatives. Too often, both professionals and citizens fall into mental traps set by narrow training and expertise — and inexperience. Usually, means are confused with ends. For example, many professionals treat "reduction of phosphorus loading" as an end when it is a means (an important one but not the only one) to the ends discussed under Step 1 — Clarifying Goals.

As shown in Figure 3-2, the Ultimate Goal, at the top of the means-ends hierarchy, is Sustainable Social Well-being. The 11 Primary Goals or Needs (Klessig, 1994; Klessig and Hagengruber, 1999) form the second tier of goals and also serve as means to the Ultimate Goal above them. Below each of the Primary Goals is a set of Sub-goals, each of which is an alternative means to that Primary Goal. Usually several Sub-goals must be used as means to achieve the Primary Goal, but unlike the Primary Goals, they are not all necessary conditions; choice is available. As the pyramid spreads out to Sub-sub-goals and further down, more and more alternative paths (different means) become available to get to the Primary Goals and Ultimate Goal.

Figure 3-3 shows the means and ends hierarchy for a northern temperate glacial lake suffering from winterkill of fish where one of the Primary Goals is to use the lake for recreational fishing.

Figure 3-2.—Top of the Means-Ends Hierarchy: Ultimate Goal and Primary Goals.

The Sub-goals are all alternative means to achieving the Primary Goal of recreational value from this lake. Similarly, the Sub-sub-goals are alternative means to each of the Sub-goals.

The means-ends hierarchy has no bottom. Each level becomes more specific and eventually the detail becomes trivial. Unfortunately, both citizens and professionals frequently get trapped in one of the lower boxes in Figure 3-3. Because of training, experience, or disposition, they have become comfortable with one means low in the pyramid. They convert that means into an end in itself and thus, have difficulty considering all the other pathways to the Sub-Goals, Primary Goals, and Ultimate Goal.

How do you overcome this natural tendency to tighten the noose of familiarity? Antidotes include broader training, diverse work experience, regular professional development, and field trips to communities using other pathways up the means-ends hierarchy.

Citizens and professionals typically start from one of the lower boxes in the hierarchy. Entry-level jobs usually confine professionals to such a box and citizens often have the solution in mind before they begin the planning process. Both groups can break out of their boxes if they are encouraged to look at the big picture — which means raising their eyes to the top of the hierarchy for the real goals and to the sides of the hierarchy for all the various means to achieve those goals.

As shown in Figure 3-3, you can take numerous pathways (means-ends sequences) to the Primary Goal of Recreational Opportunity on a lake suffering

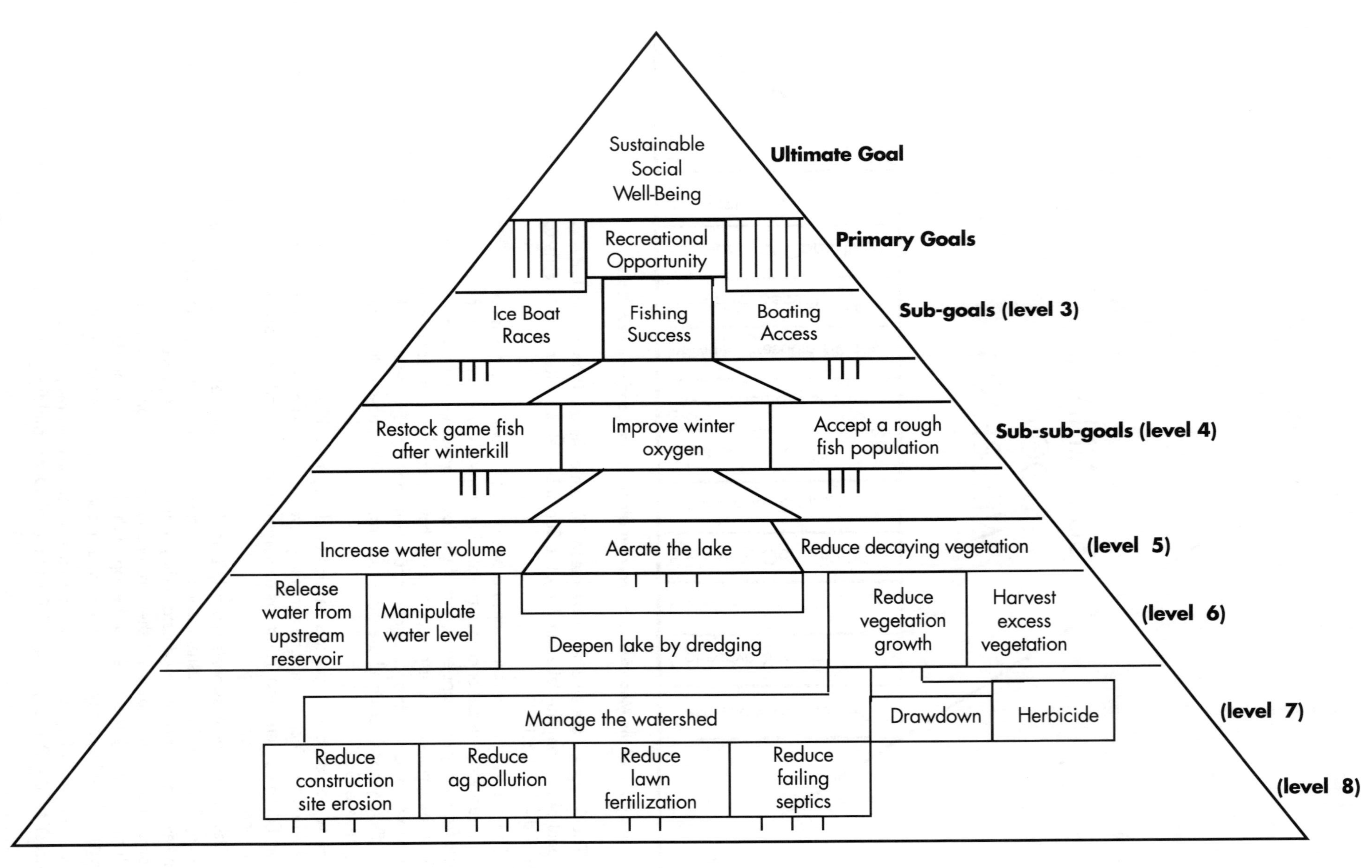

Figure 3-3.—Means-End Hierarchy for a recreational lake suffering winterkill.

from winterkill. Of course, that lake also contributes to Aesthetic Opportunity and perhaps to Economic Opportunity, Educational Opportunity, and Spiritual Dimension. Some of the pathways will overlap. A single means may lead to the fulfillment of more than one end.

Summarize in writing the alternatives from Step 3, the supporting data from Step 2, and the goals clarified in Step 1 and distribute them before the meeting where your group will make the decision on what to do (Step 4).

STEP 4: Making a Formal Decision

What We Decide to Try

The lead role played by professionals in Step 2 and Step 3 evolves to local community leaders in Step 4. The professionals' role in Step 4 is to be on tap but not on top. In addition to preparing an attractive Executive Summary of Steps 1-3, the professional(s) will probably be asked to explain the different options and the likely scenario that would follow with each of them. Professionals may also be asked to explain certain relationships and predictions.

In a democratic society, however, when you're dealing with the management of a local resource that is important to citizens in so many ways (Step 1), professional resource managers must not let themselves be dragged into making specific recommendations. A formal recommendation by a professional diminishes citizen motivation and sense of responsibility. If something goes wrong, citizens will blame the consulting firm or the government agency. Citizens will often ask professionals to tell them what to do, but such a cop-out reduces citizen ownership for their lake and their ability to make other decisions.

In a democratic society, however, when you're dealing with the management of a local resource that is important to citizens in so many ways (Step 1), professional resource managers must not let themselves be dragged into making specific recommendations. A formal recommendation by a professional diminishes citizen motivation and sense of responsibility.

If citizens are to be partners in lake management, they need to be empowered beyond taking Secchi disk readings. They can't be given an easy way out even if they ask for it.

Of course, lakes within the jurisdiction of a government agency are a different story. A government manager often relies on professionals to make the decisions for a lake on government land. But even under these conditions, lake users can be involved with some decisions.

Multiple land ownerships surround the more typical lake. Often the lake association, lake district, or municipal board will decide to pursue more than one action. A formal plan may have several elements, such as water level control, recreational use restrictions, better shoreland zoning enforcement, vegetation harvesting, fish management, wetlands purchase, etc. *The Model Lake Plan for a Local Community* (see Appendix 3-A) illustrates the selection of eight major options by the members of the Lake Hale District (Klessig et al. 1994).

If citizens are to be partners in lake management, they need to be empowered beyond taking Secchi disk readings.

STEP 5: Defining Measurable Objectives

We Need a Standard for Success

Objectives provide both the guideposts for implementation and a barometer of success. Thus, an objective should state a quantity and the completion date, e.g., "Install an aeration system with 500 feet of diffusion pipe by November 1, 2002."

A good plan has a series of objectives that follow from each of the alternatives chosen and guide the implementation process for years and perhaps decades.

A good plan has a series of objectives that follow from each of the alternatives chosen and guide the implementation process for years and perhaps decades. *The Model Lake Plan for a Local Community* (Klessig et al. 1994) includes 34 objectives with timelines ranging from less than one year to 27 years; it also serves as a generic template for a report to the community (see Appendix 3-A).

Once you've formally adopted the plan (with specific objectives) and developed the implementation strategy, put all that information into an attractive document and distribute it to all residents, local recreation and civic clubs, local elected officials, local and regional resource management professionals, and the media.

The document should serve as:

- A source of community pride;
- Recognition for the work of the citizens committee;
- Recognition for the professionals who worked with the community;
- A guide to implementation; and
- A way to keep the leaders of the local lake management organization accountable.

STEP 6: Implementing the Plan

How We Put the Plan into Operation

The plan climaxes with the actual activity — the protection and restoration of the lake. It may be getting the county board to better enforce their shoreland zoning ordinance or it may be hiring a contractor to rebuild the dam. It may be successfully protecting a wetland or a scenic bluff.

Implementation is the nitty-gritty work and the paperwork that goes with it. Many activities require permits from the state water regulatory agency. And if you're using financial assistance from agencies and other organizations, you will probably have to write special financial proposals and progress reports.

The success and efficiency of the plan's implementation will depend on how well the lead citizens and the professionals form a partnership that can orchestrate the wealth of human resources available.

Citizens will be almost entirely responsible for some elements; e.g., citizens are much more effective lobbyists than professionals are. Other activities require sophisticated engineering, limnological, fisheries, or botantical expertise.

The success and efficiency of the plan's implementation will depend on how well the lead citizens and the professionals form a partnership that can orchestrate the wealth of human resources available. Social capital is the term sociologists use for the good will that greases human organizations at every level. Social capital includes the personal networks and the institutional history of cooperation. Without that trust, people lose motivation and resort to bureaucratic and legalistic processes to protect themselves.

Making a lake plan work requires physical capital (lake and watershed), financial capital (local funds, grants, loans), human resources (citizens and professionals), and social capital (trust). Later chapters deal with the specific limnological expertise required by lake plans.

STEP 7: Evaluating the Results

Were We Successful?

Too frequently evaluations focus on the amount of money spent or number of hours invested. Both these measures of effort can reward inefficiency — a project that was more costly in terms of money and time is somehow judged to be more successful.

You can decrease the confusion of effort and results by focusing on objectives. An objective that specifies a quantity to be accomplished by a certain date is easy to evaluate. For example: the objective of "installing an aeration system with 500 feet of diffusion pipe by November 1, 2002." In the year 2003 your report to the annual meeting of the lake association can answer these specific questions:

- Was the aeration installed?
- How many feet of diffusion pipe were laid?
- Was the system operating before ice-up in the fall of 2002?

Some objectives are more difficult to quantify. Educating realtors or lot owners regarding shoreland sensitivity and development regulations occurs over time as a result of multiple exposures to information. A hierarchy of evaluation questions for educational objectives is shown in Table 3-4. They move from simple measures of effort by the educator to reported knowledge gains by the target audience to actual behavior changes by the audience to the impact of those behavioral changes.

In addition to evaluating specific objectives, the officers of the lake management organization (or a special committee) should take a comprehensive look at the whole set of objectives about every five years.

In addition to evaluating specific objectives, the officers of the lake management organization (or a special committee) should take a comprehensive look at the whole set of objectives about every five years.

Table 3-4.—Levels of achievement of educational objectives.

EVALUATION LEVEL	EVALUATION MEASURE
Input	Teacher's energy investment in preparation
Activity	Event occurrence — workshop was held
Participation	How many people were involved?
Reaction	Post -event subjective feeling about participation
Learnings	Differences between pre- and post-test of knowledge
Application	Participant used knowledge
Impact	Individual, business, family, or community changed as a result of using knowledge

STEP 8: Repeating the Process

When the Kids Grow Up, Give Them Their Turn

The End is Not the End

No plan is static. Conditions change. Scientific understanding changes. Human values and attitudes change. Plans have to change — slowly and systematically. A plan that is too responsive to changing demographics and political swings will fail to provide the stable guidance that long-term management requires. A plan that is too rigid will lose relevance and be ignored.

By the time the third general review is completed, the plan is likely to be 15 years old. If a new planning process is begun at that time, it will be completed within 20 years of its predecessors or about one human generation time.

Continuous Data Gathering

Limnological data are most useful if collected in the same places in the same way over long periods of time.

Limnological data are most useful if collected in the same places in the same way over long periods of time. While these data are used most intensely during Step 2, at least part of the collection regimen should be continuous. The lake associations around populated or large lakes can afford to hire professionals; some communities have their own lake manager. However, in most communities citizen volunteers perform continuous monitoring duties. These volunteers need training, record-keeping support, an occasional pep talk, and public recognition for work that can otherwise lose its glamour after several years.

Organizational Maintenance of Citizen Organizations

Everyone involved in a lake planning and implementation effort wants to get the job done — wants to save the lake. But like a farmer, who is so anxious to harvest the crops that s/he chases out to the field without greasing the equipment, community energy must be shared between the task of lake management and the maintenance of the lake management organization.

If community leaders are too zealous, they will forget to grease their organization. They might fail to appreciate the work of the many volunteers who do everything from water quality monitoring to organizing the Fourth of July boat parade. Sometimes zealots are so task oriented that they even publicly criticize citizens and elected officials for being "lazy" or "apathetic." Zealots often end up trying to do everything themselves because they have alienated everybody else. And, of course, they can't do everything very long. Unless a balanced leadership develops the lake management organization may actually dissolve.

Although zealots get the community focused on a lake problem, they tend to dominate a new organization. Sometimes they mellow with experience; but too often, they have to be replaced as an organization matures.

The older an organization is the more likely it will suffer from the opposite problem — lack of task orientation. It becomes too focused on having a good time with old friends and keeping everybody happy. The spring picnic, the Fourth of July boat parade, and the fall dinner become the core annual activities. Maintenance of the organization becomes an end in itself. Leaders tend to be extroverts who love to socialize. These leaders avoid conflict internally and externally and thus avoid making tough decisions about the lake.

The ideal lake management organization has a blend of leaders — some focused on the task of managing the lake and some focused on the maintenance of the organization. Such a balanced leadership team can both accomplish the plan's objectives and keep the organization strong for the long term. Best of all, under such a leadership team citizens can enjoy working with each other and be proud of their mutual accomplishments.

The ideal lake management organization has a blend of leaders – some focused on the task of managing the lake and some focused on the maintenance of the organization.

The final ingredient for the successful long-term lake management organization is new blood. Lake management organizations often have a very high percentage of retired people who, of course, provide a wealth of experience, free time, and funds. But use them wisely; encourage them to transfer their wisdom and institutional memory to the next generation. Lake management organizations should consciously and slowly turn over the organization to the next generation as the planning process is repeated.

References

Burnell, B. 1998. Strengths and weakness of major research variations. Unpubl. instructional material, Wisconsin Survey Research Laboratory, University of Wisconsin Extension, Madison.

Carman, K. and K. Keith. 1994. 69 Community Consultation Techniques. Queensland Department of Primary Industries, Brisbane, Queensland, Australia.

Delbecq, A.I., A.H. VanDeVer, and D.H. Gustafson. 1975. Group Techniques for Program Planning: A Guide to Nominal Group and Delphi Processes. Scott, Forstman, and Co., Glenview, IL.

Dillman, D. 1978. Mail and Telephone Surveys: The Total Design Method. Wiley, New York.

Klessig, L.L. 1973. Recreational Property Owners and Their Institutional Alternatives for Resource Protection. University of Wisconsin, Madison.

———. 1994. Community and Societal Sustainability: The Anti-Priority Hypothesis. R. Wolensky and E. Miller, eds. Proc. Small Cities and Regional Community Conference. Vol. 11. University of Wisconsin, Stevens Point.

Klessig, L.L. and J. Hagengruber, 1999. Eleven necessary conditions for societal sustainability. J. Human Values 5(1): 33-52.

Klessig, L.L., N.W. Bouwes, and D.A. Yanggen. 1983. The Lake In Your Community. G3216. University of Wisconsin Extension, Madison.

Klessig, L.L., B. Sorge, R. Korth, M. Dresen, and J. Bode. 1994. A Model Lake Plan for a Local Community. G3606. University of Wisconsin Extension, Madison.

Leopold, A. 1949. A Sand County Almanac. Oxford University Press, Oxford, U.K.

Roser, N.W. and W.J. Serow. 1984. Introduction to Applied Demographics: Data Sources and Estimation Techniques. Sage Univ. Paper 39. Sage Publications, Beverly Hills, CA.

Schmuck, R.A. and P.J. Runkel. 1994. The Handbook of Organizational Development in School and Colleges. Waveland Press, Prospect Heights, IL.

Shifferd, P. and S. Palmer. 1997. Being at the Lake: Problems of Organization in a Vacation Community. Report prepared at Northland College, Ashland, WI, for the Wisconsin Department of Natural Resources, Madison.

Technical Committee of the Water Resource Centers of the Western States. 1974. Water Resource and Planning, Social Goals, and Indicators: Methodological Development and Empirical Test. Utah State University, Logan.

U.S. Environmental Protection Agency. 1995. Watershed Protection: A Project Focus. EPA 841-R-95-003. Off. Wetlands, Oceans, Watersheds, Washington, DC.

G3606

APPENDIX 3-A

A model lake plan for a local community

Lowell Klessig, Buzz Sorge, Robert Korth, Michael Dresen, Jeff Bode

University of Wisconsin–Extension • Cooperative Extension
University of Wisconsin–Stevens Point • College of Natural Resources
Wisconsin Department of Natural Resources • Bureau of Water Resources Management

Contents

Preface

This publication was written for people who live on or use lakes, and for community officials involved in lake management. It sets forth a model management plan for the fictitious Lake Hale in Phantom County.

The Lake Hale model is designed to help communities care for local lakes in three ways:

1. By explaining the types of information needed to make decisions at the local level.
2. By offering a format for summarizing information and debating alternatives.
3. By furnishing an example of community commitment to a lake.

In each case, the model should be modified to fit the characteristics of the individual lake and the expectations of local residents and lake users. Keep in mind that Lake Hale is a fictitious lake with features commonly found in small to moderate-sized kettle lakes. (Large lakes and impoundments have substantially different characteristics.)

Some lake planning efforts have yielded more detailed reports than the example provided here. In fact, formal and technical documents may sometimes be necessary to meet external needs such as the Department of Natural Resources (DNR) Area-wide Water Quality Management Plans. In those cases, this model might provide a format for summarizing larger reports to distribute to community residents.

Many Wisconsin communities have received or are considering applying for a DNR Planning Grant. The information found in this model directly pertains to both the loan application process and the final report required under the grant.

The lake plan you generate will set the tone for management efforts on your lake for the next generation. You should review it every few years to determine if you are meeting your objectives on schedule and to re-evaluate your long-term goals. If you need assistance, call your DNR district inland lake coordinator or University of Wisconsin–Extension community resource development agent. (Some of these people helped produce this model by serving as reviewers.) To share ideas with other lake organizations and explore common interests, join the Wisconsin Association of Lakes.

Good luck in this important effort.

Lowell Klessig, University of Wisconsin–Extension/University of Wisconsin–Stevens Point
Buzz Sorge, Wisconsin Department of Natural Resources
Robert Korth, University of Wisconsin–Extension/University of Wisconsin–Stevens Point
Michael Dresen, University of Wisconsin-Extension/University of Wisconsin–Stevens Point
Jeff Bode, Wisconsin Department of Natural Resources

Lake Hale—a historical preface

The last glacier to visit Wisconsin created Lake Hale about 10,000 years ago. The glacier left an assortment of soil, boulders, stones and other debris as it retreated north. Occasionally, chunks of ice broke off and were buried in the glacial remains. When one large chunk melted, Lake Hale was born.

After several thousand years, small bands of Chippewa set up a summer camp at the site of the present day county park. Much later, in the 1870s, the region was extensively logged for pine. Areas that escaped the subsequent forest fires were logged again for hemlock and hardwoods. A small sawmill was built on the lakeshore between Sunset Point and Hale Creek in the early part of the 20th century. No efforts were made to protect the lake from sawdust or from the serious erosion following the forest fires.

Agricultural practices brought about additional sedimentation and nutrient enrichment of the lake. Wheat farming dominated in the 1880s, with a gradual switch to dairy farming by 1940, but sandy soils and harsh climate proved inhospitable to farming. Eight Norwegian farmers had established homesteads around the lake by 1900; by 1960, only four families remained. These farms, as well as the resort on Sunset Point, were subdivided for cottage sites in the 1960s and 1970s. The last and largest farm on the lake was owned by the Olson family; in 1989, it was developed as the Northern Heights subdivision.

The first cottages were built on the lake in the 1950s. Many were concentrated right on the lakefront—where impacts on the lake are most direct. In the 1970s and 1980s, cottages were converted to winterized homes.

A lake association was organized in 1963 after a local resident was killed in a water skiing accident. The association persuaded the town board to establish a one-way circular skiing direction and prohibit water skiing in the north lobe.

Because of increasing concern about excessive plant growth and the potential negative impact of backlot development near the lake, a lake district was established by petition to the town board in 1985. Since its formation, the district has received three planning grants from the Wisconsin Department of Natural Resources. The district contracted for aquatic plant harvesting in 1987 and 1989–91.

Overall, land use in the watershed (the land that drains toward the lake) has changed dramatically in the past 130 years. These changes have damaged the lake. This plan, and recent aquatic plant harvesting efforts, are designed to help cope with those changes and protect the lake from further degradation.

Why we care about the future— our goals for the Lake Hale community

Lake Hale and the creatures that live in and around it form an integral part of our community. We want the lake to be healthy so we can enjoy its natural beauty and use it for recreational activities. But more than that, we feel a sense of stewardship—a responsibility to protect and restore the integrity of the lake's ecosystem.

Our ultimate goal is to perpetuate the wildlife, natural beauty and recreational activities we enjoy for future generations of lake users. To meet that goal, we realize that we must develop a set of ethics for our interactions with the lake. Unless we are willing to limit the type and location of shoreline buildings we construct, the amount of shoreline we clear, the size of our boats and motors and the way we use them, the lake will no longer be the source of the natural beauty and recreational activities we enjoy today.

Specifically, we have three goals:

1. To maintain and restore the environmental integrity of the lake ecosystem.
2. To protect aesthetic opportunities for residents and visitors by preserving scenic lake views.
3. To maintain on-the-water recreational opportunities by discouraging behavior that interferes with the activities of other users.

Assessing needs and identifying problems

Concerns of lake district members

As part of the 1989 annual meeting, Jennifer Bates, our county Extension community resource development agent, conducted a "nominal group process" that rank-ordered the major concerns of lakeshore property owners. The concerns were:

1. too much noise from boats and neighbors
2. too many aquatic plants
3. crowding and lack of boating safety on weekends
4. poor fishing
5. water quality degradation
6. unattractive shoreline structures
7. litter on the lake (summer and winter)
8. harassment of waterfowl.

We included these concerns in a standard questionnaire which was sent to each family owning or renting property within the district. The newsletter carried a reminder to complete the questionnaire, and postcard reminders were sent to each address. One hundred of the 150 families responded. The information in figures 1-5 was obtained from that survey.

Most members of the lake district are more than 50 years old and are full-time residents (outnumbering seasonal residents). Of those still working, a large percentage commute to the Twin Cities.

It is clear that Lake Hale's beauty is what attracts district members; they enjoy the lake primarily for its aesthetic value and fishing. But district members also use the lake for motorized activities that often conflict with other expressed expectations. The concerns of district members reflect the increased use of the lake, as well as concern for water quality. Loss of natural vegetation on the shoreline and too much vegetation in the water are other major concerns.

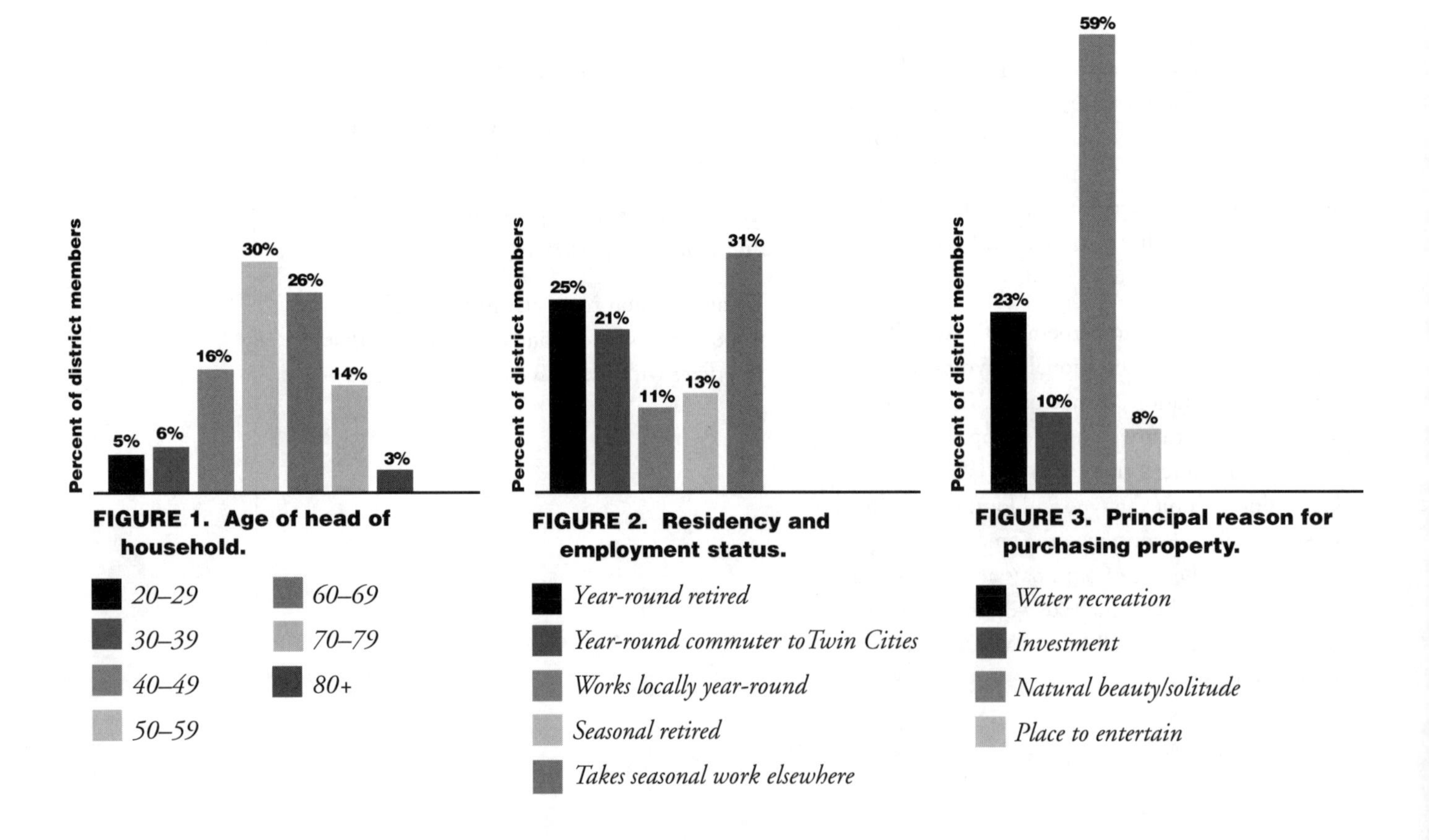

FIGURE 1. Age of head of household.

FIGURE 2. Residency and employment status.

FIGURE 3. Principal reason for purchasing property.

Additional information on the feelings, activities and characteristics of members of the Lake Hale District can be found in the report entitled "Lake Hale District Membership Survey." That report is available from Paul O'Malley, our secretary, and at the public library in Phantom City.

Public users

We hired Horace Billings, a sociologist from the University of Wisconsin–Superior to help determine the activities and preferences of people who use the public boat ramp, the public beach, the trails, the campground and the picnic area at the Lake Hale County Park. The preferences of people who stopped at the State Highway 762 overlook were also included in the study.

Billings devised a procedure to sample users at each public facility on the lake. He modified the questionnaire we used for our membership survey and hired college students to interview users on specific weekdays, weekends and holidays.

We obtained a full report on each of the types of users, how long they used the lake, and their attitudes about their experiences. (The information is

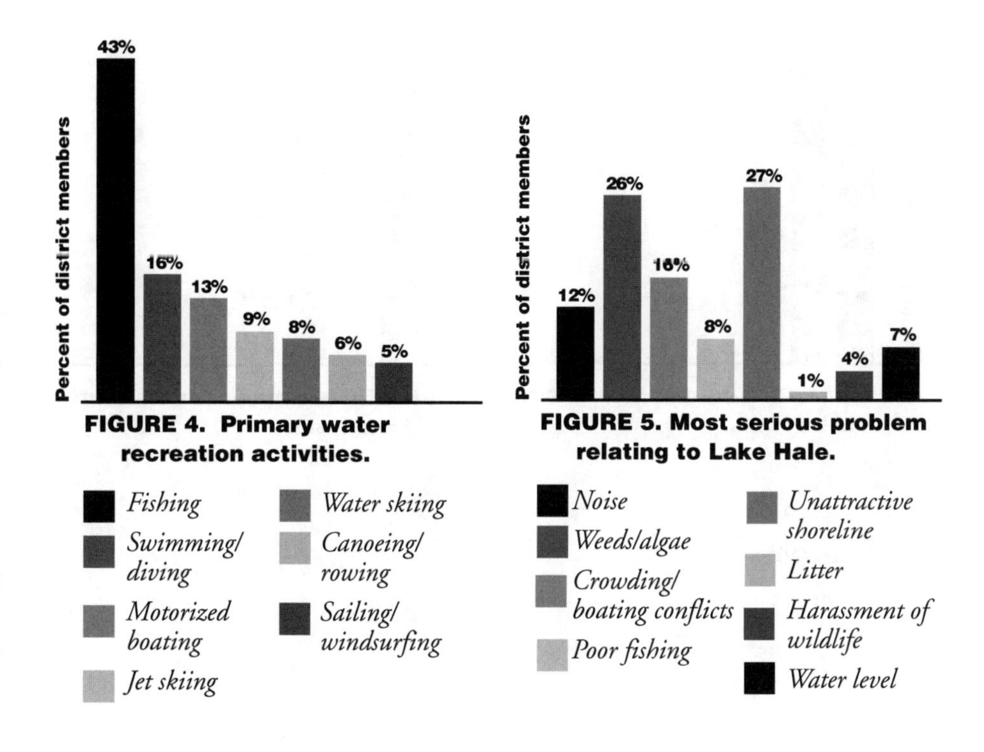

FIGURE 4. Primary water recreation activities.

FIGURE 5. Most serious problem relating to Lake Hale.

available in the "Lake Hale Public User Study.") Some summary graphs that combine all public users are provided in figures 6-10.

Lisa Manley volunteered to coordinate the counting of public users according to a schedule developed by Horace Billings. She enlisted the help of lifeguards at the county beach, who provided park, beach-user, pier fishing and boating counts. Unfortunately, the summer of 1992 was very cold, which reduced the swimming counts by approximately 50% and may have reduced the number of other users as well.

Between 3 to 5 p.m. on an average summer weekday, 107 people were at the park (76 of whom were using the beach), 9 boats were on the water, 6 people were fishing from the public pier, and 28 people stopped at the overlook on Highway 762. On the weekend, the numbers jumped to 229 people in the park with 135 using the beach, 24 boats on the water, 12 people on the public fishing pier, and 44 people at the overlook.

Bud Langley conducted monthly fishing counts on Saturdays from 7 to 9 a.m. Angler numbers were greatest on Saturday mornings in December (175), January (118), and May (62).

With the help of some students who did several all-day counts, Horace calculated the following annual estimates:

Park users	19,000*
Motorized boaters	2,700
Non-motorized boaters	1,000
Overlook users	7,000
Anglers—all types	5,200
Total users	34,900

**Beach users accounted for 9,000 of this total.*

Public access

The ramp at the County Park has parking spaces for 17 vehicles and trailers. Under the new policy for boat access, Lake Hale possesses 387 acres of open water. (Those parts of Shelter Bay and Lily Bay with emergent vegetation visible from an

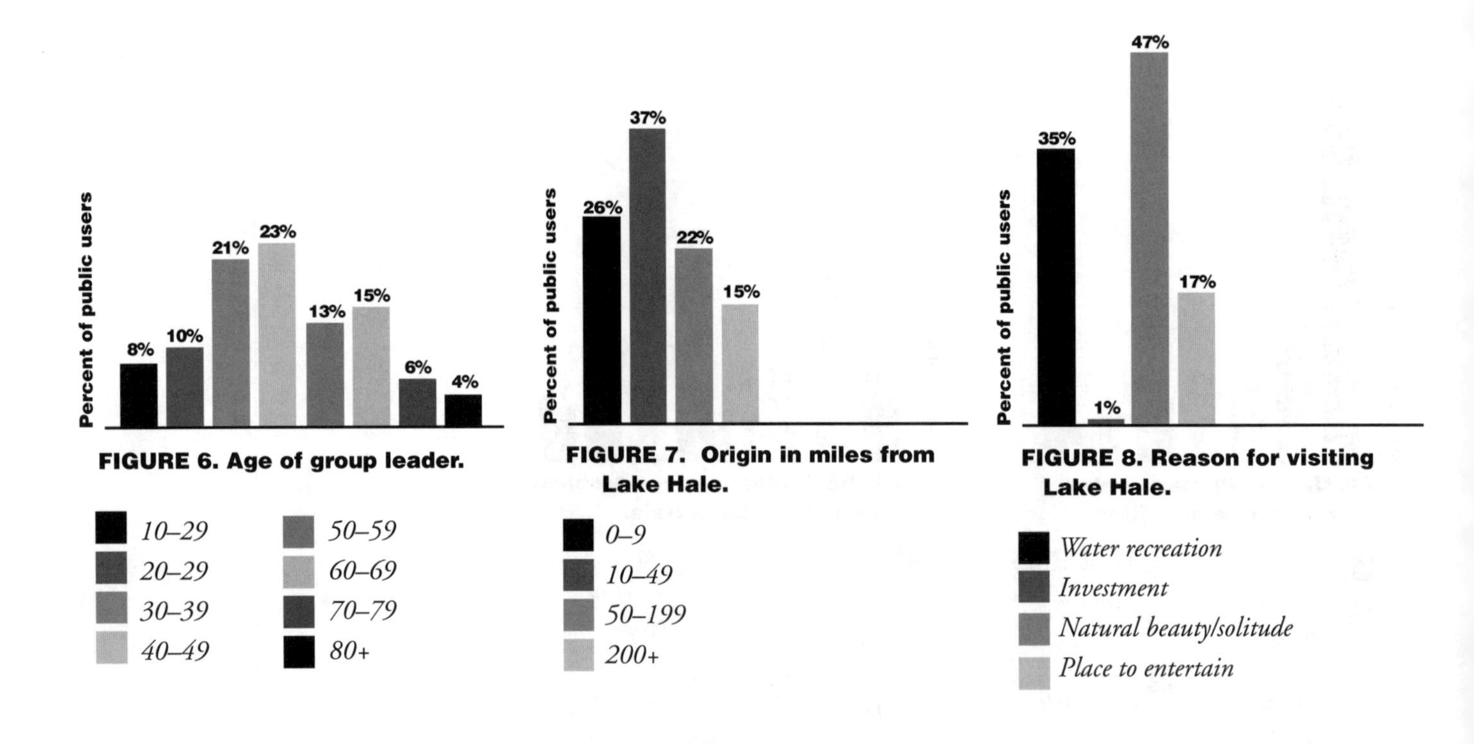

FIGURE 6. Age of group leader.

FIGURE 7. Origin in miles from Lake Hale.

FIGURE 8. Reason for visiting Lake Hale.

TABLE 1. Population levels and rates of change, 1980-1990.

	1970 pop	Change since 1960	1980 pop	Change since 1970	1990 pop	Change since 1980
Town of Meadowview	576	+6%	651	+13%	782	+20%
Phantom City	2817	+2%	2873	+2%	2959	+3%
County of Phantom	27,002	+4%	28,429	+5%	31,865	+12%

aerial photo are not counted.) The Department of Natural Resources codes indicate that a lake affords reasonable public boat access if it maintains one parking space for each 20 to 30 acres of open water. The current boating access is more than adequate under the new code.

Additional public access is provided at the Highway 762 overlook and at the county park by the 500-foot beach, 30-unit campground, fishing pier, picnic area and hiking trails.

Population changes

Based on population data that Jennifer Bates (UW–Extension) obtained, the farm population of the Town of Meadowview and Phantom County continues to decline. However, the overall population has grown consistently for three decades. As shown in table 1, 1980–90 exhibited the fastest growth. The new freeway to the Twin Cities has increased both residential development and public use of the lake. Both pressures are likely to increase.

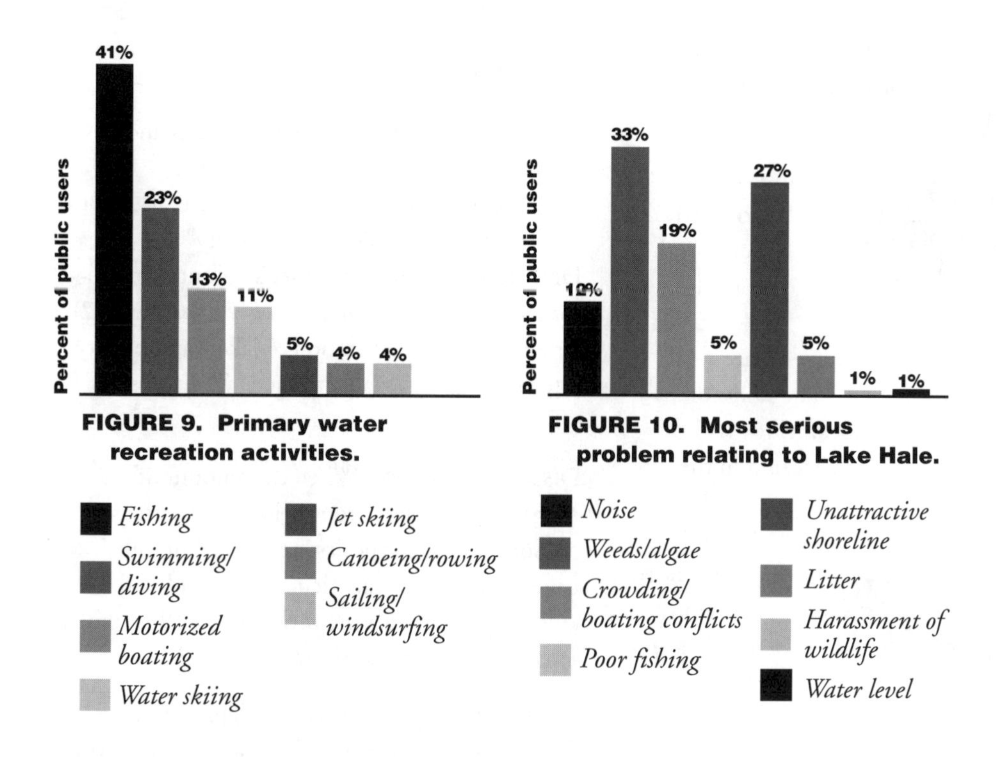

FIGURE 9. Primary water recreation activities.

FIGURE 10. Most serious problem relating to Lake Hale.

What do we know about the lake? An inventory

Water cycle

The lake's official size is 413 acres according to the *Wisconsin Lakes* bulletin. The watershed that drains to Lake Hale is 3,054 acres. Hale Creek drains a large part of the watershed to the north and east of the lake into Lily Bay. Small, intermittent streams drain the other areas of the watershed. Groundwater inflow was measured by Dartmouth Environmental Consultants as part of their water quality study. By putting gauges on the inlet creek and the outflowing river, and adjusting for precipitation and evaporation, they concluded that over 30% of the water entering the lake comes through the ground via springs.

On average, water entering the lake remains about two years before it exits through the Hale River.

Most of the north lobe is less than 10 feet deep. The south lobe is much deeper, with only a narrow band of shallow water (littoral zone) along the shore as can be seen on the hydrographic map (fig. 11). The maximum depth is 62 feet; the average depth is 14 feet. A summary of Lake Hale's physical characteristics is presented in table 2.

Most of the bottom sediments in the north lobe are decayed plant remains, or muck. Muck is also found in the center of the south lobe surrounded by sandy beaches and a few rocky areas.

TABLE 2. Lake Hale's physical characteristics.

Watershed	**3054 acres**
Lake area	**413 acres**
Maximum depth	**62 ft.**
Average depth	**14 ft.**
Volume	**5782 acre-ft. (~2 billion gal.)**
Residence time	**1.9 yrs.**
Rainfall	**30 in.**
Water budget input:	
groundwater	**32%**
Hale Creek	**35%**
other surface runoff	**16%**
direct rainfall	**17%**
Water budget exit:	
groundwater	**9%**
Hale River	**75%***
evaporation	**16%**

Hale River flow = 3 cubic ft. per second

Water quality

A summary of our water quality studies and monitoring is presented in this section. Additional detail can be found in the consultant's reports on "Lake Hale Water Quality" and the DNR reports of Kathie Jansen's volunteer monitoring efforts. *Understanding Lake Data* (G3582), a more detailed explanation of lake water quality information, is available at the UW–Extension office.

Dissolved oxygen. When the amount of oxygen in the water drops below four parts per million, some fish species are stressed or killed. Lake Hale has not experienced lake-wide fish kills. Oxygen levels were measured in 1992 and at several earlier dates as shown in table 3.

In late winter, oxygen becomes depleted in Shelter Bay and possibly other parts of the north

FIGURE 11. Hydrographic map of Lake Hale.

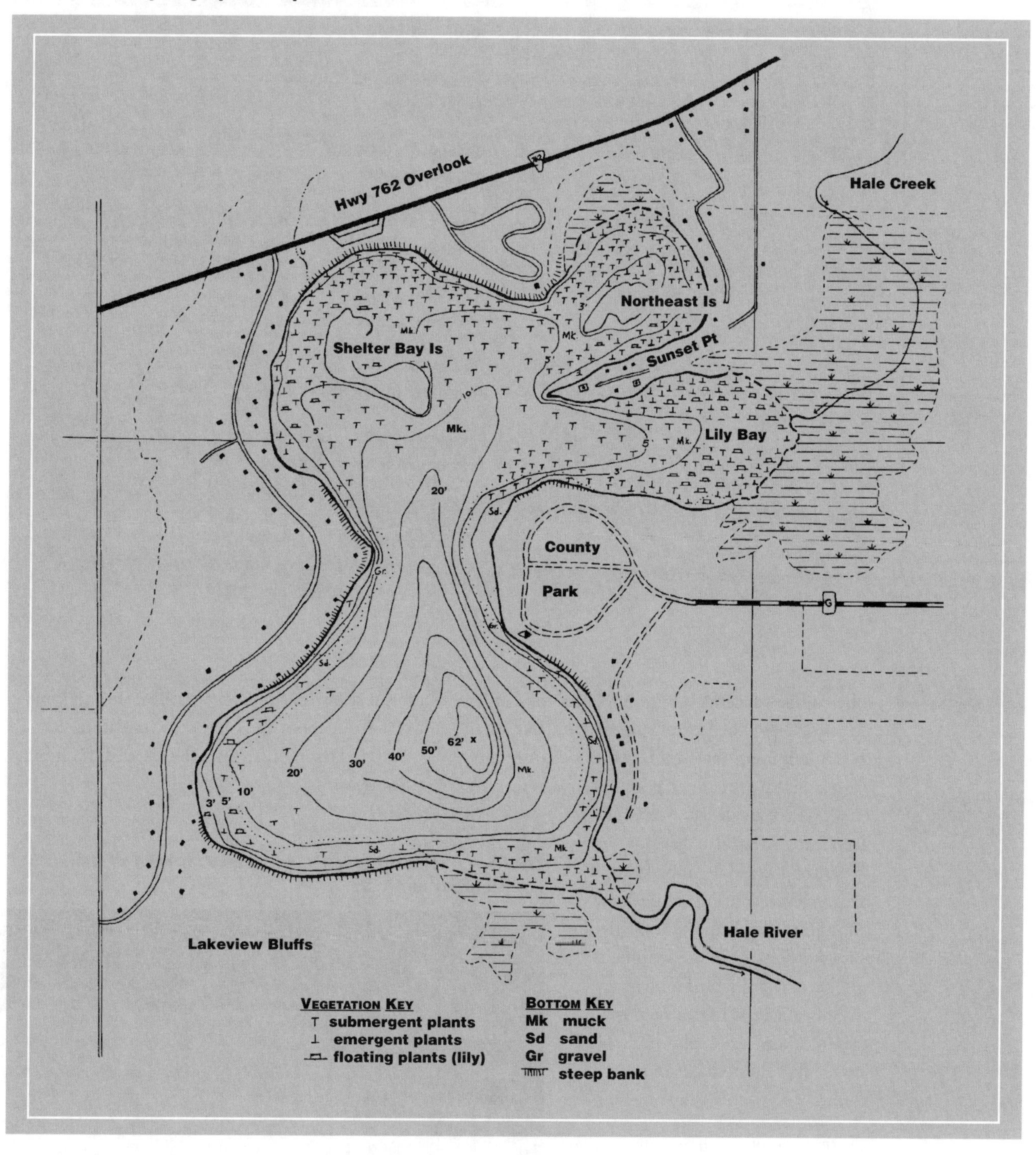

TABLE 3. Dissolved oxygen in parts per million (ppm).*

Year	Date	Organization	Location	Depth	DO level
1963	Jan 4	DNR	Mid south lobe	4 ft.	9
1981	Mar 2	DNR	Mid south lobe	4 ft	8
			Shelter Bay	4 ft.	1
1992	Jan 2	Consultant	Mid south lobe	4 ft.	10
			Shelter Bay	4 ft.	2
	Mar 4	Consultant	Mid south lobe	4 ft.	8
			Shelter Bay	4 ft.	0.5
	May 22	Self-help monitor	Mid south lobe	4 ft.	9
			Mid south lobe	40 ft.	9
	June 20	Self-help monitor	Mid south lobe	4 ft.	9
			Mid south lobe	40 ft.	7
	July 31	Self-help monitor	Mid south lobe	4 ft.	8
			Mid south lobe	40 ft.	4
	Sept 1	Self-help monitor	Mid south lobe	4 ft.	8
			Mid south lobe	40 ft.	1

**ppm = mg/liter*

lobe. While oxygen levels are adequate for our present fishery, the lack of summertime oxygen in the hypolimnion (below 25 feet) is a critical indicator that water quality has gradually deteriorated. Cisco, a fish that needs cold, deep water to survive, has not inhabited Lake Hale since 1970. As oxygen levels continue to decrease, phosphorus (an important nutrient for algae growth) in the sediments will become available for algae blooms in fall. When oxygen is present in the water, phosphorus is less soluble and remains in the sediment.

TEMPERATURE. With encouragement and training from Daryl Roberts (DNR), our self-help monitoring volunteer Kathie Jansen has taken temperature profiles of the lake at different depths. The numbers in table 4 show that the lake stratified into thermal layers from about the middle of June until October. During that time the water did not mix (turn over).

TABLE 4. Temperature (in degrees F) profile of Lake Hale (south lobe) in 1992.

DATE	DEPTH					
	3'	6'	10'	15'	20'	40'
April 16	39°	39°	39°	39°	39°	39°
May 22	47°	47°	46°	45°	45°	45°
June 20	58°	58°	49°	47°	47°	47°
July 31	69°	68°	55°	48°	47°	47°
Sept. 1	75°	75°	56°	48°	48°	48°
Oct. 1	52°	54°	54°	48°	48°	48°
Nov. 15	39°	39°	39°	39°	39°	39°

As expected, the lake had the same temperature from top to bottom as it mixed in mid-April when the ice melted, and again in mid-November just before the ice developed.

NUTRIENTS. The lake possesses moderate levels of phosphorus. Dartmouth Environmental Consultants, under their first contract in 1991, measured phosphorus levels at turnover and found 20 micrograms per liter in spring and 28 micrograms per liter in fall. These phosphorus levels will produce some algae, but are not likely to yield nuisance blooms of blue-green algae. However, if summertime oxygen loss in deep waters becomes more pronounced, the sediments will release more phosphorus and fall algae blooms can be expected to increase. If phosphorus levels increased beyond 30 micrograms per liter, Lake Hale would experience regular algae blooms.

Limnologists (scientists who study freshwater life and phenomena) use a number of indicators to classify lakes according to their nutrient richness, or level of eutrophication. On a trophic status index, Lake Hale was oligotrophic (nutrient poor) in 1850 before lumbering, farming and home building occurred around the lake. Now it is mesotrophic, or generally midway along the range as shown in table 5. This means that Lake Hale is healthy overall, but that phosphorus, which entered the lake through soil erosion, has burdened the lake with excess nutrients. Further nutrient enrichment from the watershed will trigger a release of phosphorus from the lake bottom and the lake will become eutrophic.

Dartmouth Environmental Consultants measured total nitrogen at 250 micrograms per liter. This is within the normal range and should not stimulate algae growth. However, nitrogen—from septic systems, lawns and agricultural practices in the watershed—is very soluble and can fertilize rooted aquatic plants, leading to excess vegetation. Nitrogen availability might also encourage the spread of Eurasian water milfoil, an exotic invader.

TRANSPARENCY. The Secchi disc measures water clarity by gauging the amount of algae in the water. Turbidity from soil erosion or a tan color from decaying vegetation also affect Secchi disc readings. Lake Hale has a slight tea color from natural tannic acids that drain from the wetland along Hale Creek. Art Belder, our first self-help monitor, took Secchi disc readings for four years. After he moved, Kathie Jansen continued the effort. The measurements Art and Kathie collected show a slight, but inconclusive, reduction in transparency (fig. 12).

TABLE 5. Trophic classification of Wisconsin lakes.

Trophic Class	Total phosphorus (μg/l)	Chlorophyll a (μg/l)	Secchi Disc (ft)
Oligotrophic	3	2	17
	10	5	8
Mesotrophic	18	8	6
	27	10	6
	30	11	5
Eutrophic	50	15	4

FIGURE 12. Annual average transparency.

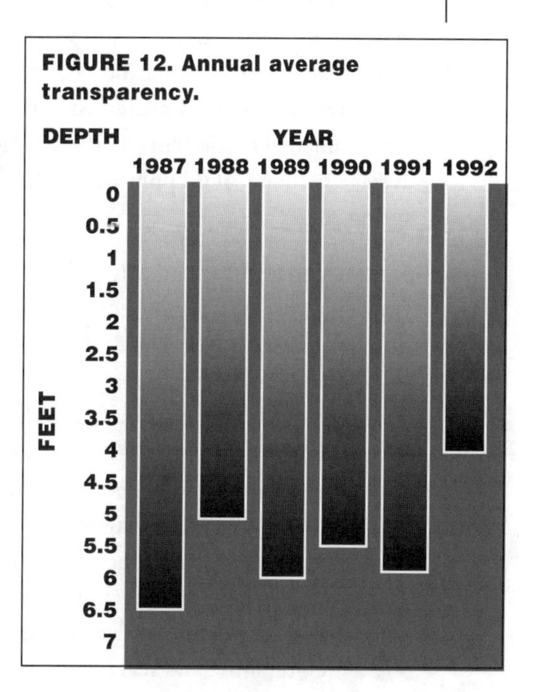

Vegetation. Excess rooted aquatic vegetation (macrophytes or weeds) currently appears to be a more serious problem than algae. According to the consultant's macrophyte survey, the most dense growth is concentrated in the north lobe. Eurasian water milfoil has become established south of the boat ramp along the east shore of the south lobe and around Sunset Point. Pond weeds are the primary species. Lily pads dominate in Lily Bay.

Aquatic plants are measured by harvesting all the plants in a square meter and weighing them after they have been dried. Our consultants found that biomass per square meter (approximately 11 square feet) was 250 grams in Lily Bay, 215 grams around Northeast Island, and 180 grams in the area around Shelter Bay Island, while the average for the south lobe was only 80 grams dry weight (454 grams = 1 pound).

Dense growths of coontail occupy several locations in the north lobe. They become a nuisance by early August and hamper fishing. But aggressive Eurasian water milfoil is likely to pose the greatest control challenge. Daryl Roberts from the DNR office in Eau Claire will advise us on how to keep milfoil in check as part of an overall aquatic plant management strategy.

Hardness. Lake Hale is not susceptible to damage from acid rain. The surrounding soils contain enough limestone and dolomite to buffer acidic precipitation. Measurements made by the environmental consultants in 1991 indicate an average hardness of 82 ppm with a dip following spring runoff.

Sedimentation. Dartmouth Environmental Consultants took a 5-foot sediment core from the deep hole in the south lobe to determine whether water quality had deteriorated over the last 250 years. (The sediment core measures the rate of sedimentation over time.) The sample indicated that water quality steadily declined from the 1870s into the 1940s. During those years, the watershed was logged and then farmed. Water quality has declined at a much more rapid rate since the 1940s because of the watershed's intensive agricultural use and real estate development along the shorelands.

Watershed

The boundary of the watershed is shown in figure 13, prepared by Ted Walinski of the County Land Conservation Department. Lake Hale drains about 3,000 acres of land; the watershed to lake area ratio is thus about 7:1. The larger the ratio, the more the watershed will have an impact on the lake through nutrient, pesticide and soil runoff. (Impoundment ratios usually average more than 100:1.) A land use survey was conducted by our original Land Use Committee with advice from Ted, using the protocol provided by Jennifer Bates.

Almost half (1,380 acres) of the watershed is used for agriculture, although none of the original shoreline farms still operate. The shoreline areas have been converted to lots. Residential areas, woodlands and wetlands each contribute over 500 acres to the total land use. The overlook and county park occupy 66 acres; the filling station, restaurant and video store cover 10 acres at the intersection of Highway 762 and the freeway.

In their land use report, the committee warned that backlot development was beginning. The report stated that several large areas of the watershed could potentially be developed with a common access (funnel) to the lake. Large numbers of homeowners without lake frontage would be able to dock boats and use a single riparian lot for lake access. Land use in 1991 is shown in figure 13 and summarized in table 6.

TABLE 6. Land use in the watershed.

	Acres
Agricultural	1380 (45%)
Forestry	555 (18%)
Residential	541 (18%)
Commercial/industrial	10 (1%)
Institutional	66 (2%)
Wetland/conservancy	502 (16%)
TOTAL	3054 (100%)

FIGURE 13. Land use within Lake Hale's watershed.

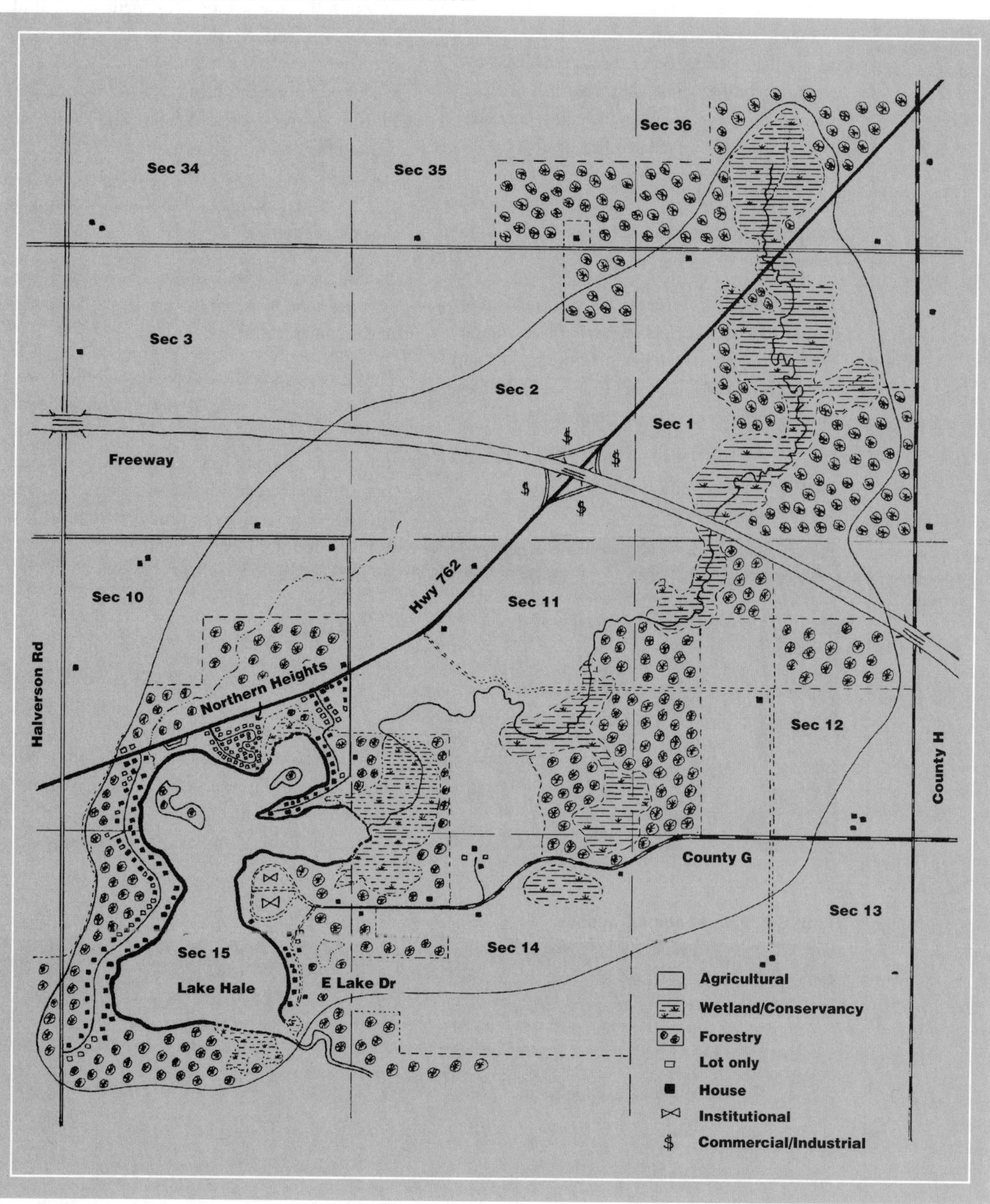

Fish

The local DNR fish manager, Betsy Olson, has met with us periodically and presented reports at the annual meeting. Betsy is concerned about excess vegetation which allows panfish to evade predators and overpopulate the lake. Increased fishing pressure on predator fish like bass and northern may also exacerbate the trend toward overpopulation of certain species. She prepared a comprehensive report for us to use in developing this plan.

The native fishery for our lake consists of largemouth bass, northern pike, cisco, black crappie, bluegill, pumpkinseed, perch, white sucker, and an assortment of forage fish (minnows). Walleye, muskie and lake trout were planted at various times. Bluegills and pumpkinseed use Shelter Bay as the primary spawning site, while northerns spawn in the the shallows and wetlands behind Northeast Island. Muskie spawn later and are effectively preyed upon by the northern fry. Walleye do not reproduce very well because the lake has few rocky areas. Walleye and muskie are planted biennially as shown in table 7. Our Fisheries Committee (Herb Latman, Albert Nice and Bo Hintz) assists the DNR with stocking and shocking activities.

Stocking rates for walleye range up to 50 fingerlings per acre. Muskie rates are 1 to 2 fish per acre on an every-other-year basis.

The fish census (table 8) taken along the shoreline in 1975, 1983 and 1990 indicated generally healthy numbers and mixes of year classes, but showed reduced panfish size. Betsy Olson noticed that the fish were much denser along the shorelines where natural vegetation was retained. Few fish were found in front of homes where the owner had "cleaned up" the shoreline.

TABLE 7. Lake Hale fingerling stocking schedule.

	Walleye	Muskie
1980	15,000	800
1982	20,000	600
1984	10,000	1,000
1986	20,000	800
1988	20,000	500
1990	10,000	800

Wildlife

Two active eagles' nests are located in other parts of the town and the occupants regularly visit Lake Hale. Osprey visits are less common. Until 1960, the lake supported a family of loons, but these birds now appear only during migration and occasionally during the summer. Lowell North, our local Loon Ranger, participates in the

TABLE 8. Fish shocking census.

Species	1980		1985		1990	
	lbs/acre	% over 6"	lbs/acre	% over 6"	lbs/acre	% over 6"
Largemouth bass	38	—	42	—	50	—
Northern pike	16	—	13	—	10	—
Walleye	11	—	7	—	10	—
Muskie	1 fish	—	1 fish	—	1 fish	—
Panfish	120	10%	130	7%	150	5%

regional Loon Watch program of the Sigurd Olson Environmental Institute at Northland College in Ashland. He believes that disturbance from boat traffic is the reason loons no longer nest on the lake.

Frog numbers seem to be down sharply; acid rain or another atmospheric pollutant is the suspected reason. Other amphibians are experiencing similar downtrends.

Boating regulations

The Town of Meadowview has prohibited power-boating at speeds greater than "slow-no-wake" in Lily Bay and behind both Northeast Island and Shelter Bay Island. Water skiing is not allowed in the north lobe and must follow a clockwise direction in the south lobe. The level of compliance with the regulations varies. Jennifer Bates helped the lake district and town prepare an application for the placement permit and state cost-sharing of the buoys. Neither the lake district nor the town has a patrol boat.

Shore development and natural beauty

Most of us bought property here so we could enjoy Lake Hale's cool blue beauty. But some of our activities have tarnished and threaten to further reduce the area's attractiveness. During its survey, our Land Use Committee noted areas that were especially beautiful or ecologically fragile and thus warranted special protection.

The committee suggested that areas shown in table 9 and figure 14 be considered for special protection. It also noted spots that looked unappealing and needed a face lift. Duane Peters, the county code (zoning) administrator, assisted in these efforts.

The Land Use Committee also wanted to register the group's concern about practices that large numbers of residents currently engage in that diminish the lake's beauty (shown in table 10). Many of these practices have been prohibited since 1970 under the Phantom County Shoreland Zoning Ordinance.

TABLE 9. Areas of special beauty or ecological significance.

Area	Distance from water	Acres	Current ownership
Sunset Point Park	*Narrow peninsula*	*1*	*Common property of people who purchased the old resort parcels*
Mouth of Hale River	*On the shore*	*40*	*County land—unmanaged and used by off-road recreation vehicles*
Hale Creek Marsh	*Extends 3 miles upstream*	*200*	*Area farmers*
Lakeview Bluffs	*Off-shore with single lake-access lot*	*100*	*Dream Estates—land development company from St. Paul*
Northeast Island	*Surrounded by water*	*7*	*Northern Heights Subdivision Assn.*
Shelter Bay Island	*Surrounded by water*	*16*	*Alice Knight from Minneapolis*

TABLE 10. Practices that diminish the beauty of Lake Hale.

Practice	Estimated number	Legal status
Constructing buildings closer than 75 feet from shore	*22*	*Prior to 1970—grandfathered. After 1970—illegal without variance.*
Improving nonconforming structures within the 75-foot setback more than 50%	*16*	*After 1970—illegal.*
Placing septic fields closer than 50 feet from shore	*21*	*Prior to 1970—grandfathered. After 1970—illegal without variance.*
Clearing more than 30% of each 100 ft. (first 35 feet from water)	*25*	*Prior to 1970—no regulation. After 1970—illegal.*
Painting buildings bright colors so that they are highly visible from or across the lake	*35*	*No regulation.*
Placing excessive night lighting on dock, lawn and buildings	*23*	*No regulation.*
Planting large lawns and other exotic vegetation that doesn't blend into the forested shoreline	*36*	*No regulation.*
Building or improving a wet boathouse	*10*	*Prior to 1979—grandfathered. After 1979—illegal.*
Maintaining general clutter on property	*41*	*No regulation except that two or more junk vehicles on a lot violates general county zoning ordinances.*

FIGURE 14. Sensitive areas in the water and on the land.

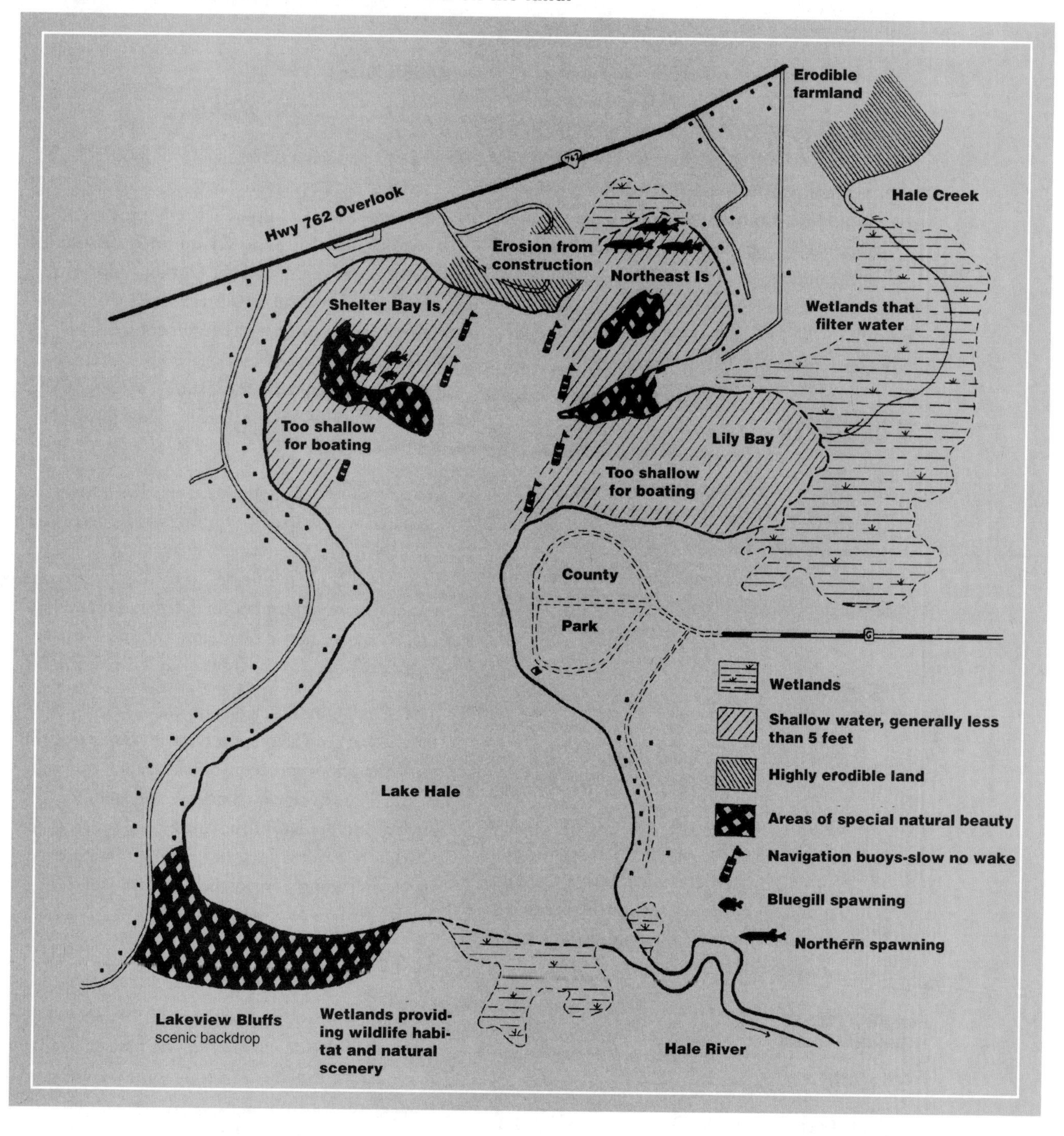

From considering options to making decisions

The following options were developed by Jerry Van Syke of Environmental Consultants after a series of meetings with the board, Jennifer Bates (UW–Extension), Daryl Roberts (DNR inland lake coordinator), Duane Peters (county code administrator) and Ted Walinski (County Land Conservation Department). The Board of Commissioners presented their recommendations to the Lake Hale District at a special meeting on April 3, 1993. **The Lake Hale Plan consists of the options adopted at that meeting.**

OPTION 1

Do nothing.

This alternative does not require spending money and, in the short run, allows us to continue to enjoy the lake rather than worry about the future. Few people voiced this opinion, and the option was not seriously debated.

OPTION 2

Dredge channels in Lily Bay.

Dredging channels would remove about 20,000 cubic yards of material at a cost of $4 to $10 per yard. While fishery habitat and boat access would be improved, the project could damage the native plant community if it was not carefully designed. A disposal area and permits would be required.
Vote to adopt: 33 yes, 106 no.

OPTION 3

Chemically treat excess plants.

With a permit from the Department of Natural Resources (Aquatic Plant Management), the district could hire a certified applicator and treat areas within 150 feet of shore. Application cost would be about $225/acre per summer.
Vote to adopt: 47 yes, 91 no.

OPTION 4

Harvest excess plants.

The excess vegetation in the north lobe could be harvested by a contractor who would charge $100 an hour, or we could purchase equipment and do the work ourselves. The Wisconsin Waterways Commission shares half the cost of equipment purchases. The estimated price of a 5-foot harvester plus conveying equipment is about $40,000. The cost of cutting 100 acres between June 20 and August 20 will run to about $10,000 per season. This option requires a feasibility analysis.

A vegetation management strategy has been prepared with assistance from Daryl Roberts of the DNR. The vegetation management strategy does not include herbicides. It focuses on physical methods to remove plant material and on better protection for native plant communities. It has a lower potential for controversy and divisiveness.

The Vegetation Management Map is shown in figure 15. A full statement of goals, objectives and implementation procedures is available from the board. The strategy includes cutting lanes for anglers and predator fish, cutting access across the north lobe twice a year, and cutting the shoreline vegetation along homeowners' property and the county park once a year.

After considering this option, the group decided to amend it to provide for three years of contract harvesting. It was then left to the 1995 annual meeting to decide whether to purchase a harvester, continue contracting, or revise the strategy for controlling excessive vegetation.
Vote to adopt: 98 yes, 39 no.

Implementation

- The Vegetation Management Committee, chaired by Sam Horsemann, will solicit a weed harvesting contractor based on the budget provided by the annual meeting. As part of a feasibility analysis, the committee will visit at least three communities that operate their own equipment and then make a recommendation to the annual meeting about purchasing equipment for Lake Hale. They will also provide information on the likelihood of the district receiving cost-sharing funds from the Wisconsin Waterways Commission.

- About 50-100 acres, primarily in the north lobe, will be cut one to three times each summer. The newsletter will provide free advertising for anyone offering or desiring shoreline clean-up services of individual lots (good employment opportunity for teens).
- The 1995 annual meeting will decide whether to purchase equipment.

FIGURE 15. Vegetation management map.

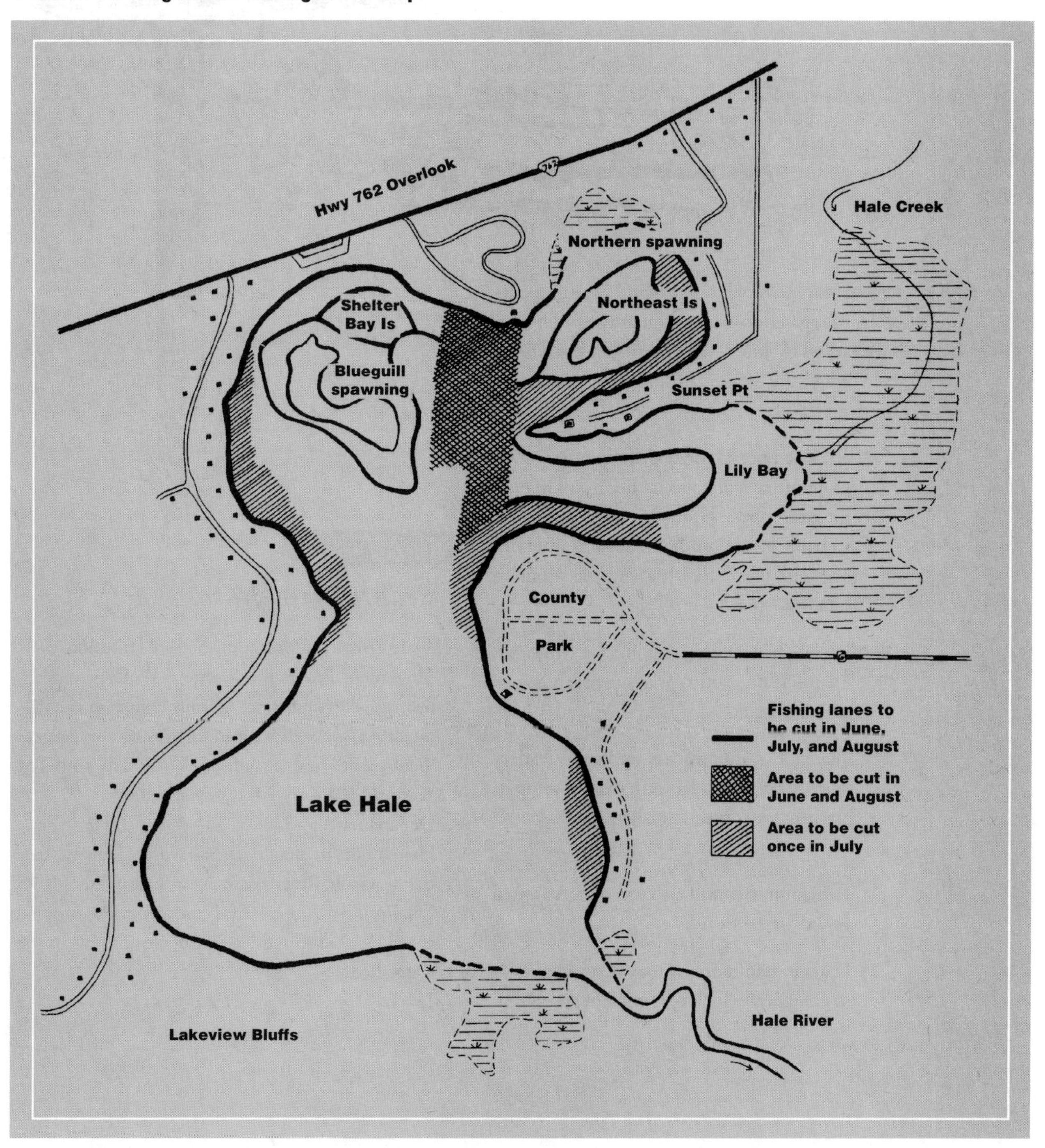

OPTION 5

Control alien species.

Eurasian water milfoil and purple loosestrife have invaded the lake and surrounding wetlands. Rusty crayfish and zebra mussels may also damage the aquatic ecosystem. The district could systematically monitor these invasions and develop prevention and control strategies as appropriate.
Vote to adopt: 139 yes, 0 no.

Implementation

- Carol Hern, who manages the county park, has agreed to erect a sign at the public access by July 1, 1993 warning boaters about the dangers of spreading exotic species.
- By December 31, 1993, the Exotics Committee will collect a library of materials on exotic species infesting lakes, subscribe to publications on the topic, and interview state and local officials about the situation in Minnesota.
- Starting in 1995, the committee will submit an article on exotic species for each issue of the newsletter.
- The Exotics Committee, under the leadership of Dr. Selma Kirkson, will develop a strategy for controlling alien species by 1996.
- The committee will also prepare an annual report to the board.
- The self-help monitor may be trained to systematically look for problem species.

OPTION 6

Reduce agricultural runoff.

The Wisconsin Nonpoint Source Pollution Abatement Program is a statewide effort to protect lakes and streams from pollution not directly associated with industries or sewage treatment plants (point sources). With help from Ted Walinski (County Land Conservation Department) and Daryl Roberts (DNR), Lake Hale might be designated a Priority Lake when the St. Croix River Basin Area-wide Water Quality Plan is updated. Funds from the program could be used to reduce runoff from farms in the watershed.
Vote to adopt: 110 yes, 21 no.

Implementation

- Commissioner Elder Tobatz has volunteered to provide leadership for this effort; as a county board member, he sits on the Land Conservation Committee. Ted Walinski will provide staff assistance.
- The district will apply for Lake Hale's designation as a Priority Lake during 1994.
- Priority Lake designation will be obtained and contracts signed by July 1, 1995.
- University of Wisconsin–Extension will provide educational support on best management practices for landowners with property in the watershed.
- Cost-sharing agricultural practices will begin by April 1, 1996.

OPTION 7

Reduce construction site erosion.

The lake district does not have the authority to regulate land use or construction practices. However, it can apply for a lake planning grant to hire legal experts to develop ordinances regulating construction site erosion control and stormwater management. The draft ordinances would then be considered by the Town of Meadowview Board and the Phantom County Board.
Vote to adopt: 126 yes, 10 no.

Implementation

- Jennifer Bates (UW–Extension) and Daryl Roberts (DNR) will help the district develop the planning grant application by August 1, 1994.
- A private attorney will be hired to draft the ordinances. Jennifer and Daryl will provide examples from other communities. Ordinances will be drafted by May 1, 1996.
- Jennifer and Duane Peters (code administrator) will hold educational sessions for local officials and interested builders and citizens.
- Duane Peters will be asked to advise the commission on the best strategy to get ordinances adopted. The ordinances will be adopted by September 30, 1996.

OPTION 8

Conduct a sanitary survey.

The Lake Hale District can request sanitary powers from the Town of Meadowview. Such powers allow the district to physically inspect septic systems, bore in drainfields, or use "snooper" equipment in the water in front of homes. Since correction orders would be issued through the Environmental Health Department, the county sanitarian would be involved; she might potentially involve the private sewage consultant from the Wisconsin Department of Industry, Labor, and Human Relations. A DNR lake planning grant could fund 75% of the effort.
Vote to adopt: 62 yes, 74 no.

OPTION 9

Purchase ecologically and aesthetically sensitive land.

Five parcels of land need protection.

1) The Sunset Point Park Association is largely defunct and the property owners are looking for a stronger organization to manage their private park.

2) Undeveloped shoreland between Sunset Point and Hale Creek and the wetlands immediately behind the shore berm are valuable fish and wildlife areas. Currently, two private owners hold these lands.

3) Alice Knight is willing to consider donating Shelter Bay Island to a responsible organization under a deed restriction that prevents the erection of any buildings on the island.

4) The Northern Heights Subdivision Association owns Northeast Island; covenants prevent any human activity on the island.

5) Lakeview Bluffs development, proposed for the southwest corner of the lake, would provide 185 homesites with a clubhouse and marina on the waterfront. The owners of these lots would overburden the lake and the construction would destroy an important vista.

Lake protection grants, available through the DNR Lake Management Program, could pay up to 50% of the appraised value of such properties. *Vote to adopt: 115 yes, 19 no.*

Implementation

- Elicia Horace, a Minneapolis attorney who knows Alice Knight, has volunteered to negotiate the donation of Shelter Bay Island to the district and expects to complete the transfer to lake district ownership by January 30, 1996.
- Bob Lark, a long-time resident of Sunset Point, will work with Elicia to obtain a consensus among the old Sunset Point Homeowners Association members to transfer the park to district ownership and management by July 1, 1997.
- Bob and Elicia will also pursue purchase or easements on the low land between Sunset Point and Hale Creek. They will also seek state lake protection grant funds for this purpose, which has a projected completion date of July 1, 1998.
- The Land Use Committee is so concerned about the potential development of backlots on Lakeview Bluff that it wants to continue fund-raising to purchase the property. A state grant and foundation support will be pursued, and the Nature Conservancy contacted. The committee hopes to complete the purchase by January 1, 1999, and pay off any mortgage by January 1, 2009. If possible, an option-to-purchase agreement will be negotiated immediately.

OPTION 10

Lobby for stronger enforcement of county zoning laws.

This option proposes that the Land Use Committee meet regularly with Duane Peters of the County Planning and Zoning Department to report illegal construction around the lake and share other concerns. Committee members could testify when variances are requested from the County Board of Adjustment or when rezoning cases go before the County Zoning Committee. For example, Dream Estates may attempt to have the Bluffs in Section 15 rezoned from a forestry to a residential area. Attempts may also be made to convert some farmland zoned A1 for "Exclusive Agriculture" to "Residential."

The Lake Hale District might join with other county lake organizations to follow up on stronger shoreland ordinances and more aggressive enforcement by the zoning office and the district attorney. Periodically, a formal zoning audit could be conducted with the assistance of the DNR. The Wisconsin Association of Lakes could advocate for stronger state legislation. *Vote to adopt: 112 yes, 25 no.*

Implementation

- If the district is unable to prevent rezoning or to purchase the Lakeview Bluff property owned by Dream Estates, the Land Use Committee will carefully monitor the development process and perhaps negotiate a development layout less damaging to the natural beauty of the bluffs and shoreline below. The committee will advocate that an independent lake capacity study be commissioned by the county and paid for by the developer.
- The Land Use Committee will serve as a Shoreland Watch and be expanded to seven members. The committee will regularly inform Duane Peters, the county code administrator, of building or remodeling that may not conform to shoreland zoning or other regulations. The committee will meet with Duane at least twice a year.

- Elder Tobatz, our county supervisor, is being encouraged to ask for a seat on the County Planning and Zoning Committee.

- The Land Use Committee will attempt to get one of its members appointed to the next vacancy on the County Board of Adjustment. At least one member of the committee will attend all County Planning and Zoning meetings and Board of Adjustment meetings.

- A zoning audit will be completed by December 31, 1995 and again by December 31, 2005. The county district attorney will prosecute at least one shoreland zoning violation each year. All wet boathouses will be removed from the lake by 2020.

OPTION 11

Operate a water safety patrol.

State funds are available to share the cost of operating a water safety patrol. If the Town of Meadowview delegated its authority to the Lake Hale District, the district could adopt its own ordinances and operate the patrol. A trained law enforcement officer, a patrol boat, and a citation system would be needed. The lake could be zoned for different uses as shown on the proposed lake use map (fig. 16). Some of these regulations already exist through the Town of Meadowview, but lack enforcement. Jennifer Bates (UW-Extension) could help the community arrive at a consensus. The ordinances would be reviewed by the DNR boating safety specialist.

Members amended this option to direct the commissioners to study the seriousness of lake use conflicts and report their findings and recommendations at the 1996 annual meeting.
Vote to adopt: 89 yes, 41 no.

Implementation

- A new Recreational Use Committee will be established to monitor the conflicts between lake users—both on the water and between water and shoreland users. The committee will provide complaint forms, summarize the results, add its own observations, and make recommendations to the board and the annual meeting. The district may recommend changes in the operations of the county park and in patrolling by the DNR warden, or the county sheriff's deputies. Subject to approval at the annual meeting, the committee may seek authority from the Town of Meadowview to adopt a more detailed lake use ordinance.

- Complaint forms to register conflicts between lake users will be available at the county park bulletin board (1995) and in the lake district newsletter (1994-95). The forms will be tabulated to document user conflict and indicate trends for discussion by the 1996 annual meeting.
- Residents are also being encouraged to videotape boating violations for review by the committee and the DNR conservation warden.

FIGURE 16. Lake use map.

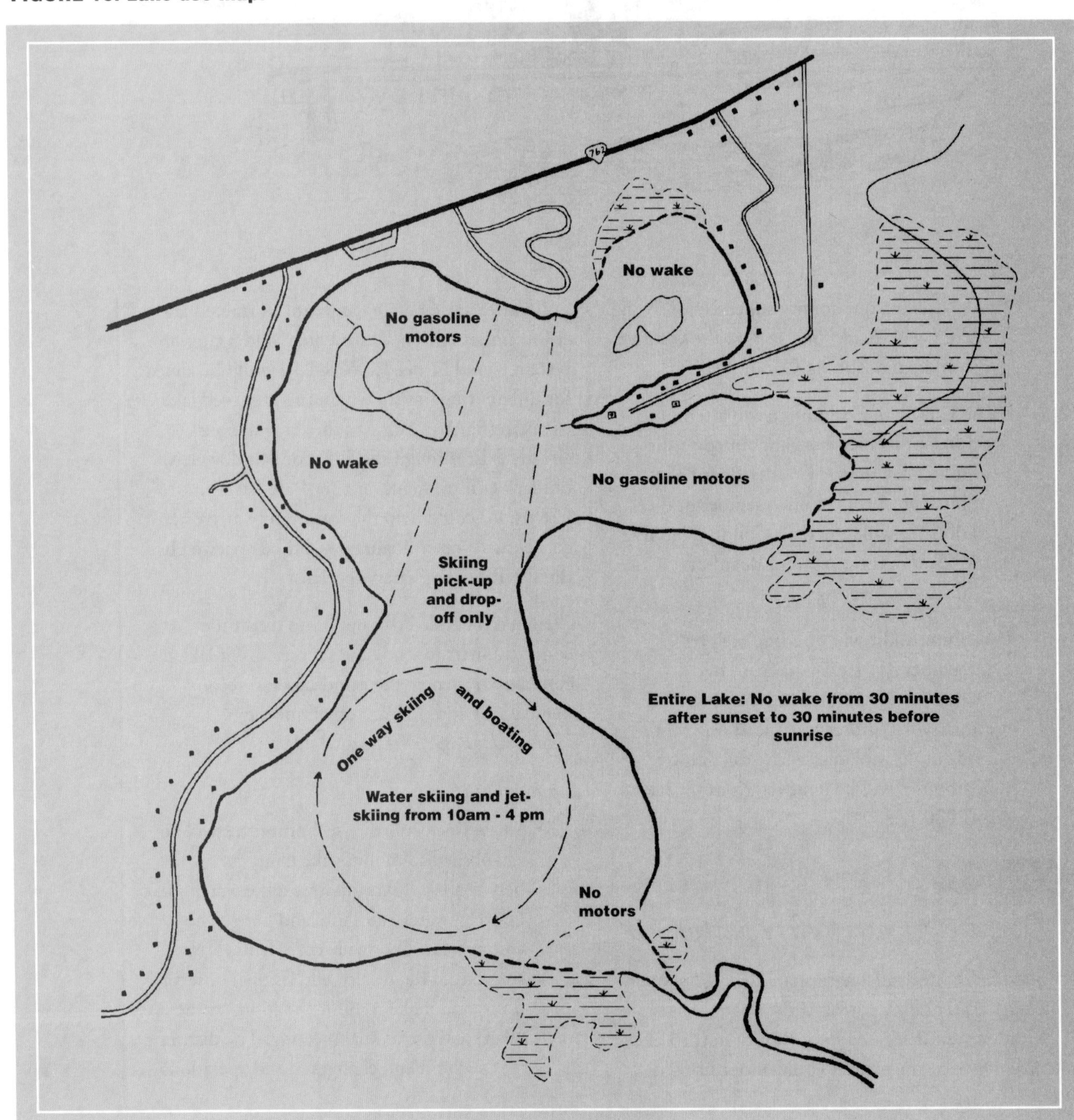

OPTION 12

Conduct an intensive educational effort.

Shoreland property owners and public users often unwittingly damage the lake ecosystem, making the recreational experiences of others less enjoyable. Many people would probably change their behavior if they realized they were harming the lake or other people. While such education does not cost very much in terms of dollars, it requires a long-term commitment and a lot of persistence. To educate our members, we could publish a newsletter, continue our monitoring efforts, include an educational element in each annual meeting agenda, and circulate videos on lake-related issues. To educate our leaders, we could require that they attend the Wisconsin Lakes Convention. To educate the general public, we could maintain an informational bulletin board and stock brochures at the county park and the Lake Hale overlook.
Vote to adopt: 136 yes, 3 no.

Implementation

- Susan Bukoltz has volunteered to edit the newsletter. She will be assisted by Harry Holtz, publisher of the *Phantom County Reporter*. Newsletters will be sent to all district property owners and residents, the town board, Elder Tobatz (County Board Supervisor), Ted Walinski (County Land Conservation Department), Duane Peters (County Planning and Zoning), Jennifer Bates (UW–Extension), Chris Harms (DNR Conservation Warden), Daryl Roberts (DNR Inland Lake Coordinator), and staff of the Extension Lake Management Program, UW-Stevens Point, in May, August and January. The problems identified in the 1989 nominal group process and the surveys that followed will receive special attention. The newsletter will direct readers to further sources of information such as the BBS-Lakes electronic bulletin board available at 800/562-5552.
- Kathie Jansen will continue as the self-help volunteer. She and Art Belder have taken more than 85 readings in the past six years. Hank Arnold has agreed to assist with the expanded program of water chemistry analysis. The DNR will continue to store the information and provide an annual report.
- The secretary will be responsible for arranging a 20- to 40-minute educational program at each annual meeting with Jennifer's assistance.
- Tiny Starr will contact public agencies and conservation groups to obtain a video library. "All Night Video" at the intersection of the freeway and Highway 762 has agreed to house and distribute the educational videos free of charge, beginning in August 1994.

- At its fall meeting, the board will provide funding for four district representatives to attend the annual spring meeting of the Wisconsin Association of Lakes held in conjunction with the Wisconsin Lake Convention. Other interested citizens will also be encouraged to attend.
- Joyce Sears and Elder Tobatz will contact the county about establishing a bulletin board at the county park. Permission to build the board will be obtained by March, 1995, and construction completed May 15, 1996. Harold Route, who just retired as district engineer with the Wisconsin Department of Transportation, will contact the department about erecting a display at the Highway 762 overlook. Permission to build a display at the overlook will be obtained by December 31, 1994, and the display will be built by July 1, 1995. Hilary Opitz has agreed to provide the commission with a draft design for the display. Wes Dirkson will draft a brochure which will be available by May 1, 1995.
- By January 1 of each year, four community leaders will have agreed to attend the Wisconsin Lake Convention.

Timelines summary

Activity	Year completed or continuing										
	1993	1994	1995	1996	1997	1998	1999	2000	2005	2010	2020
1. District leaders attend Wisconsin Lakes Convention	■	■	■	■	■	■	■	■	■	■	■
2. Three issues of newsletter published per year	■	■	■	■	■	■	■	■	■	■	■
3. Public access sign on exotic species erected	■										
4. Apply for planning grant or construction site erosion and stormwater management ordinance	■										
5. Contract harvesting	■	■	■								
6. Land Use Committee meets semi-annually with County Code administrator	■	■	■	■	■	■	■	■	■	■	■
7. Land Use Committee member attends county zoning meetings	■	■	■	■	■	■	■	■	■	■	■
8. Collect materials on exotic species	■	■	■	■	■	■	■	■	■	■	■
9. Self-help monitoring	■	■	■	■	■	■	■	■	■	■	■
10. Educational program at annual meeting	■	■	■	■	■	■	■	■	■	■	■
11. Vegetation Management Committee visits other communities	■	■									
12. County district attorney prosecutes one shoreland violation	■	■	■	■	■	■	■	■	■	■	■
13. Complaint forms available to document conflicts between lake users	■	■	■								
14. Permission to erect bulletin board in County Park		■									
15. Article on exotics in each newsletter		■	■	■	■	■	■	■	■	■	■
16. Video library of lake materials available		■	■	■	■	■	■	■	■	■	■
17. Apply for Priority Lake status		■									
18. Permission to build display at overlook		■									
19. Display at overlook built			■								
20. Bulletin board erected in county park			■								

Timelines summary (continued)

Activity	Year completed or continuing										
	1993	1994	1995	1996	1997	1998	1999	2000	2005	2010	2020
21. Construction site erosion and stormwater management ordinances drafted			■								
22. Sign contracts for Priority Lake work			■								
23. Decision on purchasing harvester			■								
24. Zoning audit conducted			■						■		
25. Transfer Shelter Bay Island to lake district				■							
26. Cost share ag practices under Priority Lake grant				■							
27. Recreational Use Committee recommendations on conflict management				■							
28. Present strategy to control exotic species				■							
29. Construction site erosion and stormwater management ordinances adopted				■							
30. Transfer Sunset Point Park to lake district					■						
31. Purchase lands or easements between Sunset Point and Hale Creek						■					
32. Purchase Lakeview Bluff							■				
33. Pay off mortgage for Lakeview Bluff										■	
34. All wet boathouses will be removed											■

Have we been successful? Evaluating our efforts

We have put forward an ambitious effort to protect Lake Hale. Our success will depend on the volunteer efforts of many people, and each of us will have a highly personalized perception of our success.

A more objective evaluation of our efforts can be made by checking the implementation boxes in the timelines summary. It will be fairly easy to determine if we have achieved these objectives. Of course, we will not meet all of them according to our timelines. We may not meet some of them at all. But such an evaluation will help us understand how well we have done. More importantly, it will help the next generation, entrusted with the stewardship of this lake, to plan for the care of Lake Hale—as we are doing on our watch.

Mini-directory

OUR COMMISSIONERS

Joyce Sears, Chair
OFFICE PHONE: 612-414-2220
LAKE ADDRESS: 2502 County G
Phone: 888-2627
HOME ADDRESS: Same
Phone: SAME

Paul O'Malley, Sec.
LAKE ADDRESS: 4271 W. Lakeshore *(Lake Hale resident)*
Phone: 888-1841
HOME ADDRESS: Same
Phone: SAME

Peter Synch, Treas.
OFFICE PHONE: 612-748-1111
LAKE ADDRESS: 18 Northern Hgts
Phone: 888-1401
HOME ADDRESS: 2711 Little John, Bloomington, MN
Phone: 612/776-4748

Sarah Robertson, Town of Meadowview
OFFICE PHONE: 721-2018
HOME ADDRESS: 1801 Halverson Rd.
Phone: 888-4678

Elder Tobatz, Phantom County
OFFICE PHONE: 721-2018
HOME ADDRESS: 1879 Halverson Rd.
Phone: 888-7172

OUR TOWN BOARD MEMBERS

Sarah Robertson, Chair
OFFICE PHONE: 721-2018
HOME ADDRESS: 1801 Halverson Rd.
Phone: 888-4678

Dave Tobatz
OFFICE PHONE: 888-1061
HOME ADDRESS: 2602 Halverson Rd.
Phone: 888-4678

Adolph (Tiny) Tonnes
HOME ADDRESS: 1890 Townline Rd.
Phone: 888-2686

OUR COUNTY BOARD REPRESENTATIVE

Elder Tobatz
HOME ADDRESS: 1879 Halverson Rd.
Phone: 888-7172

COUNTY OFFICES

Land Conservation, Ted Walinski
Rm 14 Courthouse, Phantom City
Phone: 721-1818

Planning & Zoning, Duane Peters
Rm. 180 Courthouse, Phantom City
Phone: 721-4601

Environmental Health, Tasha Holman
Rm 184 Courthouse, Phantom City
Phone: 721-4445

University of Wisconsin-Extension, Jennifer Bates
Rm 101 Courthouse, Phantom City
Phone: 721-4422

WISCONSIN DEPARTMENT OF NATURAL RESOURCES

Conservation Warden, Chris Harms
1801 Oak St., Phantom City
Phone: 721-4701

DNR Inland Lakes Coordinator, Daryl Roberts
DNR District Office, Eau Claire WI
Phone: 467-1531

Editor's note: *For the actual names, addresses and telephone numbers of community leaders and agency resource people in your locality, request a copy of the* Lake List *from your county Extension office.*

Authors: Lowell Klessig is a professor of human dimensions of natural resource management at the College of Natural Resources, University of Wisconsin–Stevens Point and a lake management specialist, University of Wisconsin–Extension, Cooperative Extension. Buzz Sorge is the inland lake coordinator for the western district of the Wisconsin Department of Natural Resources. Robert Korth and Michael Dresen are Extension lake management specialists with the College of Natural Resources, University of Wisconsin–Stevens Point. Jeff Bode is Chief, Lake Management Section of the Wisconsin Department of Natural Resources.

Illustrations: *Carol Watkins*
Maps: *Ruth King*

Issued in furtherance of Cooperative Extension work, Acts of May 8 and June 30, 1914, in cooperation with the U.S. Department of Agriculture, University of Wisconsin–Extension, Cooperative Extension.

This publication is available from your Wisconsin county Extension office or from Cooperative Extension Publications, Rm. 245, 30 N. Murray St., Madison, WI 53715, (608)262-3346.

Partial support for production and distribution provided by the Renewable Resources Extension Act, United States Department of Agriculture.

G3606 ***A model lake plan for a local community***

CHAPTER 4

Problem Identification

A lake problem is something that prevents you from using the lake the way you want to (Chapter 1 defines this as a "limitation"). You can usually identify lake problems by simply listening to lake users' complaints. When boat owners find they can't use the lake because it's choked with weeds, for example, they have clearly identified a problem.

This chapter will help you (see Table 4-1) — and all lake users, managers, and associations:

- Identify problems.
- Put problems in perspective for a specific lake.
- Understand how to diagnose the causes, not the symptoms, of problems.
- Define the causes of the lake's problems.

Common Lake Problems

A number of lakes within a region may suffer similar problems; rarely is a problem unique to a particular lake. The next few pages address the most widely occurring lake problems, ranging from algae to user conflicts.

Identifying the problem is but the first action in the process of reaching a solution; you will take a number of other steps before you learn enough to prepare a plan. This chapter also directs you to appropriate parts of this manual that will help you evaluate alternatives for solving these problems.

Algae

A source of food and energy for fish and other lake organisms, algae are a vital part of a lake ecosystem. Elevated nutrient levels can produce too many algae, resulting in noxious blooms in the water column or on the nearshore lake bottom and on rocks or aquatic plants. Large algal growths reduce water clarity and inhibit the growth of other plants; they can also deplete oxygen and cause fishkills, as well as taste and odor problems in water and fish.

Filamentous: long, thin cylindrical cells attached one to another.

But most of all, excessive algae are ugly; their blooms and tangled, filamentous masses certainly destroy the aesthetic pleasure of viewing the lake. Colonial and filamentous blue-green algae usually cause these unsightly scums, although other algae can also form blooms and mats.

Table 4-1.—Summary of Chapter 4.

Common Lake Problems	
Excess algae	Produce unsightly algal blooms
Excess attached plants	Restrict lake use
Exotic species	Reduce native species and use of lake or reservoir
Shallow water depth	Restricts boating and swimming
Turbid water	Reduces aesthetic values
Toxins	Restrict fish consumption
Acidity	Low pH causes reduction in biological community
Salinity	High salt levels restrict the biological community
Swimmer's itch	Restrict swimming
Leeches	Restrict swimming
Fecal coliforms, pathogenic bacteria, and enteric viruses	Cause illness, infections, rashes
Undesirable fishery	Increases turbidity and nutrients
User conflicts	Reduce boating and swimming
Taste and odor	Affect aesthetics and drinking water
Problem Identification	How the water quality of a lake or reservoir compares with other water bodies in the region
Obtaining Professional Advice	Selection of competent advice
Information Sources	Gather background information about your lake or reservoir
Data Collection and Analysis	
Sediment cores	Determine water quality history
Water and nutrient budgets	Determine contribution of precipitation, surface water, and ground water to lake or reservoir
Monitoring water quality	Where and when to sample a lake or reservoir
Physical Variables	
Sedimentation rate	Estimate rate water body is filling with sediment
Temperature	Amount of stratification
Transparency	Water clarity
Chemical Variables	
Dissolved oxygen	Important for fish and nutrient recycling
Nutrients	Elevated amounts cause algal and plant problems
Metals and organics	Important for fish consumption advisories
Acidification	Lowers pH; reduces fish production
Biological Variables	
Bacteria and pathogens	Important for safe swimming
Algae	Measure size and frequency of algal blooms
Macrophytes	Measure density and distribution of attached plants
Zooplankton	Important as fish food and controlling algal levels
Animal nuisances	Exotics that cause lake problems
Fish	Determine composition of fish community
Trophic State Indices	Compilation of measured parameters to assess water quality
Examples of Using Data to Manage Your Lake	
Cedar Lake	Example of a shallow lake
Mirror Lake	Example of a deep lake

Aquatic Plants (Weeds)

Aquatic plants are very beneficial to lakes: fish and macroinvertebrates live among them; zooplankton find refuge there from predatory fish; and aquatic plants stabilize sediment.

But, in excess, aquatic plants are indeed a nuisance. Too many plants limit swimming, fishing, skiing, boating, and aesthetic appreciation. An all too common problem in many lakes, excess plants are usually caused by high nutrient levels, invasions of exotic species, and a low water level.

Exotic Plants and Animals

Exotic — or non-native — species have become more of a problem in the last few decades. Most of these exotics are inadvertently brought to North America from Europe and sometimes Asia.

- One of the most dramatic historical examples was the invasive **sea lamprey** (*Petromyzon marinu*) in the Great Lakes system. This animal nearly depleted the lake trout and other salmonid fishery.

- Lakes have also suffered from exotic plants, such as **Eurasian water milfoil** (*Myriophyllum spicatum*) and **Hydrilla** (*Hydrilla verticillata*), which can limit the diversity of the lake's plant community. These plants frequently are a problem because they grow close to the lake surface. Because these plants can grow in great densities, they can obstruct boating and swimming and reduce aesthetic enjoyment. These high densities often change the fish community as well, favoring growth of panfish, such as bluegills, over larger gamefish.

- The **common carp** (*Cyprinus carpio*), introduced into North America in the 1800s, also can have detrimental effects on the lake's ecosystem. These fish are bottom feeders so they may uproot aquatic plants and stir up sediments, thereby reducing water clarity and contributing nutrients to the water-sediment interface. This is especially a problem in shallow lakes. Often these fish must be drastically reduced to improve water quality (Meijer et al. 1990).

- A more recently introduced exotic, the **zebra mussel** (*Dreissena spp.*), can have dramatic effects on lake food webs. While zebra mussels make the water much clearer, they produce far less energy for use by the higher trophic levels such as zooplankton, macroinvertebrates, and fish. The **Asiatic clam** (*Corbicula manillensis*) has a similar impact on the lake and reservoir ecosystem.

- Another invader that can have dramatic effects upon a lake is **purple loosestrife** (*Lythrum salicaria*). Although this ornamental plant blooms with pretty purple flowers, its invasive nature often excludes most other native plants that also like the water's edge. Loosestrife is also much less useful to wildlife than native species.

Shallow Water Depth

Many lakes and reservoirs lose volume — and thus, depth — as **sediment** fills in the lake, either by eroding from the watershed or originating from decaying algae and weeds in the lake itself. Increased sediment generally leads to turbid or murky water, and reduction in depth usually disrupts swimming, boating, and sailing and encourages extensive weed growth.

Dredging is one of the major lake restoration approaches used to restore depth, but it doesn't stop soil erosion in the watershed, which is the main cause of lake infilling.

Turbid Water

Turbidity: clouded water, usually because it has been stirred up or has excessive algae.

Turbid water can result from **excessive algae and/or sediment**. Sediment usually increases as a result of soil erosion in the watershed following storms; high levels of sediment are found more frequently in reservoirs and lakes with major inflowing streams. In western North America wind-blown soil can increase turbidity.

Fish can also cause turbid water, especially in shallow lakes. Benthivorous (bottom-feeding) fish such as carp and bullheads frequently stir up the water when they feed and mate. A study in the Netherlands (Meijer et al. 1990) found that reducing the carp population decreased nonalgal turbidity.

Toxins

Toxic compounds, such as pesticides or heavy metals, sometimes create problems in lakes. Toxic compounds can come from discrete sources like wastewater or industrial discharges, but can also be carried by nonpoint source runoff — and atmospheric deposition. Their principal effect is to restrict human consumption of fish.

Mercury is an example of a toxin that usually does not enter lakes from the immediate watershed, but from the atmosphere. While mercury levels in the water itself are usually not a problem, methyl mercury, an organic compound formed from mercury, can bioaccumulate in the food chain. As a result, top level predators such as gamefish can have elevated mercury levels. In nearly all cases, the mercury has been deposited from the atmosphere; thus, the sources can be far upwind. Many other toxic compounds, such as PCBs, can enter lakes through atmospheric deposition.

Acidity

Increases in lake acidity can radically change the community of fish and plant species in lakes and can also make toxic substances more soluble and magnify their adverse effects. Like some other toxins, the source of acidity usually is atmospheric. Acid precipitation can be derived from sulfur dioxide emitted from industrial sources or from nitrogen oxides in vehicle exhaust upwind from the lake. For lakes to be sensitive to acidification, they must be poorly buffered. This means the water contains low amounts of chemicals that neutralize acidity. Not all regions of North America are susceptible to acidification but there are sensitive lakes in the north central and northeastern United States, New Jersey, Florida, high elevation lakes in the western U.S. and Canada, and parts of eastern Canada.

Salinity

High salt concentrations are usually a problem in the low rainfall climates of western North America. Irrigation water used to grow crops results in leaching of high salinity groundwater into downstream water bodies. Irrigation return water also is high in salts because evaporation concentrates them. High levels of salinity can harm the aquatic ecosystem, sometimes drastically altering fish and plant communities.

Swimmer's Itch

Swimmer's itch is caused by the trematode *Schistosome dermatitis*, a parasitic flatworm that lives in birds. Snails act as intermediate hosts of the trematode (also known as a fluke), which can cause itching when it penetrates human skin. Swimmer's itch can be a problem in lakes populated by snails and waterfowl. Since normally the trematode doesn't live in humans, these organisms die in the skin and produce severe itching, but do not cause long-term effects.

Leeches

Leeches commonly live in lakes and can become a nuisance on swimming beaches because they attach to humans. They don't cause physical problems — just annoyance!

Fecal Coliforms, Pathogenic Bacteria, and Enteric Viruses

In recreational waters swimmers may contract gastrointestinal illness, skin rashes, and ear and eye infections from contact with water contaminated by fecal coliforms, e.g., Escherichia coli. Pathogenic bacteria, e.g., Salmonella, and enteric viruses may also be present. These organisms come from livestock and wildlife excrement, failed on-site wastewater disposal systems, and urban runoff (especially in areas with combined sewer overflows, i.e., communities that allow stormwater systems to accept sewage during periods of high rainfall). See American Water Works Ass. (1990) for detailed information about these organisms.

Undesirable Fishery

Three major factors can upset the balance between panfish and gamefish; these include too many nutrients, too little oxygen, and acidification.

- When **nutrients increase**, you may see larger numbers of both stunted panfish and bottom dwellers such as carp and bullheads. As noted previously, these bottom dwellers stir up sediments and thus, can greatly increase turbidity and internal loading of nutrients.

- Coldwater fishes such as trout and salmon often live in the deeper waters of a lake during the summer, preferring temperatures under 18°C, and dissolved **oxygen** levels of at least 5 mg^{-1}. To maintain a coldwater fishery, the colder waters must have sufficient oxygen.

- **Acidification** can also shift the fishery's balance, largely by retarding fish reproduction. More severely affected species may even disappear.

User Conflicts

Even though boating has always been one of the great joys of lakes — whether for water skiing, fishing, or just aesthetics — the large increase in boating in recent years (Penaloza, 1991) has produced conflicts among lake users.

To quantify this problem, Wagner (1991) reported that more than one motorboat per 25 acres can be offensive, affecting the quality of the water and sediment, the flora and fauna, even the stability of the shoreline. Asplund and Cook (1997) demonstrated that motorboats can reduce macrophyte height and density. Noise and overcrowding created by motorboats and personal watercraft (such as jet skis) can also interfere with aesthetic enjoyment.

Time and/or space zoning seems to be the most widely used answer to this problem. Time zoning restricts certain uses to specific times of the day or days of the week, while space zoning confines certain uses to specific areas (Engel, 1989a,b; Jones, 1996).

Taste and Odor

Taste and odor problems are usually related to excessive algae. Algal blooms produce odors as the algae die and decay — and they can often be quite offensive!

Taste is more noticeable when the water is used for drinking but it can also affect the taste of fish. Again, excess algae are probably the culprit.

Problem Identification

Depending on physical characteristics of the lake basin and the watershed, and the quality of incoming water, lakes are suited to different purposes.

- Reservoirs, for example, often are more turbid than natural lakes.
- Some lakes can never be crystal clear, no matter what you do.
- If the watershed is large relative to the lake surface — with highly erodible, nutrient-rich soils — your lake will always have excessive algae and weeds regardless of what you do.

Ecoregions: Comprised of relatively homogenous ecological systems delineated by geology, soils, climate, vegetation, and landform, and involving interrelationships among organisms and their environment.
—Omernik, 1987

Regional differences across the country are also important in understanding how best to manage your lake. Its quality will be determined by the ecoregion in which your lake lies (Omernik, 1987): its geology, soils, land use, and vegetation. Lakes in northern Minnesota, for example, have lower nutrient and algal concentrations and greater transparency than lakes in southern Minnesota where the soil is more naturally fertile (Heiskary et al. 1987).

Causes of Lake Problems

Your lake's problems probably resemble those of most lakes in the same ecoregion, but to identify what causes them you must understand the interactions both within your lake (among algae, macrophytes, fish, and other organisms) and between your lake and its watershed (see Chapters 2 and 6).

A natural combination of these factors may dictate that a lake will always be highly biologically productive; thus, it would be useless to try to transform it into a perfectly clear lake. If, however, people's activities have caused your lake's problems, then these effects can be reversed by combining management efforts in both the watershed and the lake itself.

To distinguish natural from people-caused problems, look at other lakes in the same region. If the water quality in some resembles yours, and they lie in relatively undisturbed watersheds, then those lakes' problems might be natural. But, if other lakes in relatively undisturbed watersheds have much better water quality, then people are probably contributing to the first lakes' problems.

If the water quality in some [lakes] resembles yours, and they lie in relatively undisturbed watersheds, then those lakes' problems might be natural. But, if other lakes in relatively undisturbed watersheds have much better water quality, then people are probably contributing to the first lakes' problems.

Other lakes in the region with relatively undisturbed watersheds make a good initial reference point for assessing the effects people may have on your lake.

You can use numerous tools to identify the causes of your lake's problems:

- Qualitative approaches, such as comparing the target lake to surrounding lakes, document subjective observations, which can reveal important patterns.
- Quantitative approaches, such as the models discussed in Chapter 5 and trophic state indices, rely on objective data.

In practice, both qualitative and quantitative approaches are usually considered.

Using these methods to identify underlying causes of problems usually requires professional assistance. An important step in defining your lake's problem, therefore, is selecting competent professional advice.

Obtaining Professional Advice

State or regional government or university personnel may be available to advise you on your lake's problems. These professionals may be county or state lake professionals, regional planning agencies, or University Extension personnel. Another source of guidance is the North American Lake Management Society (NALMS). Check out their web site (www.nalms.org) for additional information. The U.S. Environmental Protection Agency also has a good web site at www.epa.gov/owow.

Government and university professionals, however, may not actually conduct the diagnostic study. Often, this is done by a private consultant or as part of a university study.

NALMS certifies lake managers: individuals who have satisfied NALMS' requirements to possess the knowledge and experience to understand and recommend solutions for the comprehensive management of lakes, ponds, and reservoirs.

Certified Lake Manager: individual who has satisfied NALMS' requirements to possess the knowledge and experience to understand and recommend solutions for the comprehensive management of lakes, ponds, and reservoirs.

It is very important to select a competent professional with a proven track record investigating the types of problems your lake or reservoir may have.

Among the criteria to consider when selecting a consultant are:

- The candidate's (or firm's) experience in conducting lake studies, identifying the underlying causes, and formulating effective lake management plans;
- Expertise in limnology, biology, engineering, or other disciplines associated with lake management;

- Past performance in conducting similar studies or dealing with similar problems; and
- The firm's or candidate's capabilities (support staff, office facilities, equipment) to address the problems in the lake.

Information Sources

Before you begin to analyze your lake, obtain all existing information on both the watershed and the lake. Watershed districts, sanitary districts, county extension offices, county soil and water conservation districts, and city, county, and regional planning agencies usually have maps, land-use data, or aerial photographs of the watershed and lake. In addition:

- Water quality data may be available on the inflowing streams or the lake itself from state water quality agencies and federal agencies such as U.S. Geological Survey, U.S. Fish and Wildlife Service, U.S. Bureau of Reclamation, and U.S. Army Corps of Engineers.
- Fishing maps might be available that show the surface area, depth contours, location of inflowing streams, coves, and embayments, and other features of the lake that can be important in the diagnosis.
- Recent aerial photographs taken during mid- to late summer can show the extent of plant beds in the lake.
- Creel census records from state fish and game agencies can provide valuable information on historical changes in the fish community and lake productivity.
- Watershed land-use and topographic maps can help determine the location and acreage of various types of crops in the watershed and the soil types, including their potential for erosion; and the location of feedlots and barnyards, residential developments, forested and open land, and conservancy districts.
- The locations of wastewater treatment plants, industrial discharges, and storm sewers can be obtained from the sanitary district, city health department, and state natural resource or pollution control agencies.
- Discharge data and data on organic matter (for example, BOD) and nutrient concentrations in the wastewater discharge usually can be obtained from the wastewater treatment plant's discharge monitoring records (required by the U.S. Environmental Protection Agency).
- Estimates of annual runoff of water from the watershed or the amount of stream inflow to the lake might be available from the city or county planning agencies, U.S. Geological Survey, or the Natural Resources Conservation Service.
- Locations of groundwater wells in the watershed also might be available from these agencies, the local health department, or pollution control agencies.

- If the lake or reservoir supplies drinking water, source water assessments may be available from your state agency that manages drinking water programs.
- State and federal agencies may have GIS systems with watershed information in place. If the water body is a reservoir, the utility or federal agency who built it will have information.

The potential sources of nutrients, sediments, and organic matter from farms, wastewater treatment plants, urban areas, and forests can be identified. Many have been studied and some general nutrient and sediment export coefficients associated with various land uses have been published (Reckhow et al. 1980). You can combine these land-use coefficients with the annual runoff coefficients and wastewater discharge estimates to estimate the total load of material to the lake as detailed in Chapter 5.

Contact government agencies to see if load allocations (Total Maximum Daily Loads [TMDLs]) have been established for your lake or reservoir. TMDLs establish target concentrations for specific pollutants, such as nutrients; an analysis is performed to determine the maximum daily load that is allowed from various pollutant sources such as agricultural or urban. If a TMDL has been established for the lake then much of the necessary diagnostic work has already been completed. See Chapter 6 for more details.

TMDL: a target concentration (total maximum daily load) for specific pollutants.

Data Collection and Analysis

To refine the diagnosis, you will generally need more data. Your preliminary analysis — the existing information you've already collected — will tell you what to look for.

- If agricultural runoff appears to be a major contributor of nutrients and sediments, for example, then you need better estimates of loading from the various agricultural locations in the watershed to determine which ones are contributing the most to the lake.
- Wastewater discharges to a lake are usually an important source of nutrients and organic matter. Collect samples to determine the relative contribution to the lake from wastewater treatment plant effluent, stormwater sewers, and septic tanks. Estimating input from private waste disposal systems is covered later in this chapter under the groundwater section.

Sediment Cores

Often, little or no long-term data exist for a lake, making it difficult to know if its water quality has even changed, let alone how much it has deteriorated, or what has caused it. If sufficient funding is available, the best way to obtain these historical data is to conduct a paleolimnological study of the lake sediments. This will give you a record of how the lake has been disturbed by both natural and anthropogenic processes. This should be performed by a professional, e.g., a paleolimnologist.

Although only one sediment core is usually needed, if your lake is large or has multiple basins you may have to take more than one core (especially if sub-basins reflect differing perturbations). Date the cores to establish a timeline against which to measure water quality changes. Use either lead-210 or cesium-137 measurements. Lead-210 works best in lakes while cesium should be used in water bodies that are less than 130 years old. (See the section on Estimating the Sedimentation Rates.)

To evaluate changes in trophic state, examine biological remains such as algae, macrophyte, zooplankton, and insects. These parameters reflect in-lake water quality changes in nutrients, macrophyte species, anoxia in the hypolimnion, and fish predation. Changes in chemical and physical variables in the sediments document water and airshed perturbations.

For example, increased soil erosion may be traced by a higher accumulation of aluminum and zinc from urban runoff. Sediments also preserve trace metals such as lead or mercury from industrial emissions. Most important for eutrophication are the historical nutrient changes — increased deposition of phosphorus and nitrogen — that sediments preserve.

Figure 4-1 (Garrison and Wakeman, 2000) shows a paleolimnological study of Long Lake, a 1,000-acre-deep drainage lake in northwestern Wisconsin. The watershed was completely forested until logging began in the late 1880s. This initial logging

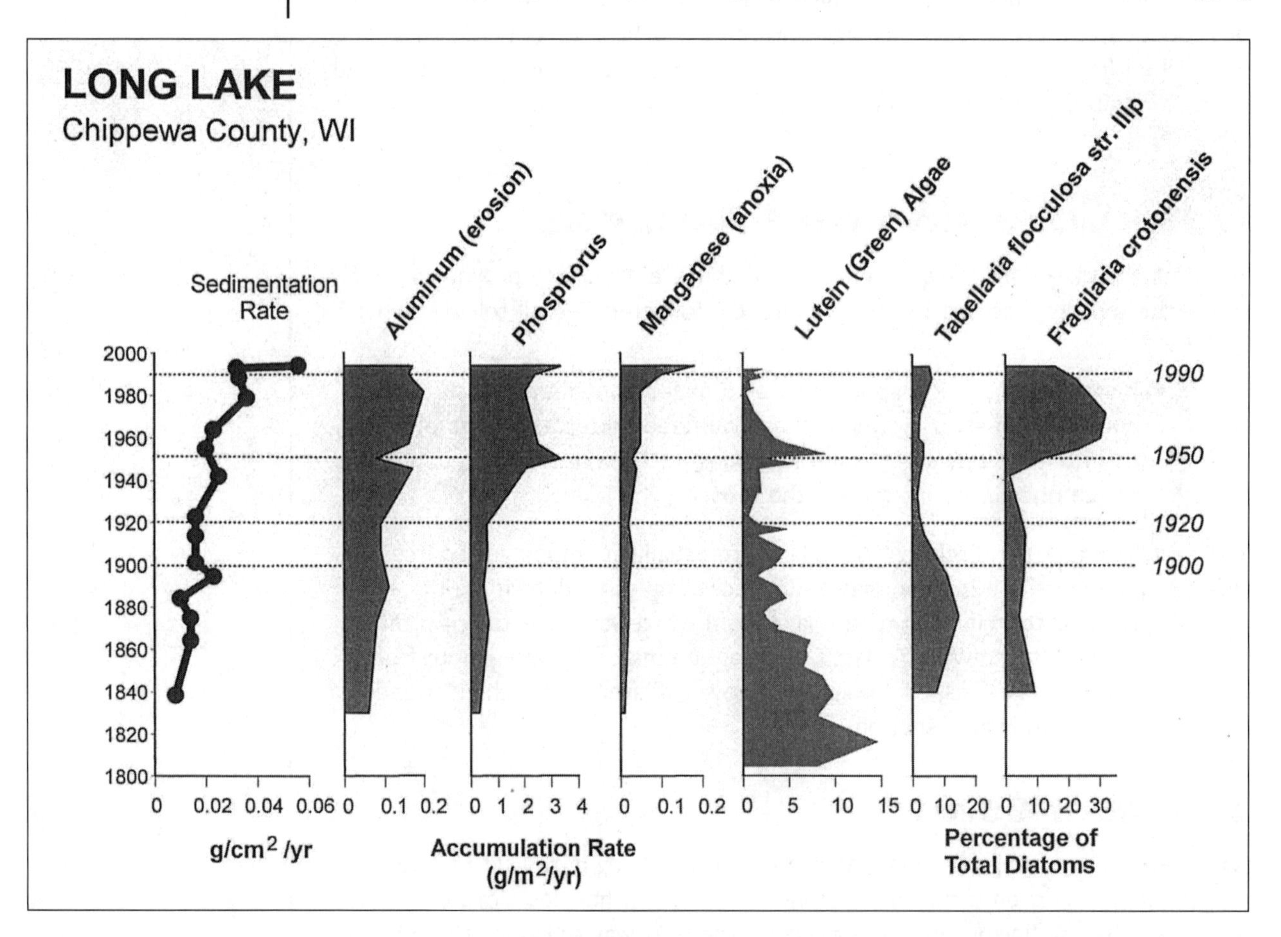

Figure 4-1.—Sediment core from a deep drainage lake indicating how watershed disturbances affect a lake's water quality (Garrison and Wakeman, 2000). Initial disturbance occurred in the late 1800s during widespread logging in the watershed. However, extensive shoreline development beginning in the 1950s had a much greater impact on the lake, resulting in increased nutrient delivery to the lake as indicated by elevated phosphorus accumulation as well as dominance of the diatom community by *Fragilaria crotonensis*. This diatom is found in surface waters with elevated nutrient levels. The algal pigment lutein is present only in green algae. Its decline during the 1950s likely reflects declining water clarity as this algal group often grows in the metalimnion. With declining water clarity, sufficient light does not penetrate to these deeper waters for growth. The high manganese in the upper portion of the core indicates increased anoxia in the hypolimnion in the last five years.

increased soil erosion (as shown by an increase in aluminum) and thus, accelerated the sedimentation rate. However, the nutrient (phosphorus) accumulation rate did not increase and the lake's water quality was only minimally affected. Shoreline development following the logging also had little effect upon the lake.

But more riparian development beginning in the 1950s had a much larger impact upon water quality by substantially increasing soil erosion, which resulted in a dramatic surge in nutrients. The algal pigment lutein, found only in green algae, also increased, and the diatom *Fragilaria crotonensis*, which indicates higher nutrient levels in surface waters, became the dominant diatom. This degraded water quality still exists; in fact, a large rise in manganese in the last five years indicates the hypolimnion is becoming more anoxic.

This core reveals that historical logging, even though it dramatically affected the landscape, had minimal effect upon the lake's water quality. Instead, shoreline development starting in the 1950s and continuing until the present has been most detrimental to the lake.

This lake could be restored most effectively by reducing nutrients contributed by riparian development.

Water and Nutrient Budgets

As described in Chapter 2, the water quality of a lake or reservoir is largely influenced by its watershed, which is usually the major source of nutrients and sediments to the lake and thus the origin of many of its problems.

This makes it highly important to know the annual nutrient load from the watershed. You can determine this by measuring the amounts of important nutrients (usually phosphorus) and water that enter the lake. The three natural sources of water are precipitation, surface water from inflowing streams, and groundwater.

Precipitation

Precipitation is much less important in the nutrient budget than in the water budget. Therefore, if funds are limited, this portion of the nutrient budget may be estimated. Precipitation is collected by various governmental agencies (e.g., NOAA, USDA, USFS, USGS) at numerous locations around North America; use the data from a site near the lake to estimate precipitation for your lake. The values for nutrient levels in rainfall throughout the country can be found in various publications.

If you need a more accurate estimate of nutrients (or other variables), collect samples on-site. Automated collectors are available but can be expensive. Other collection devices, e.g., plastic buckets, can be purchased and maintained by local citizens. If you use these, be sure to minimize the influence of wind and exclude droppings (nutrients!) from roosting birds.

NOAA: National Oceanic and Atmospheric Administration

USDA: U.S. Department of Agriculture

USFS: U.S. Forest Service, a USDA agency

USGS: U.S. Geological Survey

Surface Water

Determining water flow into and out of the lake and recording changes in lake level are essential for arriving at the annual nutrient and sediment loads to the lake. This helps establish the carrying capacity of the lake: that is, the amount of nutrients a lake or reservoir can assimilate each year without exhibiting problems.

Geodetic: Mathematical method for determining the exact position on the earth's surface.

Significant changes in lake level influence the nutrient and sediment budgets, so you must monitor the lake level during the study. To measure lake level, place a staff gauge in the lake and determine its geodetic elevation. Read it either weekly or biweekly, preferably when the water is calm.

Stream gauging stations must be placed on major tributaries near where they enter the lake and at the outlet of the lake. You don't usually have to gauge every tributary, however; the water yields from monitored sub-basins within the watershed can be substituted for unmonitored basins with similar land use. If you recognize obvious sources of pollution near a tributary stream, then you should place another gauging station near that site.

Groundwater

Estimating nutrient input from groundwater is most important in seepage lakes. No streams run into these lakes, so their only sources of water and nutrients are precipitation, groundwater, and runoff from riparian development.

When managing groundwater-dominated seepage lakes, such as those found in Florida, Minnesota, Michigan, New York, Wisconsin, and New England, the groundwater component of a nutrient budget becomes essential.

Measuring groundwater inputs is more difficult and often more expensive than determining surface water inflows. Where groundwater would be expected to contribute very little of the nutrient budget it may not be cost effective to measure it. In these instances using literature values would be better.

Defining the groundwater contribution to a lake is not as precise as for surface waters. The same general principle, however, holds true: water flows downhill. You actually define the groundwater component by measuring the elevation of the groundwater table relative to the elevation of the lake surface. Where the groundwater table is higher than the lake, the water is moving toward the lake; if the groundwater table is lower than the lake, then the lake water is moving out of the lake into the groundwater.

To define the groundwater basin around a lake, place wells on the surrounding land and then measure the water level in each well in relation to the lake level. You must also evaluate the variation of possible groundwater table slopes, soil types, bedrock types and locations, and location of permeable nearshore sediments.

Figure 4-2 shows how groundwater observation wells monitor the groundwater inflow below a septic system. In this example, three nests of wells are installed between the drain field and the lake. Their placement assumes that the groundwater along this portion of the lake shore flows toward the lake, at least for part of the year. These wells are sited to intercept the groundwater table at different levels.

If nutrients from the drain field are moving toward the lake, elevated levels will be apparent in water samples collected from the observation wells. It is also important to determine the hydraulic conductivity of the soils to estimate the rate at which the water is moving toward the lake. It is possible that the movement is so slow that the septic system contributes only a negligible amount of nutrients to the lake.

In lieu of the well system approach, several other, more focused techniques are often employed to locate specific areas within a lake where groundwater is entering or leaving. Techniques include seepage meters, small tube wells placed directly in the lake, temperature surveys, and fluorometric/conductivity measuring devices.

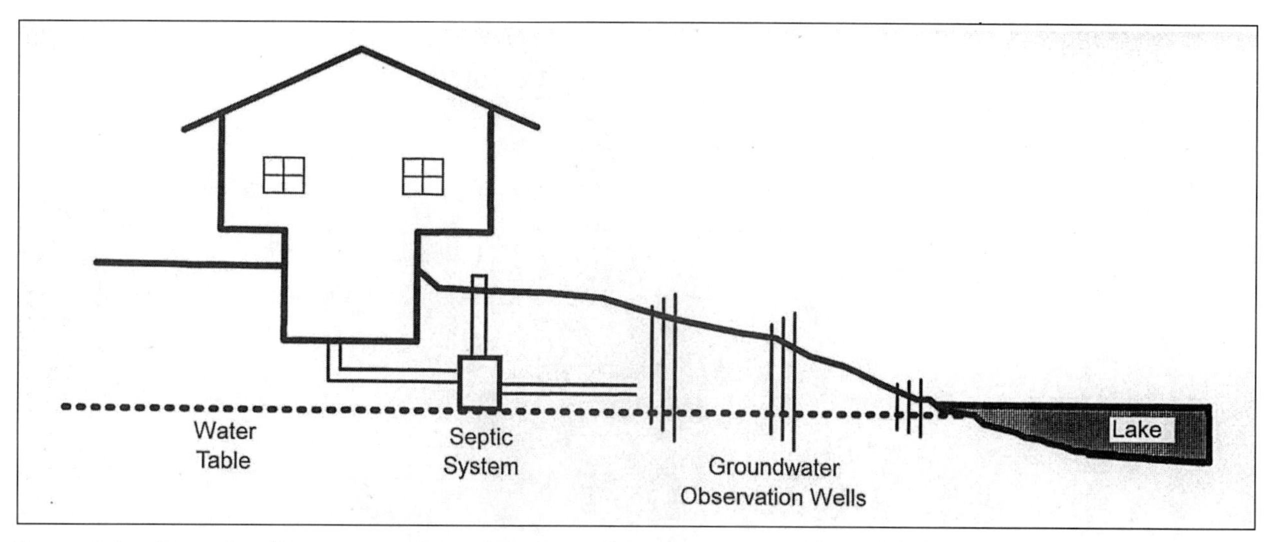

Figure 4-2.—Example of how to use groundwater wells to measure nutrients delivered from a septic tank to the lake. Determination of nutrient concentrations at the different depths and distances from the drain field indicates the contribution to the lake. Groundwater levels vary throughout the year so it is important to monitor the wells on a regular basis, especially during higher water levels.

Remember that groundwater flow into or out of a lake often varies considerably from season to season and year to year. For example, when the lake is low, groundwater often flows into the lake; when lake water levels are high the flow often reverses, with the lake contributing to the groundwater.

Be aware that groundwater flow into or out of a lake is not usually uniformly distributed around the lake. Often, groundwater only flows into a lake at a certain place and leaves around the rest of the lake.

Knowing the quantity of groundwater and the general direction of its flow can help you decide whether (or how) to sewer your lake. For example, if the soils are sandy they won't retain nutrients and they will allow septic tanks to easily seep into the groundwater, which will then carry the nutrients to the lake.

Figure 4-3 shows how to use this information in a lake. Round Lake, a seepage lake in northwestern Wisconsin, is fed solely by groundwater, and then only on the north and south sides. Only septic systems located on these portions of the lake would contribute nutrients to the lake.

Unfortunately, most lake environments are not this simple and additional evaluations often must be made to define the effects of on-site wastewater disposal systems. Most groundwater evaluations require experienced professionals, so consultants, university faculty, and state and federal agencies usually conduct them.

Monitoring Lake Water Quality

Sampling locations and depths influence the conclusions drawn from the data collected in the lake, so it is important that these stations accurately represent lake conditions.

The sampling locations and depths for physical, chemical, and biological analyses are associated directly with the properties of the lake.

- In lakes that are nearly round, a single station located over the deepest point may be adequate.
- More stations will be needed in lakes with branched, finger-like shorelines or multiple embayments, or long, narrow, natural lakes and reservoirs (Fig. 4-4).

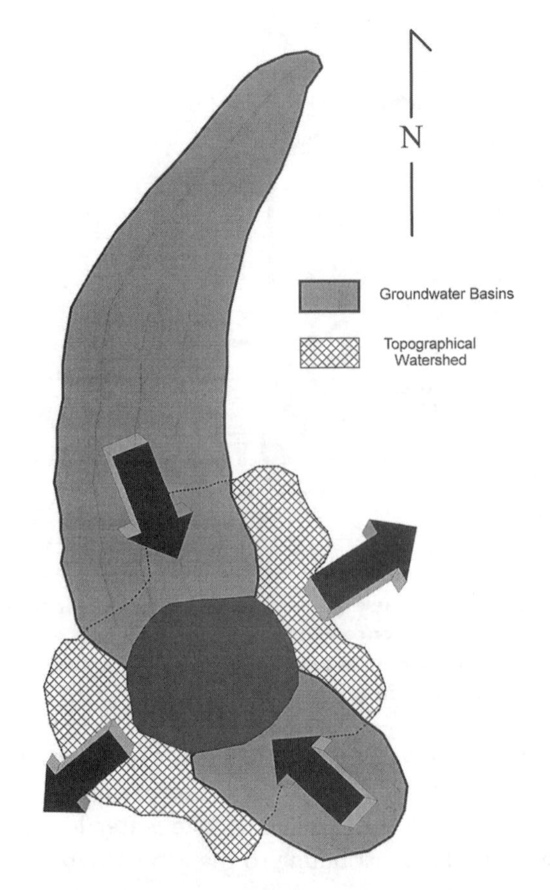

Figure 4-3.—Groundwater and topographical watershed for a seepage lake (modified from Wentz et al. 1989). Water enters the lake only from the north and south. Septic systems located in other areas of the watershed would not contribute nutrients and bacteria to the lake.

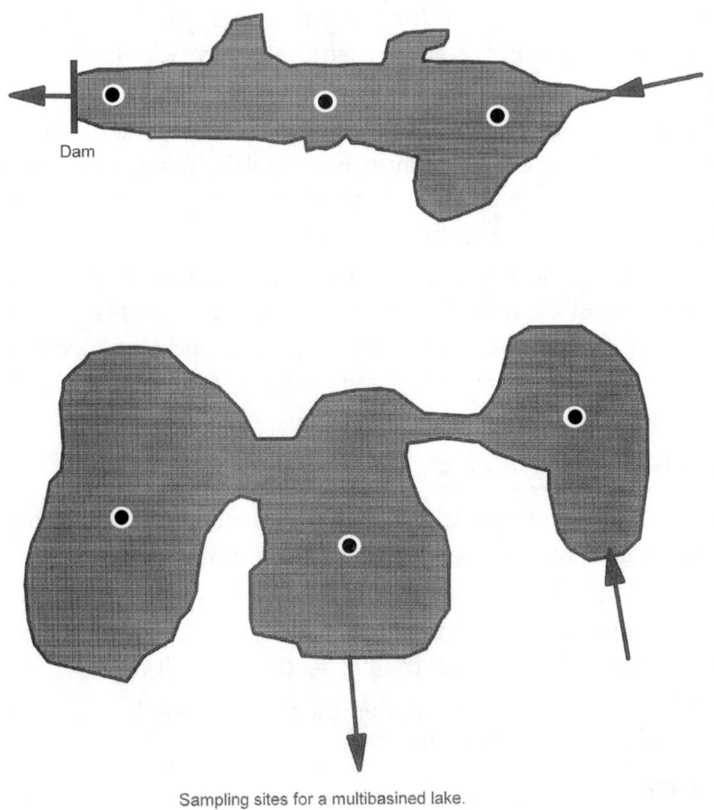

Figure 4-4.—Examples of sampling sites for reservoirs and lakes with complex shapes.

- Large lakes (e.g., >2,000 acres) should be sampled at two sites.
- In deep, stratified lakes, samples should be collected at least near the surface, in the metalimnion, near the middle of the hypolimnion, and near the bottom. One station should be in the deepest part of the lake with other stations in the shallower areas and prominent bays.
- For reservoirs, stations should be located at the river inflow, below the plunge point, perhaps near the middle, and at the deepest point near the dam.
- Fewer stations will be needed in shallow lakes that mix continuously throughout the summer, with samples taken only at the surface and bottom, but as frequently as for deeper lakes.
- See Cooke et al. (1993) for more detailed sampling descriptions.

Sampling frequency depends on what you want to know. For a general characterization of the lake, collect samples during spring turnover and monthly thereafter through early fall. You usually don't need to sample during the winter unless you're concerned about loss of oxygen during ice cover or other problems. More detailed sampling regimes will be described in the following sections to answer specific questions.

Physical Variables

Estimating the Sedimentation Rate

Although all water bodies fill with sediment over time, watershed activities — primarily construction and agriculture — can accelerate the infilling. Measuring the sedimentation rate is more important in reservoirs because they usually fill more rapidly than lakes. Occasionally, isolated areas in a lake, such as deltas where streams enter, may be infilling at an unacceptable rate.

Two methods are commonly used to determine recent sedimentation rates in lakes and reservoirs:

▾ ***Method 1 — determines the radioisotopes cesium-137 or lead-210 in the sediments.*** Although accurate, this method is relatively expensive. **Lead-210,** a naturally occurring radionuclide, is most useful since it determines the age of each sediment depth deposited in the last 150 years — but, the sediment core must be at least 130 years old. Since many reservoirs are not this old, this method will not work in those systems.

Cesium-137 works well for measuring the sedimentation rate in more recent sediments. A byproduct of atmospheric testing of nuclear weapons, Cesium-137 was deposited at its highest level during 1963 at the peak of atmospheric testing by the U.S.S.R.; thus, the average sedimentation rate since 1963 can then be calculated. Figure 4-5 shows a ^{137}Cs profile from East Twin Lake, Wisconsin. The United States began atmospheric nuclear testing in 1954 and the increase of ^{137}Cs in the core represents this date.

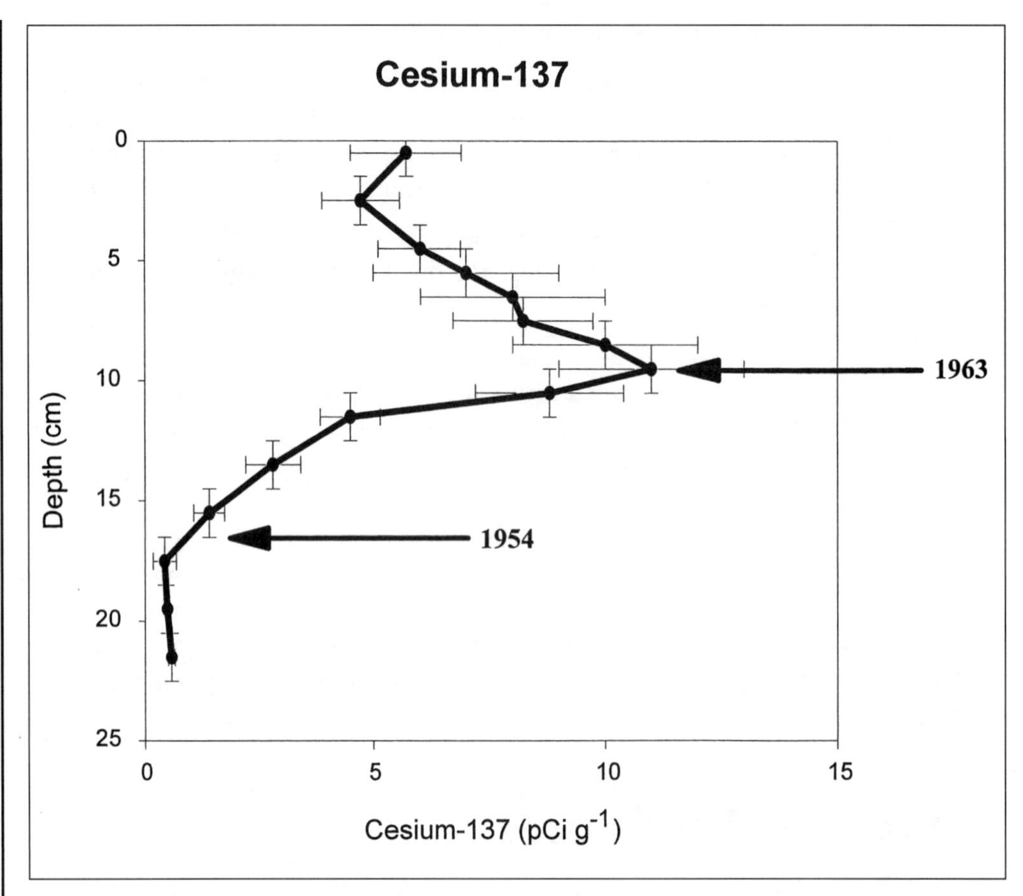

Method 1. Use the radioisotope Cs-137 to determine the sedimentation rate.

Figure 4-5.—Profile of cesium-137 indicating depths where dates 1963 and 1954 occur in a core from East Twin Lake, Wis. (Garrison, 1995).

- ***Method 2 — compares the current bottom contours (the depth to the bottom) with a similar map made several years before.*** The water level for these two surveys must be the same or the depth to the bottom must be corrected if not at the same water level. To use this method, the depth contours on both maps must have been accurately measured, which was not always the case with early maps. So, if you're not sure about the accuracy, use these maps with caution.

  Although this method is far less sensitive, it is satisfactory for natural lakes and reservoirs receiving large sediment loads, and is much less expensive than the other method.

The usefulness of these methods depends on your objective. If you're planning to dredge your lake, you must determine the rate of sedimentation before you begin. You'd waste your money dredging a reservoir that is filling in at a rate of 2 inches or more a year if you don't first control erosion from the watershed.

Temperature

Temperature patterns (thermal stratification) influence a lake's fundamental processes: the depletion of dissolved oxygen, nutrient release, and algal growth. Fish are sensitive to temperature. For example, salmonids are found in water with a summer temperature <20°C.

Use an electronic temperature meter to take readings, and take them every meter from the top to the bottom of the lake. Although sampling is usually done either monthly or biweekly, how often you sample will depend on what you want to know.

Temperature measurements can help you decide, for example, whether a shallow lake briefly stratifies and then mixes periodically throughout the summer. Take weekly measurements during the summer, in conjunction with measuring for dissolved oxygen (see the section on dissolved oxygen).

Deeper lakes that remain stratified throughout the summer may require less frequent measurements to understand general temperature patterns.

Transparency

Transparency is based on the transmission of light through water and is related, in part, to the natural light attenuation of the water being measured, the amount of suspended solids in the water, and the natural color of the water.

Secchi depth is probably the most frequently used variable in limnology: it measures the clarity (transparency) of the water by lowering a 20-cm plastic or metal disk divided into alternating black and white quadrants — the Secchi disk — into the water until it can no longer be seen. The depth is first recorded at that point, then again after raising the disk until it just becomes visible.

The average of these two depth measurements is recorded as the Secchi depth, referred to as the "Secchi transparency" of the lake. The greater the Secchi depth, the clearer the lake.

Secchi depth can be correlated with phosphorus and chlorophyll *a* to determine how eutrophic the lake is by using the Trophic State Index developed by Carlson (1977). See Trophic State Indices section for details.

Use the Secchi disk on the shady side of the boat — and don't wear sunglasses since they may allow you to see deeper. Avoid taking Secchi measurements early in the morning or late in the day as the low angle of the sun precludes accurate measurements of water clarity.

Volunteers in many states and provinces collect Secchi measurements on their lakes during the summer, usually submitting the data to a government agency. These data can be used to compare the water clarity of lakes both regionally and statewide.

The Great American Secchi Dip-In — coordinated by the creator of the Trophic State Index, Dr. Bob Carlson — provides a national perspective of water clarity. Volunteers make Secchi measurements in the U.S. and around the world for the week around July 4 and Canada Day. These data give scientists and volunteers a sense of how transparency varies according to water type, regional geology, and land use. Even more important, these annual Dip-In snapshots can be put together to form a changing picture of transparency over time. You can find the data from the Dip-Ins at www.dipin.Kent.edu.

Other less subjective measurements are available. These involve using a photometer that accurately measures photosynthetically available light at different depths.

Chemical Variables

Dissolved Oxygen

In shallow lakes that mix periodically during the summer, dissolved oxygen should be measured at the same time as temperature. Stagnant periods when dissolved oxygen in the bottom drops to zero followed by mixing periods can result in phosphorus being released from the bottom and redistributed throughout the lake — this means algal blooms.

Deeper lakes that remain stratified during the summer need to be frequently sampled for dissolved oxygen and temperature. To track the oxygen levels, sample every 2 meters from the top of the water column to the lake bottom every two weeks from the onset of stratification until mixing in the fall. When dissolved oxygen begins to decline in the bottom waters, sample every meter until the DO approaches zero (less than 1 mg L^{-1}).

These data can also be used to determine the anoxic factor, which indicates the lake's health. The anoxic factor, which has been developed by Nürnberg (1995a) (Chapter 5, Table 5-1), estimates the number of days in a year (or season) that a sediment area equal to the whole-lake surface area is covered by anoxic water.

The extent and duration of anoxia is very important for fish and macroinvertebrate habitat; so is determining the amount of nutrients coming from the bottom sediments. Figure 4-6 shows an anoxic lake.

Low dissolved oxygen may cause both summer and winter fishkills. During summer months, the dissolved oxygen in shallow eutrophic lakes may decline following a rapid algal die-off. Natural causes can severely deplete dissolved oxygen, but so can unwise management; for instance, treating an algal bloom in the entire lake with herbicides can drastically reduce the dissolved oxygen and cause a fishkill. Also, lakes that freeze during the winter can lose enough dissolved oxygen

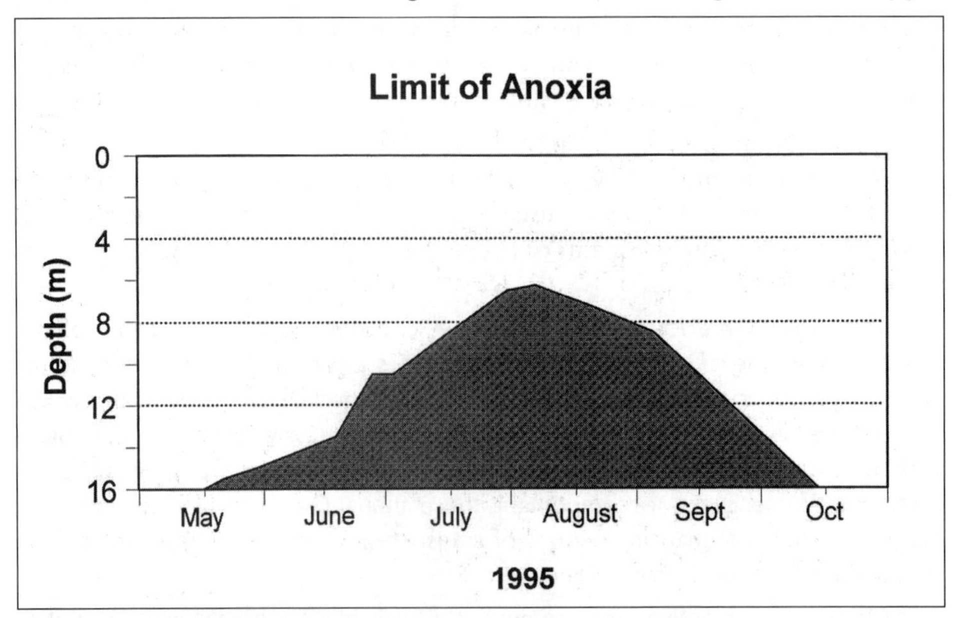

Figure 4-6.—Extent and upper limit of bottom waters that don't have oxygen during the summer in Lake Delavan, Wis. (Garrison, 1998). The seasonal limit of anoxia is determined by temperature stratification and the nutrient status of the lake as well as the lake's morphometry.

to kill fish. To determine the potential for winterkill, DO should be measured at least monthly. Nürnberg's winter anoxic factor (Nürnberg, 1995b) can also predict winterkill.

Nutrients

A lake study usually focuses on nutrients critical to plant growth: principally phosphorus and nitrogen. Chemical analyses may include total soluble phosphorus, soluble reactive phosphorus, total Kjeldahl nitrogen, nitrate nitrogen, ammonium nitrogen, total and dissolved solids — and, occasionally, chloride or potassium — as indicators of agricultural or urban source problems.

For eutrophication studies, total phosphorus is generally the single most important nutrient to determine in the incoming and outgoing streams. Phosphorus is often the key nutrient in determining the quantity of algae in the lake (see Chapter 2). Controlling it is usually the only practical solution to algal growth in a lake.

Although total phosphorus is the most commonly measured form of this nutrient, other forms are also present. The bioavailable forms are dissolved or loosely bound P that is readily available for algal uptake. Concentrations of these forms (e.g., soluble reactive P and total dissolved P) are necessary to run some trophic models, e.g., BATHTUB. To determine the amount of these forms of P present, filter the sample soon after collection to remove particulate matter.

Many lake management decisions will be made based on the total phosphorus coming into a lake. Modeling efforts (see Chapter 5) to predict water quality changes resulting from a project are based on the total phosphorus loading.

Phosphorus measurements are especially important during spring overturn as this value is used in predicting summer trophic status. If the lake is completely mixed from top to bottom (DO and temperature profile), take only one sample just below the water surface. Sample for phosphorus at 1 meter (39.37 inches) below the surface biweekly or monthly. If the lake stratifies, extend your sampling throughout the metalimnion and hypolimnion. Samples taken below the mixed zone help estimate the amount of internal loading from the bottom sediments during anoxic conditions (see Chapter 5, Table 5-2 for ways to estimate internal phosphorus loading).

Spring turnover: Process of water layers in a lake reversing position, usually in the spring.

In relatively large lakes (over 1,000 acres) with very little oxygen in their bottom waters, internal loading can occur throughout the summer. When conditions are calm, the anoxia may extend into the metalimnion, further encouraging phosphorus release into the water column.

Strong winds, e.g., a frontal passage, can cause thermocline oscillations; the resulting water movement pumps this high phosphorus water into the mixed zone (Stauffer and Lee, 1973). To determine if this is a problem, sample for DO and phosphorus prior to and immediately following these winds. Since such storms are difficult to predict, you should collect samples weekly and store; discard them if the winds fail to materialize.

In shallow lakes, elevated pH can also cause internal phosphorus loading (also see Chapter 5). When algae are actively growing during an algal bloom, they remove carbon dioxide (a weak acid) from the water column, causing pH to increase. James et al. (1996) have shown that at pH values above 9.5, phosphorus can be released from the sediments at rates equal to or exceeding release rates under anoxia. In shallow lakes that experience algal blooms, measure pH weekly or biweekly during the summer. If pH levels exceed 9.5, high levels of phosphorus may be entering the water column.

The total nitrogen (N) to total phosphorus (P) ratio (N:P) in the lake water can help determine which algae dominate. For example, look for nitrogen-fixing blue-green algae when nitrogen is low. This can happen where phosphorus levels are naturally abnormally high, e.g., central Florida or hypereutrophic lakes.

Metals and Organics

Human health concerns are growing over increasing concentrations of trace metals (e.g., mercury and cadmium) and organics (such as PCBs) in lakes. Although usually deposited by the atmosphere at relatively low levels, these toxins bioaccumulate in the food chain. So, as higher and higher levels of organisms consume food containing these toxins, their bodies concentrate them. Top-level predators such as game fish like northern pike and walleye, which people eat, may contain harmful levels of these toxins even though water level concentrations are very low.

Sampling lakes and reservoirs for these compounds is very difficult and requires highly specialized — and very expensive — ultraclean techniques. Instead, sample organisms that people eat, e.g., game fish. These are much easier to sample and they provide more relevant information as to whether these toxins are a problem. This sampling should be conducted by professionals as specific sampling methods are required to prevent contamination. Many state agencies routinely sample such foods. See Chapter 2 for more information on airsheds.

Acidification

An indication of acidity in lake water, pH is measured on a scale of 0 to 14.

An indication of acidity in lake water, pH is measured on a scale of 0 to 14. The lower the pH, the higher the concentration of hydrogen ions (H^+) and the more acidic the water. A reading of less than 7 means the water is acidic; if the pH is greater than 7, it is basic (alkaline). Because the pH scale is logarithmic, each whole number increase or decrease on the scale represents a 10-fold change in the hydrogen ion concentration.

Many lakes are naturally alkaline and not the least bit sensitive to acid precipitation. Those that have pH values less than 7 can be sensitive depending upon their location. Lakes in the Upper Midwest are not sensitive unless their pH values are less than 6.0. In the northeastern U.S., and eastern Canada, lakes with pH values less than 7.0 can be sensitive to acidification, especially during spring runoff.

Acid rain typically has a pH of 4.0 to 4.5 while pure rainwater would have a pH of 5.7 as a result of atmospheric carbon dioxide. In contrast, most lakes have a natural pH of about 6 to 9.

Alkalinity: Measure of water's ability to resist changes in pH by neutralizing acid input.

Alkalinity is a measure of the acid neutralizing capacity of water; that is, its ability to resist changes in pH by neutralizing acid input. In most lakes, alkalinity is a complex interaction of bicarbonates, carbonates, and hydroxides in the water. The higher the alkalinity, the greater the ability of water to neutralize acids.

Low alkalinity lakes typically have pH values below 7. When alkalinities are less than 20 mg L^{-1}, the Gran analysis method should be used. The Gran method (U.S. EPA, 1989) provides information that is referred to as "acid neutralizing capacity" because in addition to alkalinity, it includes the dissociated organic acids and other compounds that help buffer (increase alkalinity) the water.

Biological Variables

Biological indicators are the best symptom of problems associated with high nutrients in a lake or reservoir. These usually include algae and macrophytes (aquatic weeds), and, at times, the fishery, bacteria, and exotics such as zebra mussels. Depending on the perceived problems, you will probably have to examine only some of these.

Bacteria and Pathogens

Animal waste from barnyards and feedlots can carry harmful bacteria and pathogens into a lake. They can also enter water bodies from combined sewer overflows during storms.

Ducks and geese can be a bigger problem. In addition to adding nutrients to your lake, their waste can contribute bacteria such as *E. coli*. Since it is very difficult to get rid of waterfowl, you should periodically test the water at swimming beaches and any other areas where people may drink water. Additional information is available in APHA (1992) and AWWA (1990).

The best source of information about bacteria or other pathogens is your local health department. They usually test waters and deal with these problems.

Algae

Chlorophyll *a* is the most common measure of the amount of algae in the water column. The average summer chlorophyll a concentrations can tell you how severe your lake's algal problems are. Peak chlorophyll a concentrations in an oligotrophic lake may range from 1.5 to 10.5 $\mu g\ L^{-1}$ — in a eutrophic lake from 20 to over 200 $\mu g\ L^{-1}$.

Photic zone: Upper portion of the water column that receives sunlight. This is roughly two to three times the Secchi depth.

To measure chlorophyll *a*, collect an integrated water sample from the mixed portion of the lake either by taking water samples from several depths and mixing them together, or by using a tube that extends through the photic zone. Sample on a biweekly or monthly basis during the spring and summer.

Microscopic examination can help you identify the types of algae in your lake and thus, understand the lake's problems. Blue-green algae most frequently cause aesthetic problems since they can float at the surface, leave a paint-like film on the shores, and cause taste and odor problems. Other algae can also change the color of the water; *Synura* turns it red.

Eutrophic Cedar Lake experiences seasonal phytoplankton changes. June finds algal levels low in this shallow lake, but even then blue-green algae dominate (Fig. 4-7). Blue-greens, especially *Lyngbya* and *Aphanizomenon*, continue to dominate through July but become less important in August and eventually are replaced by the eutrophic diatom *Aulacoseira*. The algal community reaches its highest level in September when diatoms dominate. Because diatoms do not float as well as other algae their dominance indicates the importance of wind, in addition to nutrients, in structuring the algal community in this large shallow lake.

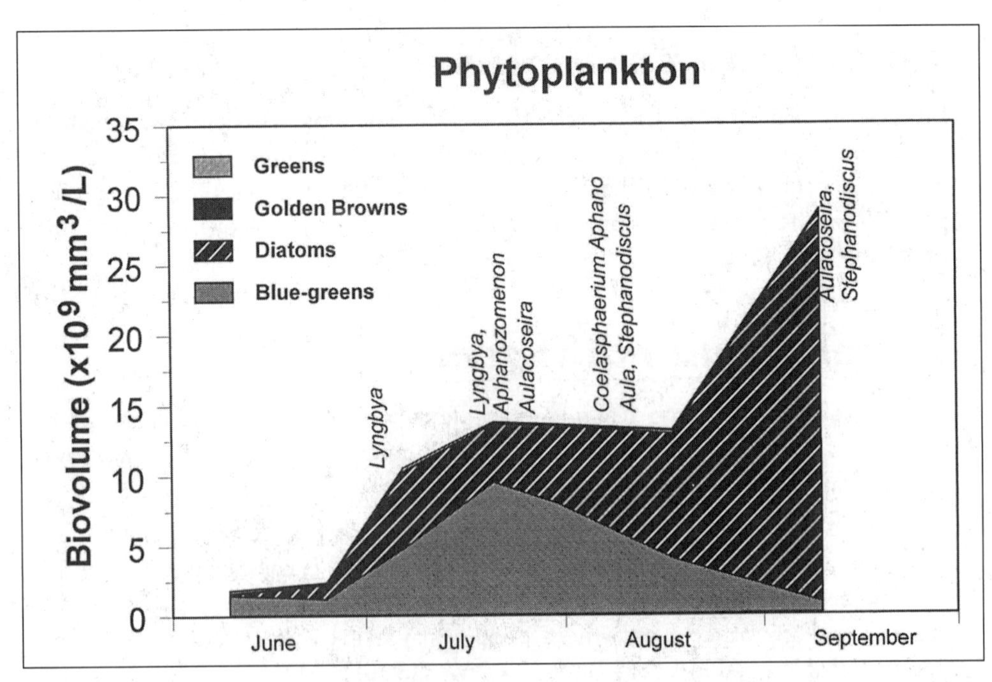

Figure 4-7.—Example of seasonal phytoplankton changes in eutrophic Cedar Lake, Wis. (Garrison, unpublished data).

Macrophytes

Aquatic plant communities range from completely submerged macro (large) algae (e.g., *Chara* or *Cladophora*) to rooted plants with floating leaves (e.g., water lilies) to completely submersed plants (e.g., pondweeds) to free-floating plants (e.g., duckweed or water hyacinth). See Figure 2-19 for a description.

Macrophyte: A plant large enough to be seen without magnification.

Macrophyte densities vary seasonally between lakes in an area and among regions. In the Upper Midwest, macrophytes might average several hundred pounds per acre, while in Florida several tons per acre are common.

While plants may be very dense in eutrophic lakes, the community is usually fairly simple, often containing large amounts of exotic plants, e.g., Eurasian watermilfoil and/or hydrilla. An excellent source of information on aquatic plants is the Aquatic Plant Information Retrieval System (Aquatic Plant Management Society at www.apms.org).

Survey your plant community once or twice during the growing season. Run a transect perpendicular to the shoreline toward the deep area of the lake, then collect samples along this transect either using a modified lawn rake (Deppe and Lathrop, 1992) or by snorkeling or scuba diving.

Observe the water depth, height of plant growth, species composition, and density. If you're using the rake method, measure density on a scale of 1-5 with 5 being the densest. The rake method will not work on low growing plants, which must be sampled by hand.

Samples collected by hand focus on a known area (e.g., 0.1 m^2) to determine the actual biomass of the plants. You can also estimate plant quantity on a subjective scale such as: A = abundant, B = common, S = sparse.

Use this information to create a map of the lake's macrophyte community that shows their distribution: e.g., emergents, floating leaves, and submergents. Note the sensitive and especially valuable species. Changes in sensitive species over time can be a good indicator of changes in a lake's condition.

Comparing macrophyte studies between different years can yield important information about a lake. Fox Lake, Wisconsin, used the rake method in 1994 and 1995 to assess the plant community. The second year, plants were much denser and grew deeper (Fig. 4-8). The increased density was especially evident at sites represented by transects 2 and 17. In 1995, the lake was much clearer than in 1994, thus encouraging macrophyte growth.

This macrophyte information can also help you:

- Decide where to concentrate control efforts such as harvesting or dredging, and
- Predict how deep plants might grow if the water clarity improved.

Since macrophytes provide habitat for fish and wildlife, you probably want to protect certain areas that contain sensitive or endangered species.

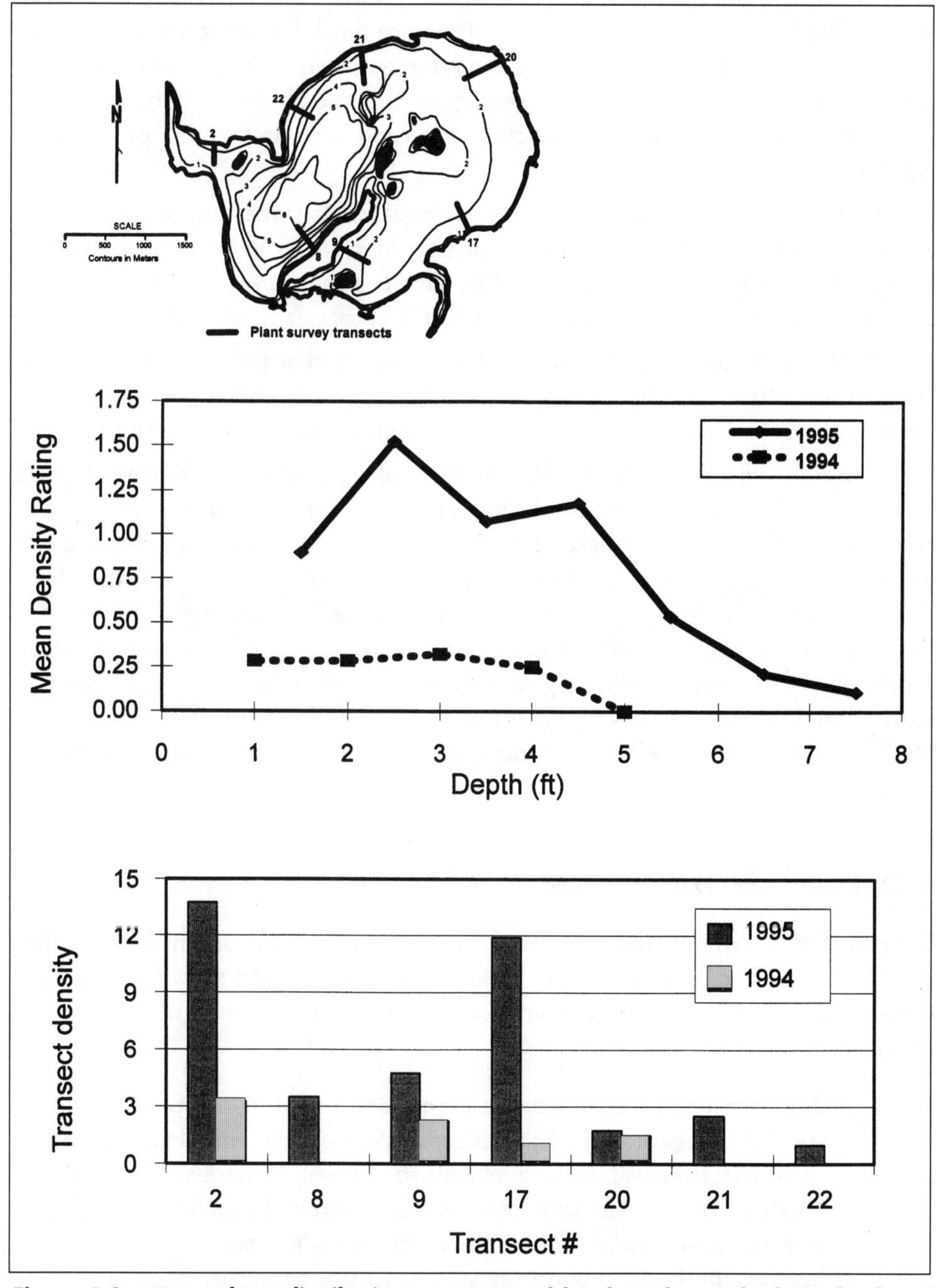

Figure 4-8.—Macrophyte distribution as measured by the rake method (Asplund and Johnson, 1996). Transects sampled are shown on the map of Fox Lake, Wis. Macrophytes were denser and grew deeper in 1995 because of the greater water clarity.

Zooplankton

Zooplankton are microscopic animals that are an essential part of the food chain. Because they feed on algae, these animals can significantly increase water clarity even if the nutrient level is moderate. But zooplankton are also important to the fish diet: some species such as crappies and perch depend upon them as adults; and many game fish feed on zooplankton during their early life stages.

Zooplankton should be sampled if you're considering using a biomanipulation technique (e.g., increasing zooplankton to increase water clarity). Collect samples by pulling a plankton net vertically from the bottom of the lake to the water surface. Sample on a weekly or biweekly basis following spring turnover through the fall.

Biomanipulation: Using biological (usually predator/prey) relationships to achieve desired results (see Chapter 7 for a full discussion).

Determine the dominant species and their average length. Some species such as large *Daphnia*, commonly called water fleas, are much better at controlling algal levels. But they're also the zooplankton preferred by fish. You do want to know whether (and how much) fish are feeding on the zooplankton. If you find large zooplankters, especially *Daphnia pulicaria*, you can conclude that fish predation is not important.

Large numbers of *Daphnia* can reduce algal levels since they are large consumers of algae (smaller zooplankton consume much less). In many lakes this biomanipulation occurs in late spring when edible algae abound following spring turnover, but there are few planktivorous fish (fish that eat zooplankton).

Even in eutrophic lakes the water is very clear at that time; but this so-called "clear water phase" is generally short-lived. As the water warms, fish begin to actively feed on the zooplankton, thus reducing their consumption of algae.

Planktivory: consumption of zooplankton by fish and other predators.

The importance of this can be seen by looking at changes in large *Daphnia* and their reproductive potential, which is determined by the number of eggs per adult female. If they're being eaten by fish, the number of these large *Daphnia* will decline even though their reproductive potential increases, indicating they have abundant food (Fig. 4-9a). In the example from Lake Delavan (Fig. 4-9b) fish are not eating a significant number of large *Daphnia*. In September and October even though food resources are limiting zooplankton growth, *Daphnia* are still present in the lake. The reason planktivory is low in this lake is when it was restocked in 1990 following a complete fish eradication, very few fish that eat zooplankton were added.

Animal Nuisances

Certain animals can become real pests in lakes. Many of these are introduced species that have few natural predators — like zebra mussels and rusty crayfish. Even if you have only a few of these animals, they can soon become a nuisance if not controlled.

- **Zebra mussels and Asiatic clams** are a recently introduced exotic species that can dramatically affect a lake's ecosystem. Once they're in your lake, they will be very difficult to control. To determine whether they've entered your lake, you can install artificial substrates in your lake and periodically examine them for zebra mussels. Many states have established a monitoring system to track the spread of this nuisance.

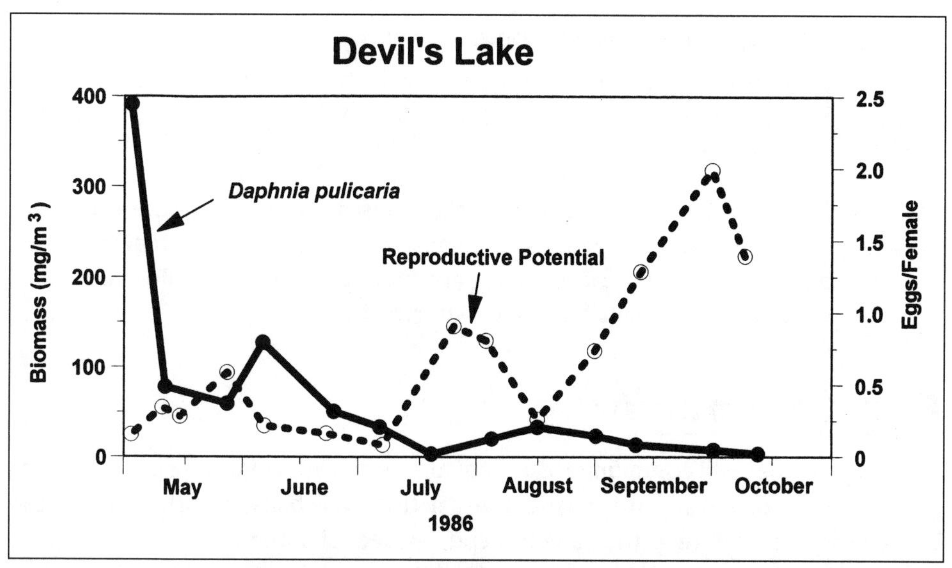

Figure 4-9a.—Devil's Lake, Wisconsin: Seasonal changes of the large herbivorous zooplankton *Daphnia pulicaria* (Wis. Dep. Nat. Res. 1988). Because of its large size it consumes large amounts of phytoplankton but fish also eat it. Numbers were high in spring with an abundance of food and little fish predation. By July, fish predation reduced its numbers even though sufficient food was available, as indicated by the increased reproductive potential.

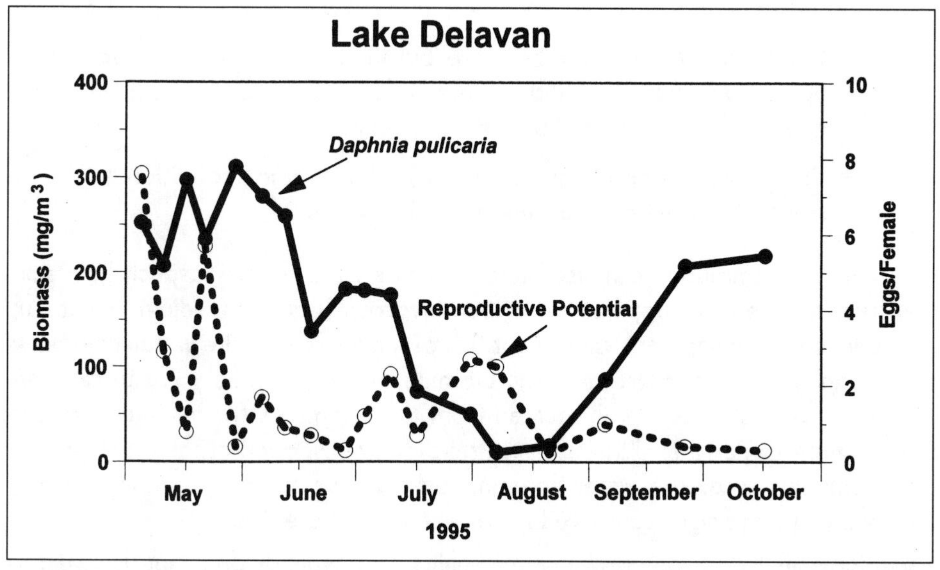

Figure 4-9b.—Lake Delavan, Wisconsin: In this lake nutrients are higher so levels of *Daphnia pulicaria* are higher than in Devil's Lake. Unlike Devil's Lake, fish predation is not a problem as *Daphnia* numbers remain high in September and October despite limited food resources. *Daphnia* decline in July and August is caused by invertebrate predators such as the phantom midge *Chaoborus* and the zooplankter *Mesocyclops edax* (Garrison, 1998).

- **Rusty crayfish** — another introduced species — can have profound effects upon the fish community and the lake ecosystem in general. Rusty crayfish are especially aggressive, often wiping out native species. They feed voraciously on aquatic plants and in some instances have severely depleted a lake's macrophyte community. To find them, place baited traps around the shallow water of the lake and examine them periodically.

Two native species can also become problems:

- **Leeches** can become so abundant as to be a problem for swimmers. Sample them by placing baited traps in swimming areas and periodically examining them.
- **Blood flukes** — the organism that causes swimmer's itch — are difficult to detect. You'll usually know they're around when swimmers begin to complain. Blood flukes are most common in shallow water, especially near plant beds and waterfowl areas.

Fish Community

A survey of the fish community can provide useful information on the species present, their size distribution, and the relative availability of fish prey to the larger fish predators (e.g., the gamefish species, see Chapter 2).

If fishing is poor, then a survey of the fish community should tell you why. You may find that:

- The fish species people aren't catching does not even live in the lake. Lake conditions may not be suitable for its habitat or survival; conditions could have changed and thus eliminated it; or, it could have been wiped out by a combination of overfishing and poor reproduction.
- Or, the species may still be there, but in very low numbers because of poor reproduction — either because of unsuitable habitat or intense competition for food with another predator.
- The gamefish population may be large, but in poor condition or stunted in size because they lack suitable prey.

The fish community can also dictate a lake's water quality, especially in shallow lakes, where large populations of benthivorous (bottom-feeding) fish stir up the sediments, moving significant quantities of nutrients into the water column. In such cases, you need to estimate the biomass of these fish so you can calculate how much phosphorus is translocated from the sediments to the water column. For example, Lamarra (1975) estimated that for every pound of carp 0.11 pounds of phosphorus enter the water column. You may not be able to improve water quality if you don't significantly reduce or eliminate these fish.

Biomanipulation to improve water quality also depends on properly structuring the fish community. Fish (such as perch and bluegills) that feed primarily on zooplankton can decimate the large zooplankton that normally feed on algae. When fish that feed on other fish such as pike and walleye are dominant, the zooplankton community is free to feed on algae and thus, decrease algal levels.

If biomanipulation is to succeed, you must know the number of planktivorous (plankton-eating) and piscivorous (fish-eating) fish in the lake. Many shallow lakes have not been successfully restored until the number of bottom-feeding fish has been significantly reduced.

Fishery management practices can be applied to solve most of these problems, but only if the problem is first identified. The state fish and game agency can often be enlisted to conduct the fish community survey, to help interpret its results, and to suggest a fishery management strategy.

Additional information about fishery science can be found in Baker et al. (1993) and on the web site of the American Fisheries Society (www.fisheries.org).

Trophic State Indices

Several indices may be used to compare the severity of a lake's problems with other lakes in the area. Often referred to as "trophic state indices," they simplify complicated environmental measurements and can quantify lake changes following implementation of protection and restoration practices (Carlson, 1977; Kratzer and Brezonik, 1981; Walker, 1984).

The trophic state index concept is based on the belief that, in many lakes, the degree of eutrophication is primarily related to increased nutrient concentrations — phosphorus, in particular. An increase in phosphorus concentration is expected to increase the amount of algae (see Chapter 2) as measured by chlorophyll *a*. Simultaneously, water transparency declines (as measured by Secchi disk).

Carlson's Trophic State Index (Carlson, 1977) is the most widely used. It compares chlorophyll *a*, Secchi transparency, and total phosphorus concentration. High index numbers indicate increased eutrophy; low numbers, oligotrophy (low levels of nutrients and algae, clear water). TSI = 0 represents a Secchi transparency of 64 meters. Each halving of transparency increases the TSI by 10 units. A TSI of 50, thus, represents a transparency of 6.6 feet (2 meters), the approximate demarcation between oligotrophic and eutrophic lakes. See Chapter 5 for the exact Carlson TSI formulas.

Suppose that a lake had a transparency index of 60 before it was restored. Two years later, as a result of a watershed project that reduced phosphorus loading, the index was 40, indicating an improvement in water quality. A TSI of 40 might be common to undeveloped lakes in the area, perhaps indicating that the lake has improved about as far as it can. Significant upward movement of the index in later years would indicate the lake has returned to its previous condition. The index, therefore, is a useful tool for assessing the lake's current condition and for monitoring change over time.

The Carlson TSI works well in most lakes that are phosphorus-limited but poorly in lakes that are nitrogen-limited, suffering turbidity from erosion, or experiencing extensive macrophyte problems (Brezonik, 1984).

Figure 4-10a shows TSI plots for a northern lake of moderate water quality. By scanning the TSI plots, the lake professional can begin to understand the patterns in a specific lake and appreciate the seasonal variations without having to analyze phytoplankton and phosphorus concentrations and interpret their relationships.

TSI values can also be used to detect unusual conditions in a lake. TSI values calculated for phosphorus, for example, may not be the same as simultaneous calculations of TSI from Secchi disk or chlorophyll *a* measurements. To understand this situation, you will have to examine the database in greater detail. At Lake Delavan (Fig. 4-10b), a sizable population of large zooplankton suppressed the algae, thus lowering chlorophyll *a* and producing clearer water than would be expected given the phosphorus concentrations.

Another scenario: If the TSI for phosphorus and chlorophyll *a* are similar but the TSI for Secchi depth is less, this may indicate that inorganic turbidity or water color is significantly reducing clarity.

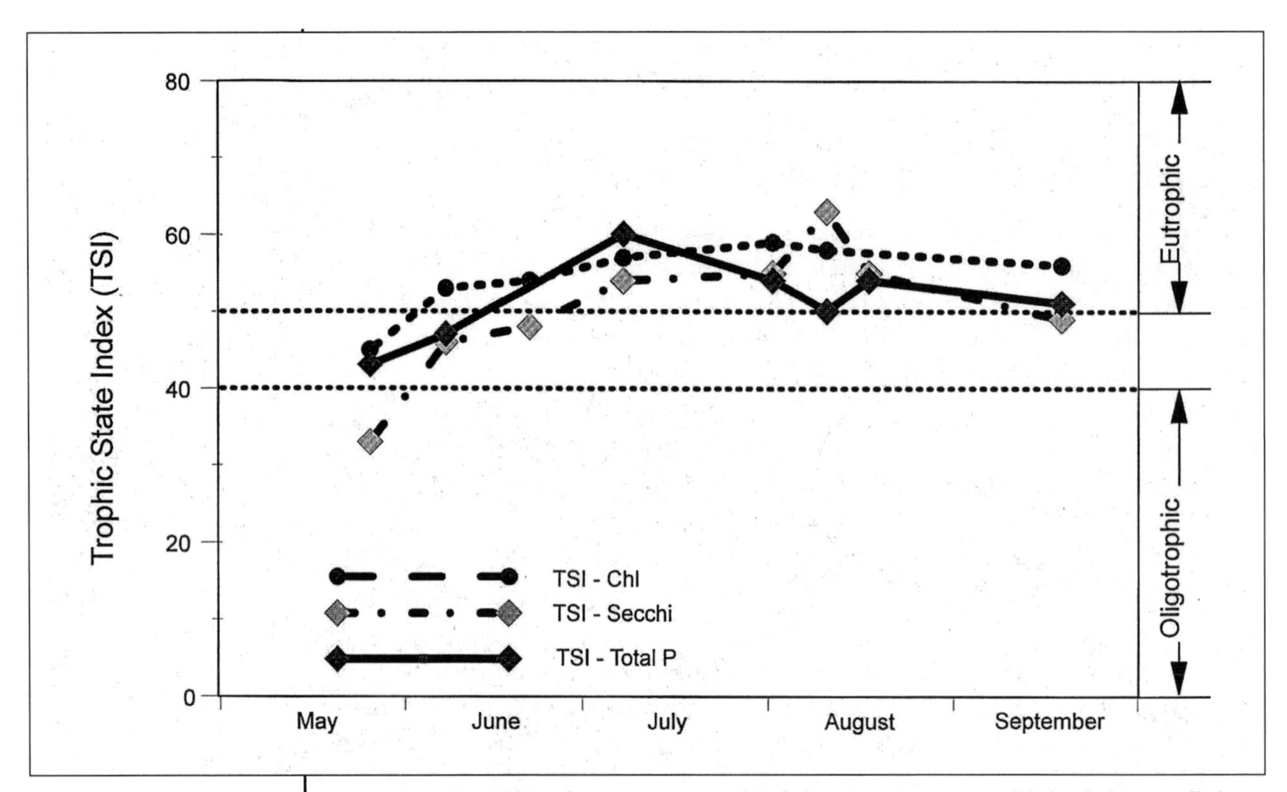

Figure 4-10a.—Plot of TSI in a eutrophic lake (Garrison, unpublished data). All three indicators are in good agreement.

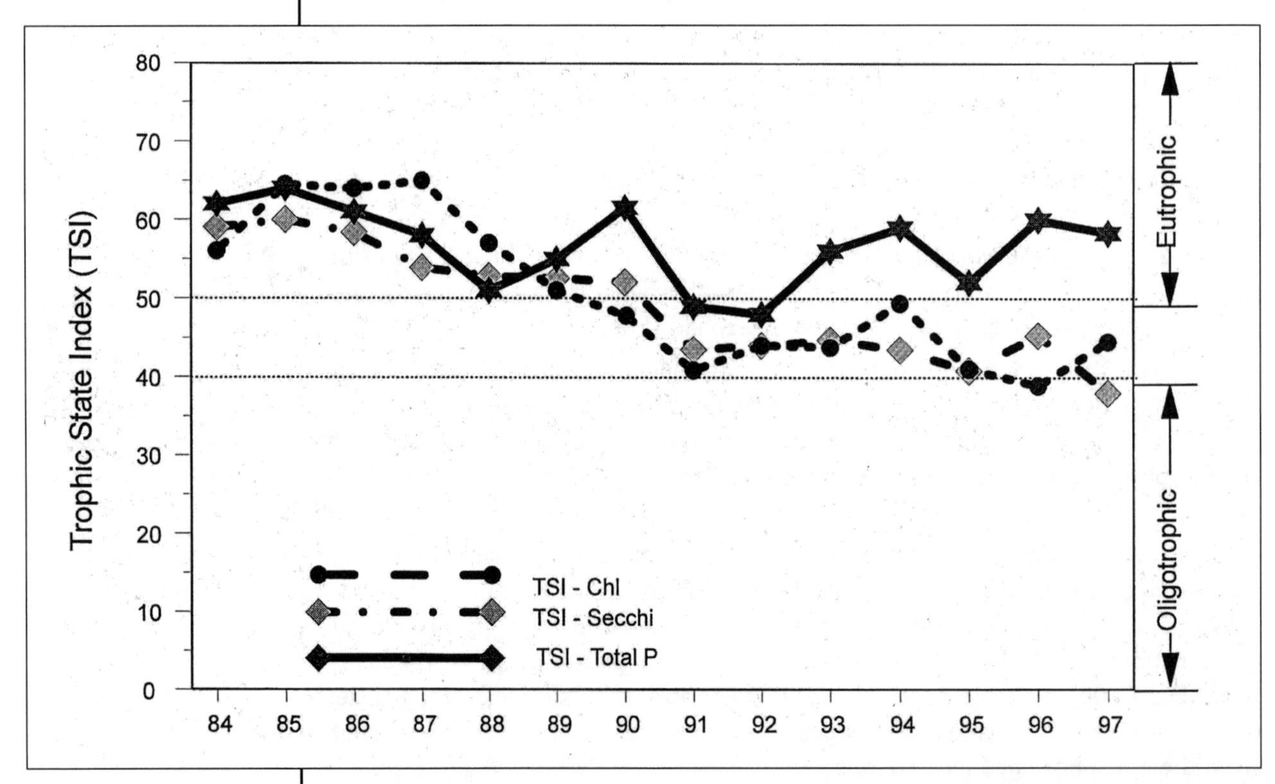

Figure 4-10b.—Plot of TSI for Lake Delavan, Wis., over a period of 14 years (Robertson, unpublished data). All the indicators are in agreement from 1984 through 1989. At this time a number of restoration measures were performed, including stocking large numbers of piscivorous fish. Since 1993 this has resulted in much better chlorophyll *a* and Secchi values than would be predicted from the phosphorus values.

Other indices are more appropriate for specific applications:

- Walker (1984) developed an index for reservoirs.
- Brezonik (1984) developed an index that fits the needs of Florida lakes and includes situations where nitrogen rather than phosphorus may be limiting algal growth.
- Porcella et al. (1979) included a term in their Lake Evaluation Index that represents the amount of lake surface covered by macrophytes.

Using the Data to Manage Your Lake

Amassing all these data is critical to the diagnostic study of a lake; it enables the lake manager or consultant to understand the severity of the lake's problems and figure out what has caused them. Only then can a lake management plan be formulated.

To succeed, a lake management plan may first have to concentrate on the watershed to address the sources of the lake's problems. Correcting the cause of the problem such as limiting nutrient runoff from fields or streets is always preferable to simply addressing the symptoms in the lake.

If the watershed's contribution of nutrients is not reduced, treating the problem within the lake is more expensive than necessary and may not succeed. For example, treating bottom sediments with alum improves a lake's water quality for only a short time if nutrients continue to pour in from the watershed (Garrison and Knauer, 1984b).

Two examples follow that will help you understand how to use a diagnostic study to formulate and implement a lake management plan; one is from a shallow lake, the other from a deep lake.

Cedar Lake, Wisconsin

A large, relatively shallow lake in western Wisconsin, Cedar Lake covers 1,100 acres with a maximum depth of 28 feet. It had experienced large and extensive blue-green algal blooms, especially in late summer, and was periodically treated with copper sulfate.

Lake residents requested a diagnostic study that determined phosphorus loading from the largely agricultural watershed to be about 1,500 kg per year. Phosphorus loading models (see Chapter 5) based only on watershed inputs suggested that the in-lake phosphorus value should be much lower than was actually measured. The mean summer phosphorus concentration was well within the eutrophic range at 60 µg L^{-1}.

The data, however, showed that both phosphorus and chlorophyll increased dramatically in late summer. Phosphorus values exceeded 150 µg L^{-1}, causing a large algal bloom. Temperature and dissolved oxygen profiles taken semi-weekly in 1987, revealed that stratification began in late May and the bottom waters quickly became anoxic. When that happened, phosphorus levels became very high as phosphorus was released from the sediments (Fig. 4-11). When the lake mixed in late summer, this phosphorus mixed into the surface waters, fueling the large algal bloom.

Although the lake had an excellent sport fishery, a fish survey also found a sizable carp population. Of course, these bottom dwellers (as discussed earlier in

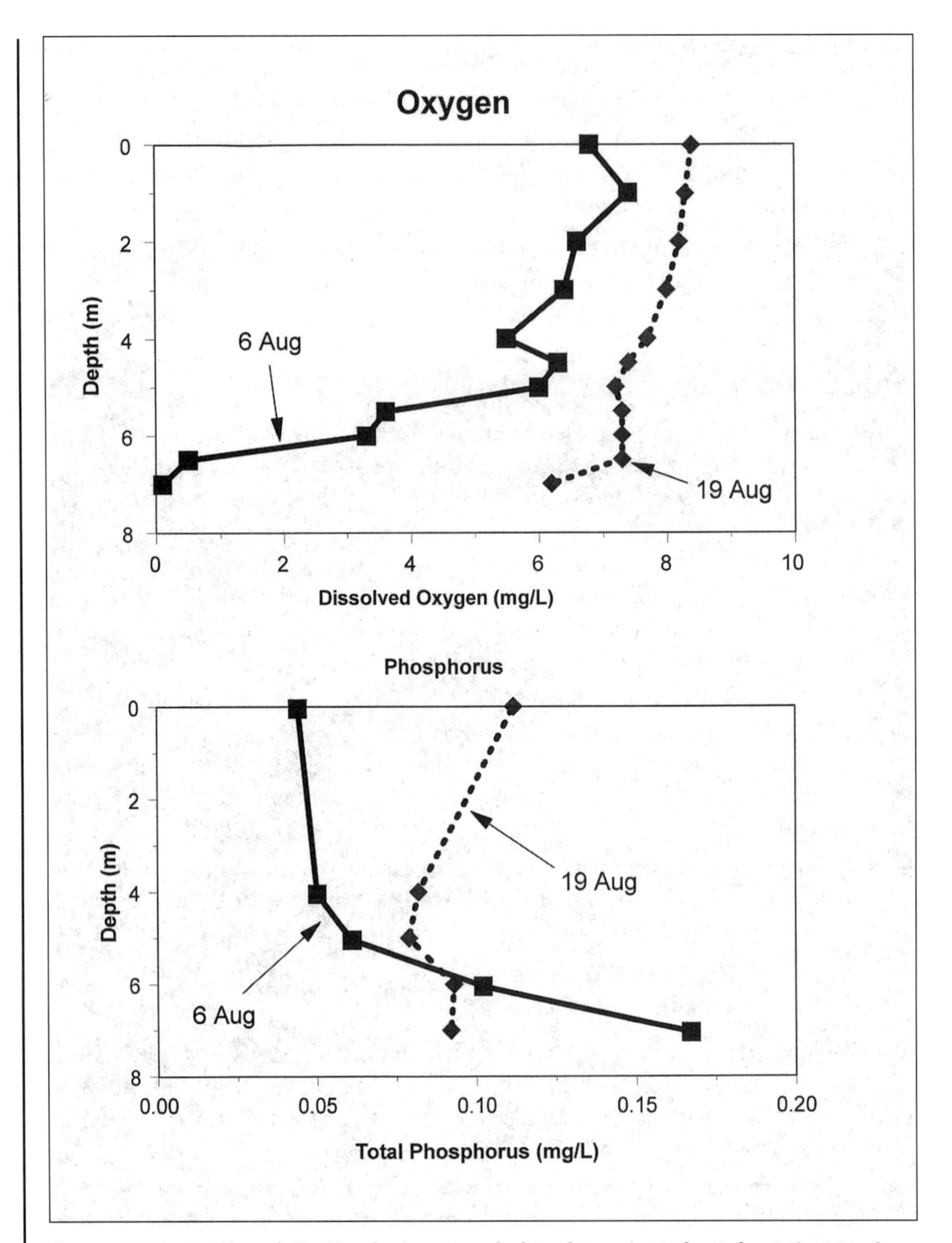

Figure 4-11.—Profiles of dissolved oxygen and phosphorus in Cedar Lake, Wis. (Garrison, unpublished data). During stratification, P is released from the bottom sediments. Following mixing, this P is moved into the surface waters where it is available to cause algal blooms.

this chapter) move phosphorus from the sediments into the water column as they feed (Lamarra, 1975), raising the phosphorus level throughout the lake.

When the phosphorus budget was revised to include internal as well as external loads (Fig. 4-12), in-lake phosphorus concentrations were accurately predicted. And the major targets for restoring the lake became the internal loading of phosphorus from carp activity in the sediment.

Because the sport fishery was so good, using rotenone to eradicate the fishery was not an option (see Chapter 7). Instead, a commercial fishery attempted to reduce the carp population. Although a large number of carp were removed,

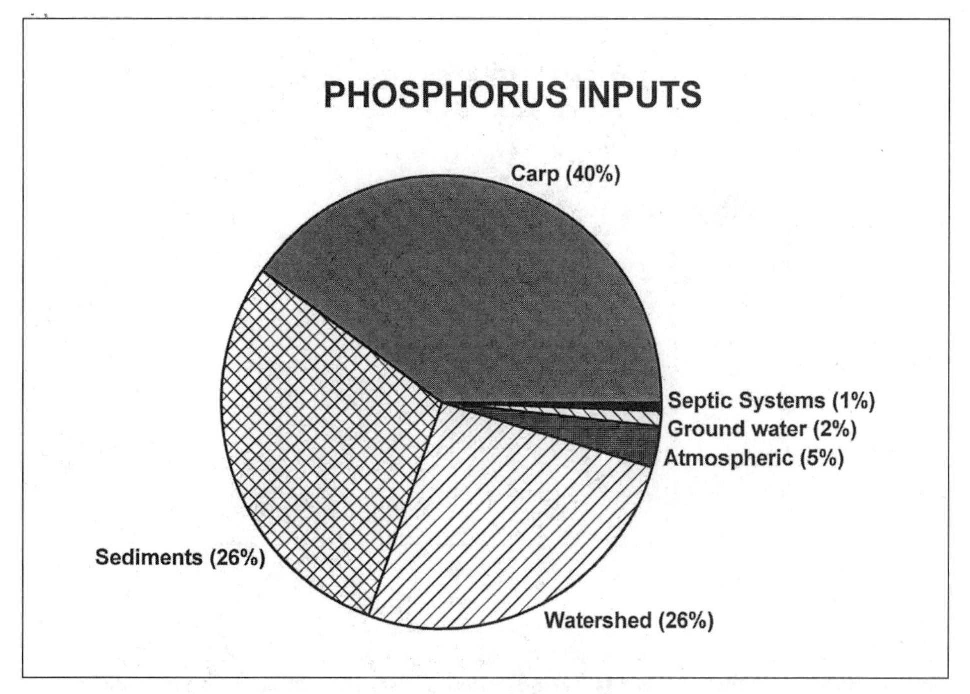

Figure 4-12.—Phosphorus load for Cedar Lake (Garrison, unpublished data). Nearly two-thirds of the load comes from internal sources. Therefore, the restoration effort was targeted toward reducing P input from these sources.

enough remained to contribute a significant amount of phosphorus and summer algal blooms continued.

To reduce phosphorus release from anoxic sediments, a destratification system was installed (Chapter 7). This system consisted of a blower that discharged 300 cfm of air with a manifold system located near the center of the lake that distributed the air into the water column. The idea was not to directly introduce air into the water column but instead reduce lake stability so that less wind would be needed to mix the lake. The blower was turned on before stratification began (late May) and operated continuously until early September.

This system succeeded in reducing phosphorus release from the sediments by 70 percent (Fig. 4-13). The lake was still anoxic but for a much shorter period of time and over a far smaller area.

The destratification system also largely eliminated the algal bloom. Previous chlorophyll *a* concentrations exceeding 100 µg L^{-1} (Fig. 4-14a,b) also declined. Algal levels are still in the eutrophic range with peak summer chlorophyll *a* values of about 50 µg L^{-1}. The only way to further lower these levels is to decrease nutrients entering from the watershed and reduce the carp population still more.

Mirror Lake, Wisconsin

Mirror Lake is a small urban lake in central Wisconsin with a surface area of 13 acres and a maximum depth of 43 feet. Mirror Lake had experienced repeated blue-green algal blooms and winter fishkills (Knauer, 1975), so the city commissioned a diagnostic study to determine the annual incomes of water and total phosphorus and to examine the lake's water quality.

Although Mirror Lake is classified as a seepage lake with no permanent inflowing streams from the watershed, urban stormwater discharged into the lake,

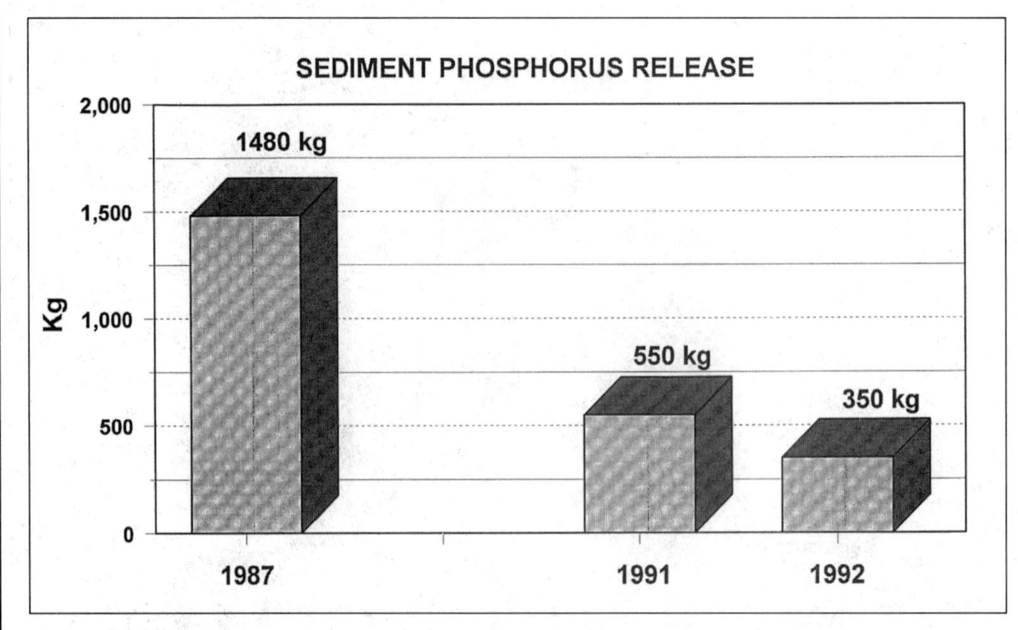

Figure 4-13.—Prior to operation of the destratification system, a large amount of P was released from the sediments during anoxia. The restoration reduced sediment release by about 70 percent (Garrison, unpublished data).

making it a drainage lake. Studies of water and external nutrient loads during 1972 and 1973 (Table 4-2) revealed that city storm sewers contributed more than half the phosphorus income to Mirror Lake and thus, should be targeted by the lake restoration project. The study demonstrated that phosphorus loading peaked during spring showers and intense late summer rainfalls (Knauer, 1978).

Table 4-2.—Annual phosphorus loads for Mirror Lake, Wisconsin, 1972 and 1973.

SOURCE	1972 (%)	1973 (%)
Storm sewers	50	57
Overland flow	16	21
Groundwater	21	18
Precipitation	13	4
TOTAL	**100**	**100**

Total phosphorus concentration in the lake averaged 90 $\mu g\ L^{-1}$, a very high value, with extremely high concentrations in the hypolimnion, particularly near the sediments. This indicated that the sediments contributed a substantial amount of internal loading.

The algae in the lake during the summer were unlike those found in many other eutrophic lakes. Massive blooms of a blue-green alga *Oscillatoria agardhii* characterized the spring and fall, but the summer season saw this species confined to the metalimnion (see Chapter 2), while blue-green algae dominated the upper waters following storms.

A sediment core indicated that storm sewers caused the lake's poor water quality. Among other things, the core was analyzed for chlorophyll pigments common in *Oscillatoria*. The first algal bloom, as recorded by pigments in the sediments, occurred in the early 1940s, just a few years after storm drainage was diverted into the lake.

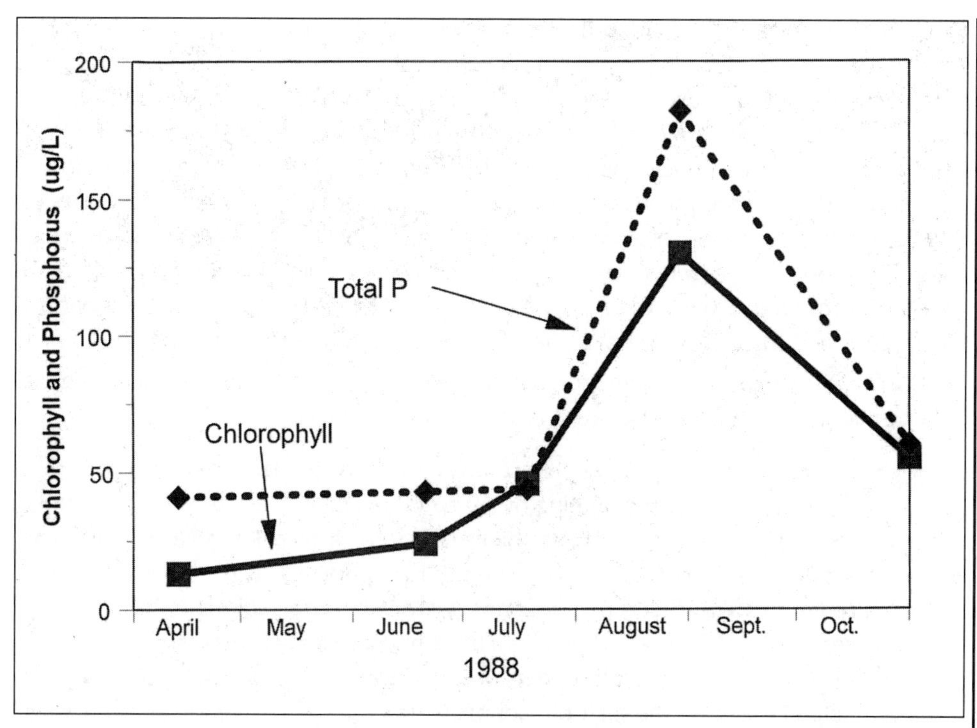

Figure 4-14a.—Trophic state indicators in Cedar Lake prior to the installation of the destratification system (Garrison, unpublished data).

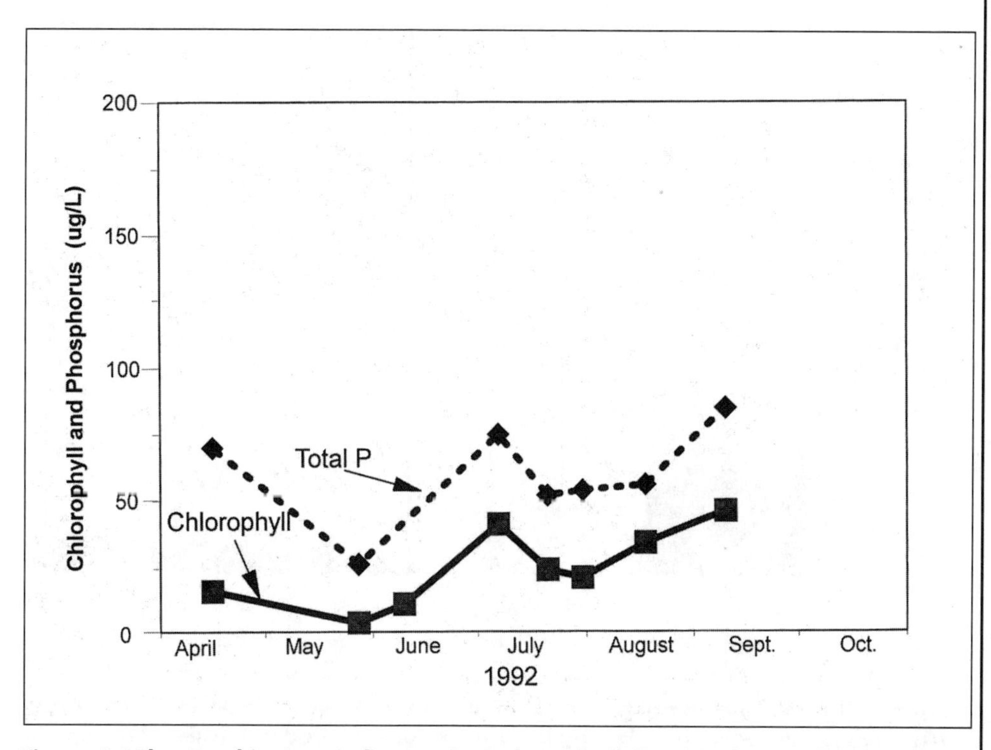

Figure 4-14b.—Trophic state indicators in Cedar Lake following the installation of the destratification system (Garrison, unpublished data). This system greatly reduced the extent of anoxia thus reducing P sediment release that caused the late summer algal bloom.

The diagnostic study also demonstrated that dissolved oxygen was very low during the winter. An analysis of the data revealed that this problem was caused by poor lake mixing during fall months before ice developed. Because of the lake's small size but relatively deep depth, fall turnover didn't happen until late in the fall. Much of the lake's volume is in the hypolimnion, and during stratification the bottom waters accumulate a large quantity of reducing chemical substances. When the lake mixes, these reducing chemicals consume oxygen, often decreasing dissolved oxygen concentrations to 4 mg L^{-1} throughout the water column. If ice forms before enough oxygen is exchanged with the atmosphere, not enough oxygen is left when it freezes to keep fish alive throughout the winter.

The data from the diagnostic study were used to formulate a lake management plan and lake protection and restoration strategies:

- In 1976, the storm sewers were diverted from the lake, reducing the external phosphorus loading by 50 to 60 percent. Despite this reduction, lake phosphorus concentrations remained high (Fig. 4-15), very similar to the prediversion average of 90 µg L^{-1}. Phosphorus released from the sediments was recycling phosphorus stored in the sediments from 35 years of storm sewer drainage. These phosphorus-rich waters moved from the bottom to the upper waters during spring and fall mixing, which helped maintain the high phosphorus levels.

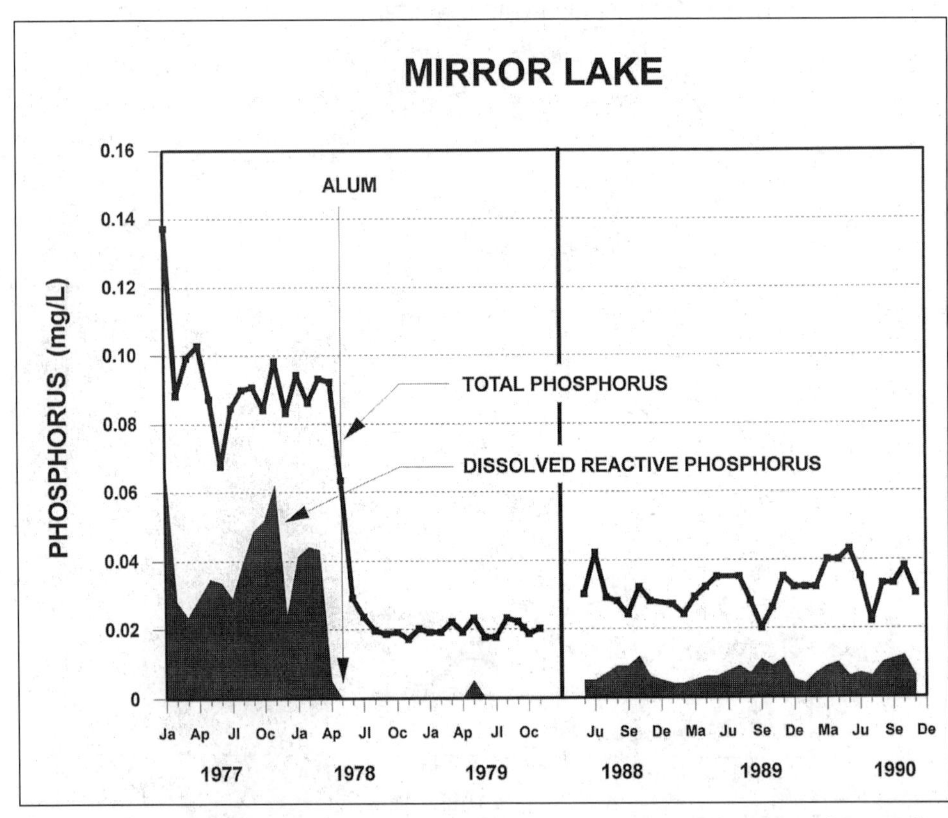

Figure 4-15.—Volume weighted mean phosphorus concentrations in Mirror Lake, Wis., before, immediately following the alum treatment, and a decade following the treatment (Garrison and Ihm, 1991). Mean total P concentration prior to the alum treatment was 90 µg/L and 20 µg/L following the treatment. By 1988 the P levels had increased to 32 µg/L but this was entirely a result of increased concentrations in the hypolimnion. Deep water concentrations a decade following the alum treatment were still considerably reduced from pre-treatment levels and epilimnetic concentrations during the period 1988–90 were similar to those experienced following the alum treatment.

- Aluminum sulfate (alum: Chapter 7) was applied to Mirror Lake in May 1978 to inactivate the sediment phosphorus release. As shown in Figure 4-15, total phosphorus declined to about 20 µg L^{-1} and dissolved reactive phosphorus was virtually undetectable for two years following the treatment.

- Ten years following the alum treatment, the lake was reexamined for three years. The volume weighted mean P concentration had increased from 20 µg L^{-1} to 32 µg L^{-1} in the hypolimnion, indicating recurrence of some internal loading — probably because the alum layer was now buried. By the late 1980s the alum layer was 2 inches (5 cm) below the sediment surface (Garrison and Ihm, 1991). While the alum prevented the release of phosphorus from the deeper sediments, phosphorus in the sediments above 2 inches would be released into the bottom waters during anoxic conditions. Although sediment phosphorus release was higher (0.20 mg m^2 day^{-1}) than immediately following the alum treatment (0.07 mg m^2 day^{-1}), it was considerably less than the pretreatment rate of 1.30 mg m^2 day^{-1} (Table 4-3). And even though the mean phosphorus concentration was higher a decade following the alum treatment, the summer epilimnetic concentration was similar to that immediately following the alum treatment (Garrison and Ihm, 1991).

Table 4-3.—Release rates of phosphorus and ammonium measured by *in situ* nutrient regeneration chambers for Mirror Lake, Wis. Since ammonium release rates are not affected by the alum treatment, these figures provide a further indication of the effectiveness of the alum treatment (Garrison and Ihm, 1991).

TREATMENT	PHOSPHORUS (mg m^2 day^{-1})	AMMONIUM (mg m^2 day^{-1})
Pre-alum (1978)	1.30	6.6
Alum (1978–81)	0.07	5.1
Post-alum (1990)	0.20	5.4

- The storm sewer diversion and alum application also reduced the size of the algal blooms. Blue-green algae no longer bloomed following summer storms and the quantity of the alga *Oscillatoria agardhii* was considerably smaller.

- To solve the problem with low dissolved oxygen under the ice, an artificial circulation device was used in the fall to thoroughly mix the lake. The circulation unit was turned on at the beginning of November and operated until the beginning of ice cover (usually early December). This extended fall mixing ensured that when the lake iced over, enough oxygen remained in the water column to prevent winter fishkills (Fig. 4-16).

These case histories represent real and highly successful uses of the diagnosis-feasibility-implementation approach to lake protection and restoration. Once the causes of the problems were identified, money was directed to long-term solutions instead of wasting it on temporarily effective treatments (e.g., copper sulfate treatments). The lesson here is that lake management should proceed along step-by-step approaches that are based upon a knowledge of both the watershed

The lesson here is that lake management should proceed along step-by-step approaches that are based upon a knowledge of both the watershed and the lake and are directed at the causes, not the symptoms of the problems.

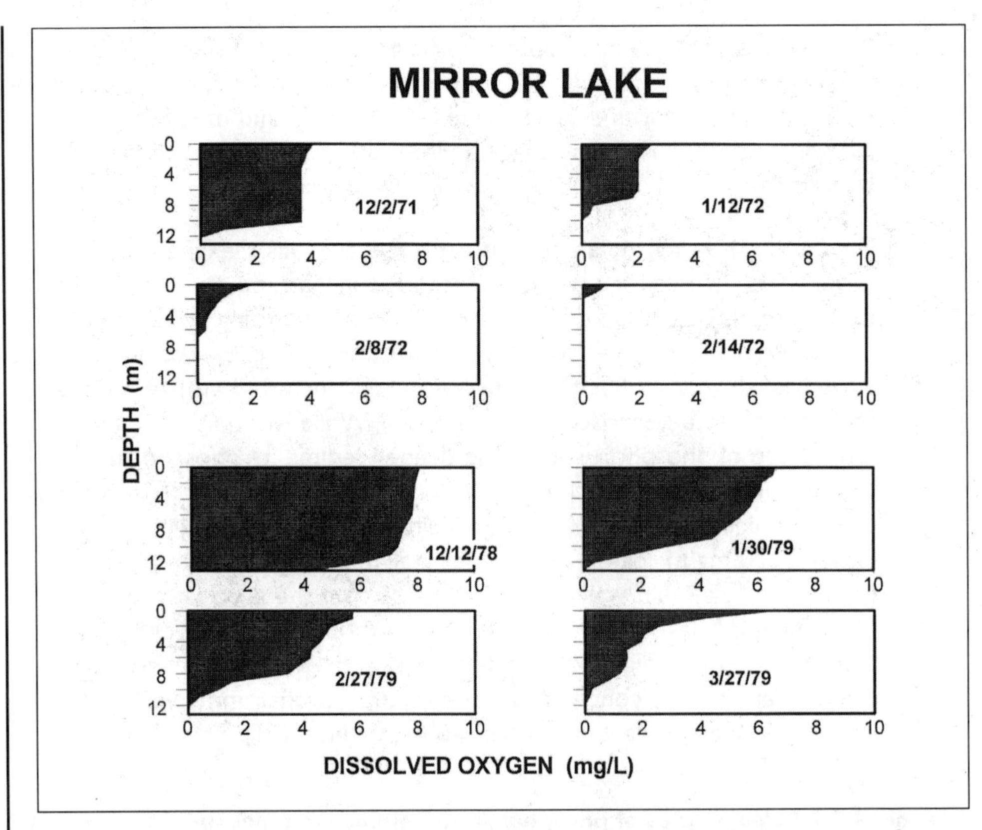

Figure 4-16.—Winter dissolved oxygen profiles for Mirror Lake demonstrating the success of the artificial destratification system (modified from Garrison and Ihm, 1991). Without this system the lake frequently mixed in the fall just prior to freeze up. Because of the high oxygen demand from reduced substances from the bottom waters, the DO concentration throughout the water column was low. The destratification system prolonged fall turnover allowing sufficient oxygen to become dissolved throughout the water column to prevent winterkill conditions.

and the lake and are directed at the causes, not the symptoms of the problems. Effective lake management plans (Chapter 8) integrate watershed management practices (Chapter 6) with in-lake restoration procedures (Chapter 7).

References

American Public Health Association. 1992. Standard Methods of Water and Wastewater. 18th ed. Am. Pub. Health Ass., Am. Water Works Ass., Water Environment Federation, Washington DC.

American Water Works Association. 1990. Water Quality and Treatment: A Handbook of Community Water Supplies. 4th ed. McGraw-Hill, New York.

Asplund, T.A. and C.M. Cook. 1997. Effects of motor boats on submerged aquatic macrophytes. Lake Reserv. Manage. 13:1-12.

Asplund, T.R. and J.A. Johnson. 1996. Alternative Stable States in Fox Lake, Dodge County, Wis. Results of 1995 plankton and water quality monitoring. Wis. Dep. Nat. Resour., Madison.

Baker, J.P., H. Olem, C.S. Creager, M.D. Marcus, and B.R. Parkhurst. 1993. Fish and Fisheries Management in Lakes and Reservoirs. EPA 841-R-93-002. Terrene Institute and U.S. Environ. Prot. Agency, Washington, DC.

Brezonik, P.L. 1984. Trophic state indices: rationale for multivariate approaches. Pages 441-5 *in* Lake and Reservoir Management. Proc. 3rd Annu. Conf. N. Am. Lake Manage. Soc., Knoxville, TN. EPA 4401/5-84-001. U.S. Environ. Prot. Agency, Washington, DC.

Carlson, R.E. 1977. A trophic state index for lakes. Limnol. Oceanogr. 22:361-9.

Cooke, G.D., E.B. Welch, A.B. Martin, D.G. Fulmer, and G.C. Schrieve. 1993. Effectiveness of Al, Ca and Fe salts for control of internal phosphorus loading in shallow and deep lakes. Hydrobiol. 253:323-35.

Deppe, E. and R.E. Lathrop. 1992. A comparison of two rake sampling techniques for sampling aquatic macrophytes. Find. No. 32. Wis. Dep. Nat. Res. Manage., Madison.

Engel, S. 1989a. The restructuring of littoral zones. Lake Reserv. Manage. 2:235-42.

———-. 1989b. Lake use planning in local efforts to manage lakes. Pages 101-5 *in* Proc. Natl. Conf. Enhancing States Lake Management Programs, May 1988. Northeast. Ill. Plann. Commiss., Chicago, IL.

Garrison, P.J. 1995. Paleoecological analysis of East Twin Lake, St. Croix County. Wis. Dep. Nat. Resour., Madison.

———-. 1998. Final Report for Delavan Lake, Walworth County Zooplankton Study for the period of 1995-97. Wis. Dep. Nat. Resour., Madison.

Garrison, P.J. and D.M. Ihm. 1991. First Annual Report of Long-term Evaluation of Wisconsin Clean Lake Projects: Part B Lake Assessment. U.S. Environ. Prot. Agency, Washington, DC.

Garrison, P.J. and D.R. Knauer. 1984a. Lake restoration: a five year evaluation of the Mirror and Shadow lakes project, Waupaca, Wis. EPA 440/5-81-010. U.S. Environ. Prot. Agency, Washington, DC.

———. 1984b. Long-term evaluation of three alum-treated lakes. Pages 513-17 *in* Lake Reserv. Manage. EPA 440/5-84-001. U.S. Environ. Prot. Agency, Washington, DC.

Garrison, P.J. and R.S. Wakeman. 2000. Use of paleolimnology to document the effect of lake shoreland development on water quality. J. Paleolim. 24: 369-93.

Heiskary, S.A., C.B. Wilson, and D.P. Larsen. 1987. Analysis of regional patterns in lake water quality: using ecoregions for lake management in Minnesota. Lake Reserv. Manage. 3:337-44.

James, W.F., J.W. Barko, and S.J. Field. 1996. Phosphorus mobilization from littoral sediments of an inlet region in Lake Delavan, Wis. Arch. Hydro. Biol. 138: 245-57.

Jones, W.W. 1996. Balancing recreational user demands and conflicts on multiple use public waters. Am. Fish. Soc. Symp. 16:179-85.

Knauer, D.R. 1975. The effect of urban runoff on phytoplankton ecology. Verh. Int. Verein. Limnol. 19:893-903.

Kratzner, C.R. and P.L. Brezonik. 1981. A Carlson-type trophic state index for nitrogen in Florida lakes. Wat. Resour. Bull. 17:713-15.

Lamarra, V.J. Jr. 1975. Digestive activities of carp as a major contributor to the nutrient loading of lakes. Verh. Int. Verein. Limnol. 19:2461-8.

Meijer, M.L., M.W. de Haan, A.W. Breukelaar, and H. Buiteveld. 1990. Is reduction of the benthivorous fish an important cause of high transparency following biomanipulation in shallow lakes? Hydrobiol. 200/201: 303-15.

Nürnberg , G.K. 1995a. Quantifying anoxia in lakes. Limnol. Oceanogr. 40:110-11.

————-. 1995b. The anoxic factor, a quantitative measure of anoxia and fish species richness in central Ontario lakes. Trans. Am. Fish. Soc. 124:677-86.

Omernick, J.M. 1987. Ecoregions of the conterminous United States. Freshw. Ann. Ass. Am. Geog. 77:118-25.

Penaloza, L. 1991.Boating Pressure on Wisconsin's Lakes and Rivers. Tech. Bull. No. 174. Wis. Dep. Nat. Resour., Madison.

Porcella, D.B., S.A. Peterson, and D.P. Larsen. 1979. Proposed method for evaluating the effects of restoring lakes. Pages 265-310 *in* Limnological and Socioeconomical Evaluation of Lake Restoration Projects: Approaches and Preliminary Results. EPA 600/3-79-005. U.S. Environ. Prot. Agency, Washington, DC.

Reckhow, K.H., M.N. Beaulac, and J.T. Simpson. 1980. Modeling phosphorus loading and lake response under uncertainty: A Manual and Compilation of Export Coefficients. EPA-440/5-80-011. U.S. Environ. Prot. Agency, Washington, DC.

Stauffer, R.E. and G.F. Lee. 1973. The role of thermocline migration in regulating algal blooms. Pages 73-82 *in* E.J. Middlebrooks, ed. Modeling the Eutrophication Process. Utah State Univ. Logan. Republ. by Ann Arbor Science, Ann Arbor, MI.

U.S. Environmental Protection Agency. 1989. Handbook of Methods for Acid Deposition Studies Field Operations for Surface Water Chemistry. EPA/600/8-84/023. Research Triangle Park, NC.

Wagner, K.J. 1991. Assessing the impacts of motorized watercraft on lakes: Issues and perceptions. Pages 77-93 *in* Proc. Natl. Conf. on Enhancing States' Lake Management Programs. May 1990. Northeast. Ill. Plann. Commiss., Chicago.

Walker, W.W. 1984. Trophic state indicies for reservoirs. Pages 435-40 *in* Lake Reserv. Manage. Proc. 3rd Annu. Conf. N. Am. Lake Manage. Soc., Knoxville, TN. EPA 4401/5-84-001. U.S. Environ. Prot. Agency, Washington, DC.

Wisconsin Department of Natural Resources. 1988. A Two-year Study of Devil's Lake: Results and Management Implications. Bur. Res., Madison.

Wentz, D.A., W.A. Rose, and J.T. Krohelski. 1989. Hydrologic component. Pages 5-1 to 5-77 *in* D.R. Knauer and S.A. Brower, eds. The Wisconsin Regional Integrated Lake-Watershed Acidification Study: 1981-83. EPRI EA-6214. Electric Power Res. Inst., Palo Alto, CA.

CHAPTER 5

Predicting Lake Water Quality

Models

Mathematical models express in quantitative terms the cause-effect relationships that control lake water quality, and can be used both to diagnose lake problems and evaluate possible solutions. Model formulas are derived from scientific theories combined with observations of conditions in real lakes. In lake studies, models are employed in two basic situations:

1. **Diagnostic**: What is going on in the lake? What is the present water quality? Models provide a frame of reference for interpreting lake and watershed monitoring data. They tell the user what to expect in a lake with given morphometric, hydrological, and watershed characteristics. These expectations may not always be met, however, because of natural variation in the observations and unique features of the lake that are not represented by the model. This result, in turn, can help clarify important cause and effect relationships.

2. **Predictive**: What will happen to the lake if we take certain actions? Models can be used to predict how lake water quality will change in response to changes in nutrient inputs or other factors. Once the model is calibrated and verified with baseline conditions in your lake, you can perform experiments on paper or computer instead of engaging in full-scale experimentation with the lake itself (a usually infeasible and inefficient process).

Calibrate: To obtain a best fit between model predictions and observed data by adjusting model parameters.

Mathematical lake models can address many questions:

- What did the lake look like before anyone arrived? (pre-development scenario)
- What is the lake's present water quality? (existing scenario, lake assessment)
- How will future watershed development affect the lake's water quality? (post-development scenario)
- What are the most important sources of nutrients to the lake?
- What level of nutrient loading can the lake tolerate before it develops algae problems? (goal setting, TMDLs [total maximum daily loads])

- How much must nutrients be reduced to eliminate nuisance algal blooms?
- How long will it take for lake water quality to improve, once controls are in place? (restoration scenario)
- How successful will restoration be? (based on a water quality management goal such as target levels for lake phosphorus, chlorophyll, or transparency)
- Are proposed lake management goals realistic? Are they cost effective?

Use modeling only for evaluating those types of problems you understand well enough to express them in concise, quantitative terms. Some situations, like exotic species introductions, unique accidents, or unusual weather conditions are not predictable; therefore, modeling is not possible. Sometimes it's even unnecessary, especially if the lake or reservoir is well studied and no future changes are expected.

Models are just tools used by lake management professionals to develop their assessments and recommendations. The lake manager decides whether and which type of models to use, what supporting data should be collected, how the models should be implemented and tested, and how the model results should be interpreted.

Validate or Verify: To compare model predictions with observed data, using data that are different from those used to calibrate the model, and to conclude that the results of the comparison are acceptable.

Mass Balance: The mass of a substance in a fixed volume is neither created nor destroyed. Rather, it changes only according to the input and output fluxes across the volume's boundary, which in the case of a lake may be its bottom, surface, or shoreline.

Ideally, a model is applied only to situations and conditions for which it has been tested and verified. If no validated models are available or it is not possible to validate a chosen model (perhaps for financial reasons), the model's results should be interpreted with caution. The lake manager's choice of appropriate models for a given lake or reservoir should be based on regional experience, limnological knowledge, and the types of predictions desired (e.g., detailed versus spatial-temporal averages). The lake manager should consider how closely the lake characteristics (e.g., morphometry, hydrology, natural lake versus reservoir) reflect the characteristics of the lakes that were used to develop and verify a model.

The many types of lake models differ in complexity, assumptions, data requirements, and methods of calculating results. Most are based on the mass balance concept where all the fluxes of mass to and from specific compartments must be accounted for over time. Complex dynamic simulation models such as "CE-QUAL-ICM" (U.S. EPA, 1997) can make detailed predictions of a lake's response to pollution over space and time by using fine spatial grids for compartmentalization and small time steps. Such models, however, require a lot of input and field data for calibration and testing.

Complex models can usually be simplified by making certain assumptions. In some cases, spatial variation of nutrients in a lake may be unimportant, so you can eliminate those compartments in the mass balance model. For example, Chapra and Canale (1991) developed a spatially homogeneous lake model with a separate sediment compartment that helps them predict the long-term temporal dynamics of average lake phosphorus concentrations.

You might further simplify a model by assuming the lake is in temporal steady-state: i.e., its nutrient concentration is no longer changing over time. A lake's mass balance then reduces to simple algebraic equations that a spreadsheet can handle for the average lake nutrient concentration. Under this steady-state assumption, you can compare different scenarios (e.g., pre- and post-development), but you can't determine the time it will take to reach the second scenario, nor predict temporal trends. Walker's (1987) BATHTUB model for reservoirs assumes a temporal steady-state but allows for some spatial variation by dividing the reservoir into basins.

Models should be only as complicated as necessary for the task at hand. Unnecessary detail built into a model makes it more difficult and expensive to construct and verify. A simple model can be more robust and useful for management decisions than one requiring difficult to obtain detailed input data and then making many detailed predictions that will never be compared to observations (see U.S. EPA, 1999). Because spatial homogeneity, temporal steady-state, and other assumptions make for simpler lake models that still make useful predictions, such models are frequently used — and are emphasized in this chapter. A review of other useful lake models can be found in U.S. EPA (1997).

Robust: Description of a model that predicts well over widely varying conditions.

Spatial Homogeneity: evenly distributed (well-mixed) throughout a fixed volume.

Temporal Steady State: In the context of a mass balance, there may be fluxes in and out of the system, but combined inputs equal combined outputs so that the mass in a fixed volume is constant over time.

Modeling eutrophication is the primary purpose of this chapter. The general concepts of eutrophication (see Chapters 2 and 4) are based on the observation that nutrient availability (phosphorus in particular), algal production, and fish production are strongly correlated. Therefore, increasing or decreasing the phosphorus loading to a lake will generally have the same effect on nutrient availability, algal growth, and fish production.

As will be shown, these relationships can be modeled step by step, starting with a lake's phosphorus budget, followed by observed relationships between phosphorus concentration and water quality, and concluding with predictions of a lake's response to future development or restoration (e.g., through goal-setting with TMDLs). But lakes and reservoirs have other water quality problems besides those based on phosphorus, and thus we will also assess the value of establishing mass balance models for the nutrient nitrogen, total suspended solids (causing turbidity), and acids.

Eutrophication: The Problem

Excessive nutrients that promote aquatic growth, especially algae, were identified as the most important problem in 44 percent of all U.S. lakes surveyed in 1998 (U.S. EPA, 2000a). Nutrients were also deemed excessive in more than half of the lakes with impaired water quality. Therefore, models are frequently used to evaluate eutrophication problems related to algae. Eutrophication modeling is based on several general observations made in many temperate lakes and reservoirs over many years:

1. Lake algal growth is usually determined (limited) by the supply of phosphorus. Even if other factors, e.g., nitrogen or light, become limiting, algae biomass and blooms usually increase with increasing lake water phosphorus concentrations (Nürnberg, 1996).

2. Increasing or decreasing the amount of phosphorus entering the lake over an annual or seasonal period will increase or decrease the average concentrations of phosphorus and consequently of algae (as in 1).

3. A lake's capacity to absorb increased phosphorus loading without experiencing higher phosphorus concentrations and consequent algal blooms increases with volume, depth, flushing, and sedimentation rates.

Phosphorus: Phosphorus is usually measured in units of weight of phosphorus per volume of water; for example: mg/m^3 = µg/L = ppb.

• **Total Phosphorus** represents all phosphorus in a sample.

• **Particulate Phosphorus** represents only a certain fraction that is greater than a defined size (often 0.45 micron).

• **Total Dissolved Phosphorus** is all phosphorus smaller than the defined size.

• **DRP or SRP (Dissolved or Soluble Reactive Phosphorus)** is the part of Total Dissolved Phosphorus that is readily available and is often used as an approximation to the chemically pure form of orthophosphate.

In other words, algal growth is mainly controlled by phosphorus. Thus, the condition of your lake water depends on the balance between the amount of incoming and outgoing phosphorus and the volume of water available for dilution. A large, deep lake with a high flow will be able to handle a larger phosphorus load better than a small, shallow, or stagnant lake.

Most eutrophication models simply summarize these relationships in mathematical terms. In particular, steady-state phosphorus mass balance models use estimated lake inputs, outputs, and lake morphometry to predict long-term average lake water phosphorus concentration, which in turn is empirically related to variables indicating algal biomass (chlorophyll), water transparency (Secchi disk depth), hypolimnetic anoxia (lack of oxygen), and/or fish species abundance and production (Fig. 5-1). These relationships are called "empirical" because they are statistically based on the responses of water quality to average phosphorus concentration observed in a large number of lakes and reservoirs.

Total phosphorus functions as the currency of water quality, since it is the primary controlling variable for water quality and is correlated to the others, like chlorophyll, hypolimnetic anoxia, and transparency that all help define a lake's trophic state (see Chapter 4). It also is more stable and easier to determine. For example, a one unit increase of phosphorus yields a certain amount of algae, which in turn decreases Secchi disk transparency by a certain depth and increases hypolimnetic anoxia by a certain amount of time and volume.

Modeling Eutrophication

Eutrophication modeling is a step-by-step process:

▼ STEP 1. ***Development of hydrological (water) and nutrient budgets.*** A lake model is only as good as the data used in its construction. Good hydrological and nutrient lake budgets provide that data. If no or very little inflow and outflow stream data are available, these flows can be modeled separately with various degrees of detail. At one extreme, a watershed simulation model might use 20 years of daily precipitation data and numerous sub-watershed areas to provide long-term average water flow. Alternatively and more simply, data derived from other comparable studies can be used to determine approximate water and phosphorus loads for a steady-state lake model.

▼ STEP 2. ***Calculation of lake phosphorus concentrations from external and internal phosphorus loading*** as determined in Step 1. Phosphorus models can simply describe the lake as a steady-state, completely mixed water body and determine annual average lake total phosphorus, or they can use a simulation model to specifically model various lake compartments (e.g., bays and basins, different layers) or seasons separately, or even detailed phosphorus changes in time and space. Sometimes a combination of simple and more elaborate models is useful: the simple model can verify the overall input and output of a more detailed model.

Figure 5-1.—Eutrophication modeling concepts.

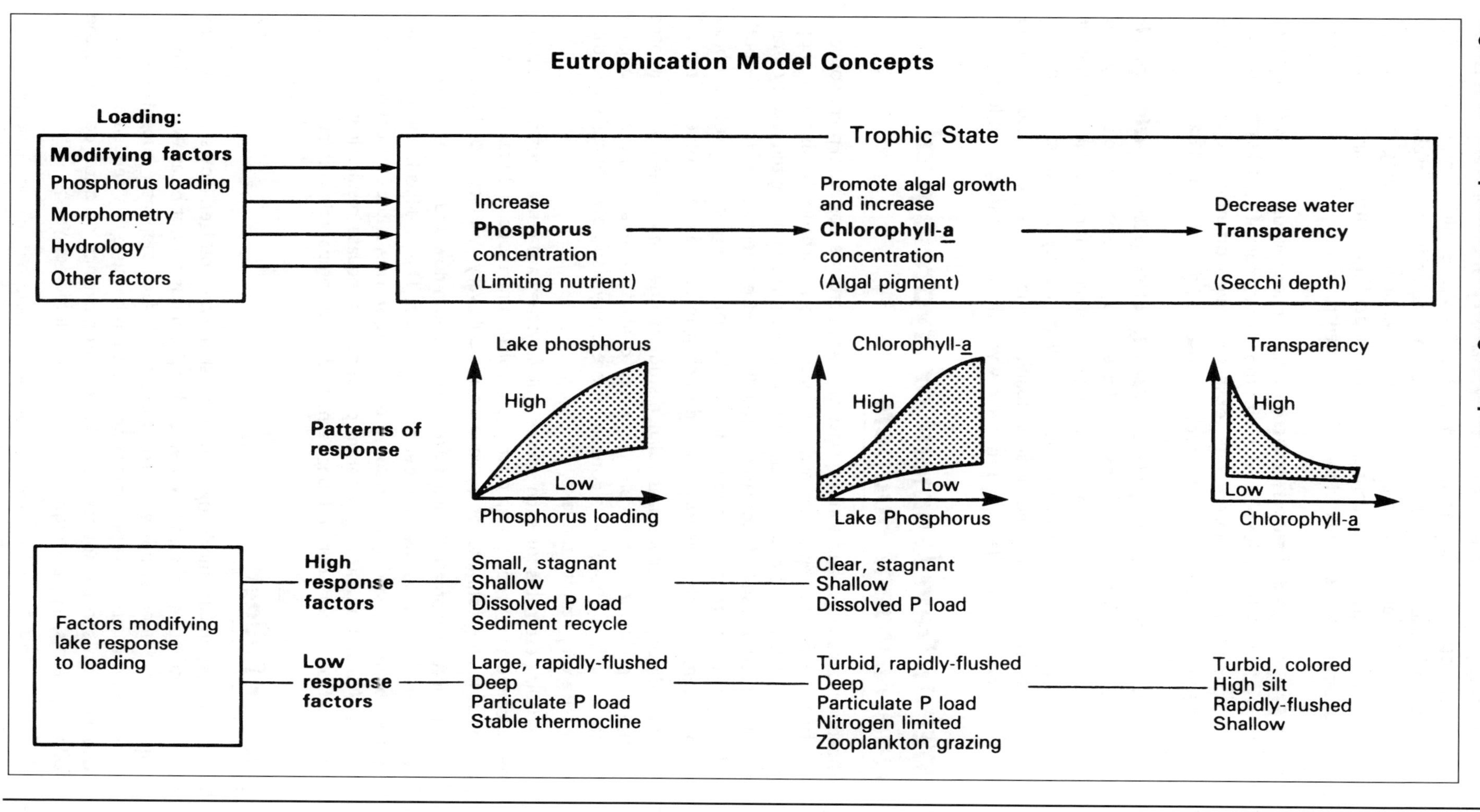

▼ STEP 3. ***Prediction of water quality from lake phosphorus concentration*** determined in Step 2. Empirical models are frequently used to predict water quality variables from phosphorus. These models often require only one predictor variable, e.g., average summer epilimnetic phosphorus concentration, to model the water quality variable in question, e.g., summer chlorophyll concentration.

▼ STEP 4. ***Model verification*** of lake phosphorus and water quality predictions from Step 2 and 3 with monitoring data, if available. Model predictions are compared with observed lake conditions to assess the sources and degrees of variation of observed and predicted data.

▼ STEP 5. ***Forecasting and tracking changes in water quality.*** Combinations of Steps 1 to 3 are used to predict possible changes in water quality. Once the model has been validated for your lake (Step 4), it can be used to predict the changes in water quality likely to result from future changes in phosphorus loading. The model can also remove anthropogenic effects and thus show you the lake's natural condition.

STEP 1: Development of Hydrological and Nutrient Budgets

Phosphorus loading changes in response to season, storms, upstream point sources, and land use. For example, converting a forest into an urban subdivision or shopping center usually increases the amount of phosphorus entering the lake by a factor of 5 to 20, a result of increased water flow and nutrient concentration – both related to runoff from impervious surfaces. Evaluating the loading gives us a basis for projecting how the lake will respond to changes in land use or other factors.

In addition, a detailed mass balance study can determine the relative impact of various pollutant sources on lake water quality. For example, even though a stream has a high phosphorus concentration, it may also flow minimally, and thus contribute very little to the lake.

Loading estimates for each source will vary; they are ranges, not fixed quantities. Depending on monitoring intensity, calculation methods, and natural variation, an annual loading estimate for a given stream could vary by a factor of 2 or more.

To directly estimate loading from a source, both flow and concentration must be properly quantified over the period in question. This is difficult and expensive because both vary widely in response to season, storms, and other unpredictable factors. When a monitoring program produces inadequate results, it is better to use data from a more detailed long-term study of a comparable watershed.

Water Budget

The first step in lake modeling is to establish a water budget. Flows carry pollutants into and out of lakes, and water quality problems cannot be analyzed without a quantitative understanding of lake hydrology. In fact, the water budget is as important as evaluating the pollutant concentration because it is needed to quantify loadings from specific sources. The basic water balance equation considers the following terms, typically expressed as water volume per year:

Inflow + precipitation = outflow + evaporation + change in storage

Figure 5-2 illustrates possible water flows that contribute to the total budget.

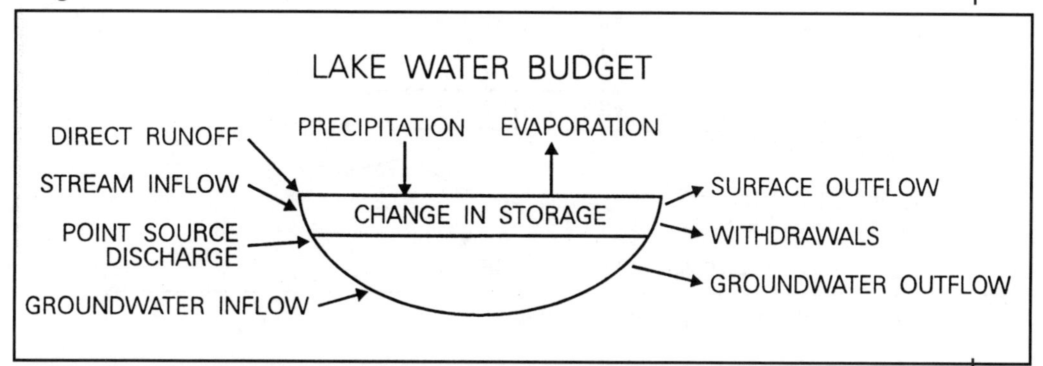

Figure 5-2.—Schematic water budget.

- **Inflows** may come from tributary streams, point source discharges, shoreline runoff, and groundwater springs.
- **Outflows** may include the lake outlet, groundwater discharges, and withdrawals for water supply, irrigation, or other purposes.

Major inflow and outflow streams should be gauged directly and continuously over the long term. If only short-term data are available, use them in conjunction with long-term regional climate data (**precipitation, evapotranspiration**) to calibrate a model that will give you long-term average stream hydrology. Use estimates (for example, runoff coefficients) to quantify smaller streams.

The **change in storage** accounts for fluctuations in the elevation of the lake surface — whether it's "high" or "low" — over the study period; this is sometimes significant in reservoirs.

Once all flow terms have been determined, check the water budget for balance. Major discrepancies may indicate an error in an important source of inflow, outflow, or storage (such as unknown or poorly defined stream or groundwater flow). It is relatively difficult to establish water balances in seepage lakes because of the problems and expense of monitoring groundwater flows. Significant errors in the water balance may indicate a need for further study of the lake's hydrology.

Phosphorus Budget

The cornerstone for evaluating eutrophication problems, the lake phosphorus budget evaluates and ranks phosphorus sources that may contribute to an algal problem. The basic concept and mathematics are relatively simple, although estimating individual budget items often requires considerable time, monitoring data, and expertise.

The following terms are evaluated and typically expressed as phosphorus mass per year (if an areal load, as phosphorus mass per lake surface area per year):

External load = outflow load + sedimentation - internal load + change in storage

Figure 5-3 illustrates external and internal phosphorus sources that may contribute to the total budget.

The term **external load** is the amount of phosphorus per year that enters the lake from all external sources. It may come from tributaries, sewage treatment

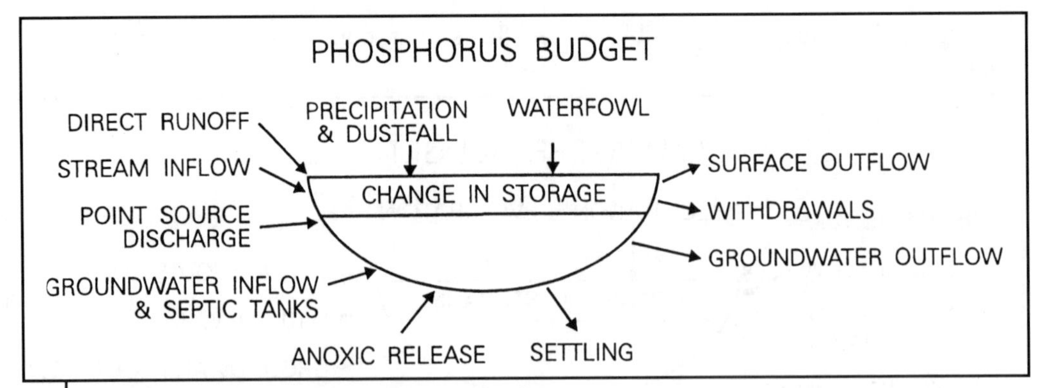

Figure 5-3.—Schematic phosphorus budget.

plants discharging directly to the lake, precipitation and dust fall, septic tanks, groundwater, runoff, and waterfowl. Estimating these loads is the most important and generally most expensive step in the modeling process.

Intensive monitoring programs to define and quantify at least some major loading sources can pay off in good, reliable results. Lake water is usually monitored for all parameters at the same time so that loadings can be related to lake responses.

Stream loading, often the largest source, is usually estimated from stream flow and phosphorus concentrations monitored periodically (weekly, monthly) and supplemented with samples taken during storms. Storm sampling is very important, particularly in small streams that flash flood, because a very high percentage of the annual loading may occur during short, intense storms. If these events are not sampled, it will be difficult to develop reliable loading estimates. You might also try combining monitored hydrology with empirically predicted phosphorus concentration to arrive at the stream load.

A complete monitored external load budget is so costly and labor intensive — and takes so long — that, instead, you may want to make indirect estimates based on the characteristics of your watershed. This method is based on the concept that two watersheds in the same region and with similar land-use patterns and geology will tend to contribute the same amount of phosphorus per unit area, and thus data can be extrapolated from one or more monitored watersheds to others.

Whether this method will work for your lake depends largely on the availability of good data on regional export coefficients (mass of phosphorus per watershed area, per year) for the land uses and watersheds in your area. Export coefficients have been compiled for a number of land uses (see Chapter 2, Table 2-1). This approach is much less costly than direct monitoring and can be as good or better for long-term predictions, especially when you monitor inflow infrequently or for only a few years.

The term **outflow load** relates to phosphorus leaving the lake in surface outlet(s); withdrawals for water supply, irrigation, or other purposes; and groundwater seepage. You can usually measure these flows and concentration, although if groundwater seepage dominates the outflow, it will be difficult to directly determine the outflow loading.

The term internal load refers to all internally derived phosphorus. The most important source comes from bottom sediments that release phosphorus when their surface goes anoxic. This happens frequently in the summer in eutrophic stratified lakes. To determine the internal load from anoxic sediments, use either your lake's phosphorus budget, the phosphorus increase in the hypolimnion, the

Table 5-1.—Anoxic Factor.

The anoxic factor (AF) (Nürnberg, 1995) quantitatively summarizes the extent and duration of anoxia (lack of oxygen) in stratified lakes. It is based on a series of measured oxygen profiles and morphometric data and can be computed for any lake or reservoir. To render this index comparable across lakes of different sizes, AF is corrected for lake surface area by simple division. Expressed this way, AF is a ratio that represents the number of days in a year or season that a sediment area equal to the lake surface area is anoxic. Hence its units are d/yr or d/season; i.e., summer or winter. Anoxic factors can be predicted from average phosphorus concentration and lake morphometry when oxygen profiles are not available.

To compute AF, first inspect the oxygen profiles and determine the depth at which DO concentration is 1 mg/L. When this concentration is found about 1 m (3 feet) above the bottom, the sediment-water interface is likely anoxic, and processes requiring reduced sediment surfaces, like phosphorus and iron release, will commence. Next, the period of anoxia (t_i in days) must be multiplied by the corresponding hypolimnetic area (a_i in m^2) and divided by the lake surface area (A_o, m^2) corresponding to the average elevation for that period. These terms of n, numbers of periods at different oxyclines are then added up. In this way, AF is comparable between lakes, like other areal measures, e.g., areal nutrient loads and fish yield.

When classified with respect to trophic state, below 20 d/yr indicate oligotrophic conditions, 20 to 40 d/yr are usually found in mesotrophic lakes, 40 to 60 d/yr represent eutrophic conditions and above 60 d/yr is typical for hypereutrophic conditions.

$$AF = \sum_{i=1}^{n} \frac{ti \times ai}{Ao}$$

increase in phosphorus concentration at fall turnover, or laboratory estimates of sediment release rates coupled with the anoxic factor as a variable to describe the duration and extent of hypolimnetic anoxia (Tables 5-1, 5-2).

Internal loads can be quite substantial in eutrophic lakes, and, when good management has reduced external loading, even more than the phosphorus entering the lake from the watershed (Fig. 5-4).

Other internal sources that are usually less important include:

- Phosphorus transport from the bottom of the lake to the epilimnion (the top) via algal migration.
- Release from shoreline sediments resulting from resuspension and turbulence.
- Release from sediments caused by changes in pH, and sediment disturbance by bottom-dwelling fish and macrobenthos (Welch, 1992).

Macrobenthos: Insect larvae and other small invertebrates living at the bottom of a pond or lake (these are actually large compared to plankton).

Sedimentation is the downward flux of phosphorus to the sediment. Lake water quality is usually better with higher sedimentation, because less phosphorus remains in the water column to stimulate algal growth.

The term **change in storage** accounts for changes in the total mass of phosphorus stored in the water column between the beginning and end of a study period. Such changes reflect differences in lake volume, average phosphorus concentration, or both. This term is positive if the phosphorus increases and negative if it decreases.

Water and phosphorus budgets supply the information necessary to compare the individual loading terms and thus rank sources and identify candidates for watershed management or in-lake restoration techniques. For example, the Lake Wilcox phosphorus budget clearly indicates that treating the tributary

Table 5-2.—Determination of internal phosphorus load in stratified lakes.

Internal phosphorus load is the phosphorus released from lake sediments into lake water. Often it accumulates in the hypolimnion during stratified seasons and is primarily released from anoxic sediment surfaces. Internal load (just like external) can be returned to the sediments leading to various estimates ranging from "gross," which most closely approximates the total amount of internal load, to various "net" estimates which include some degree of re-sedimentation of internal load. Five different ways of estimating internal phosphorus load are listed below:

1. **Net estimates from complete phosphorus budgets:** In a complete mass balance that includes all external P inputs and the total P loss in the outflow, internal phosphorus load is implicitly considered. A net internal load can then be computed from the increase in outflow mass over that predicted from a retention model that assumes no internal load:

$$\text{net internal load} = \text{external load} \times (R_{pred} - R_{meas})$$

where retention (R_{pred} and R_{meas}) is explained in Table 5-3. This estimate of internal load is the smallest and deviates the most, because it includes sedimentation and precipitation of internal load. This net estimate is related to internal load by using the same retention model as used for external loads (Nürnberg, 1998):

$$\text{net internal load} = \text{internal load} \times (1 - R_{pred})$$

2. **Partially net estimates from in situ phosphorus increases in the summer hypolimnion:** Internal load estimated from phosphorus increases in the hypolimnion is higher than that based on an annual budget (1), because of some sedimentation during that period.

3. **Partially net estimates from in situ phosphorus increases at fall turnover:** Internal load can also be estimated at fall turnover, when the surface water concentrations increase due to mixing of hypolimnetic phosphorus-rich water with epilimnetic water. Some additional sedimentation might already have happened before and during turnover, so this value will probably be slightly below the estimate determined from summer increases in the hypolimnion (2).

4. **Gross estimates from phosphorus release rates and anoxia:** The highest estimate of internal load, which should be closest to the amount of phosphorus actually released from the anoxic sediment surfaces, can be computed from release rates and the anoxic area and time, i.e. the anoxic factor. This value should be highest, since it does not incorporate any sedimentation. When there are no measured release rates available, a measured phosphorus concentration in the profundal sediment can be used to predict a range for the phosphorus release rate (Nürnberg, 1988).

5. **Gross estimates from complete phosphorus budgets:** The equations in 1 can be combined so that internal load can be estimated as:

$$\text{internal load} = \text{external load} \times (R_{pred} - R_{meas}) \;/\; (1 - R_{pred})$$

For example, long-term averages of internal load (kg/yr) estimates for Lake Wilcox, Ontario were estimated as follows:

1. Phosphorus budget (net) 39
2. Hypolimnetic increase 135
3. Fall turnover 122
4. Release rate 157
5. Phosphorus budget (gross) 169

The net estimate of 39 kg/yr (estimate 1, that considers a predicted retention of 0.77) corresponds to an internal load of 169 kg/yr (estimate 5). This is similar to estimate 4 which does not consider retention of internal load either.

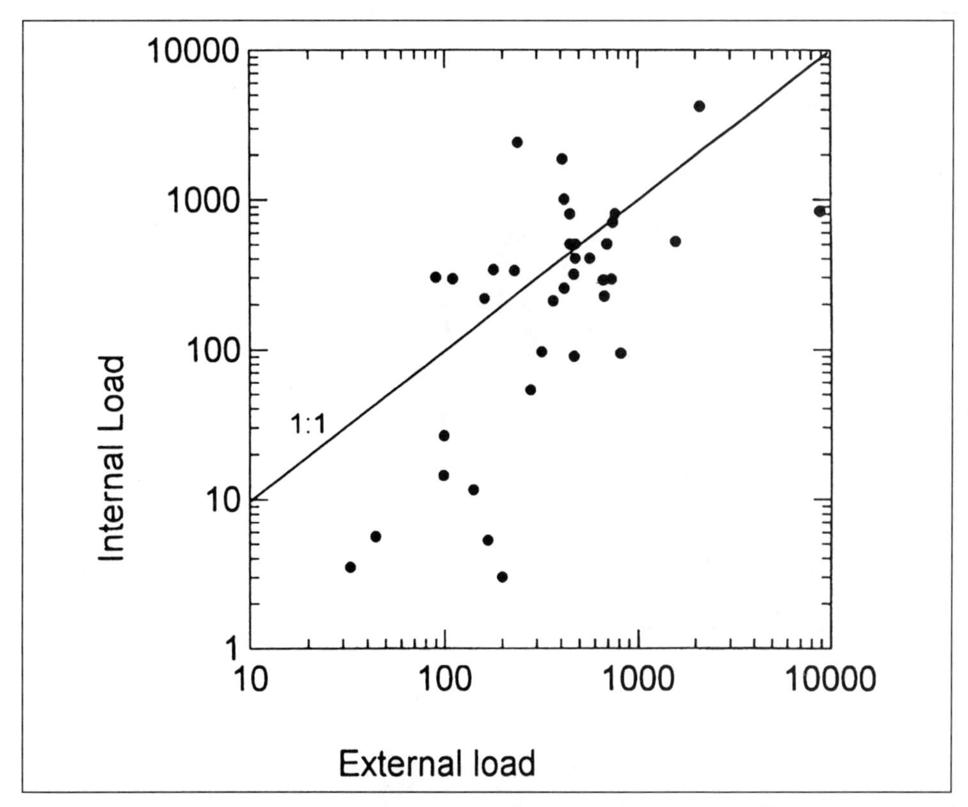

Figure 5-4.—Internal versus external phosphorus loading ($mg/m^2/yr$) in stratified lakes with anoxic hypolimnia.

would not reduce eutrophication because it accounts for less than 10 percent of external and less than 3 percent of total load (Fig. 5-5). And the only way to eliminate late-summer algal blooms in Cedar Lake was by treating the high internal phosphorus load (Chapter 4).

STEP 2: *Predicting Phosphorus Concentration*

Steady-state phosphorus mass balance models usually assume that the lake is spatially homogeneous. They are driven by four fundamental variables calculated from lake morphometry, and water and phosphorus budgets:

1. The average **input phosphorus concentration** is the sum of all external phosphorus loads to the lake weighted by the outflow of water.

   Average input phosphorus concentration = external phosphorus load/outflow

   This equation represents the "real" average inflow concentration only when total inflow equals outflow. Where these flows differ, e.g., in reservoirs, this term has a more theoretical meaning. Typically, outflows and loads are calculated on an annual basis to even out seasonal variation.

   This basic measure of inflow quality is important in determining how watershed point and nonpoint sources affect eutrophication of the lake. Thus, long-term management (as in the TMDL process) frequently focuses on reducing average input phosphorus concentrations.

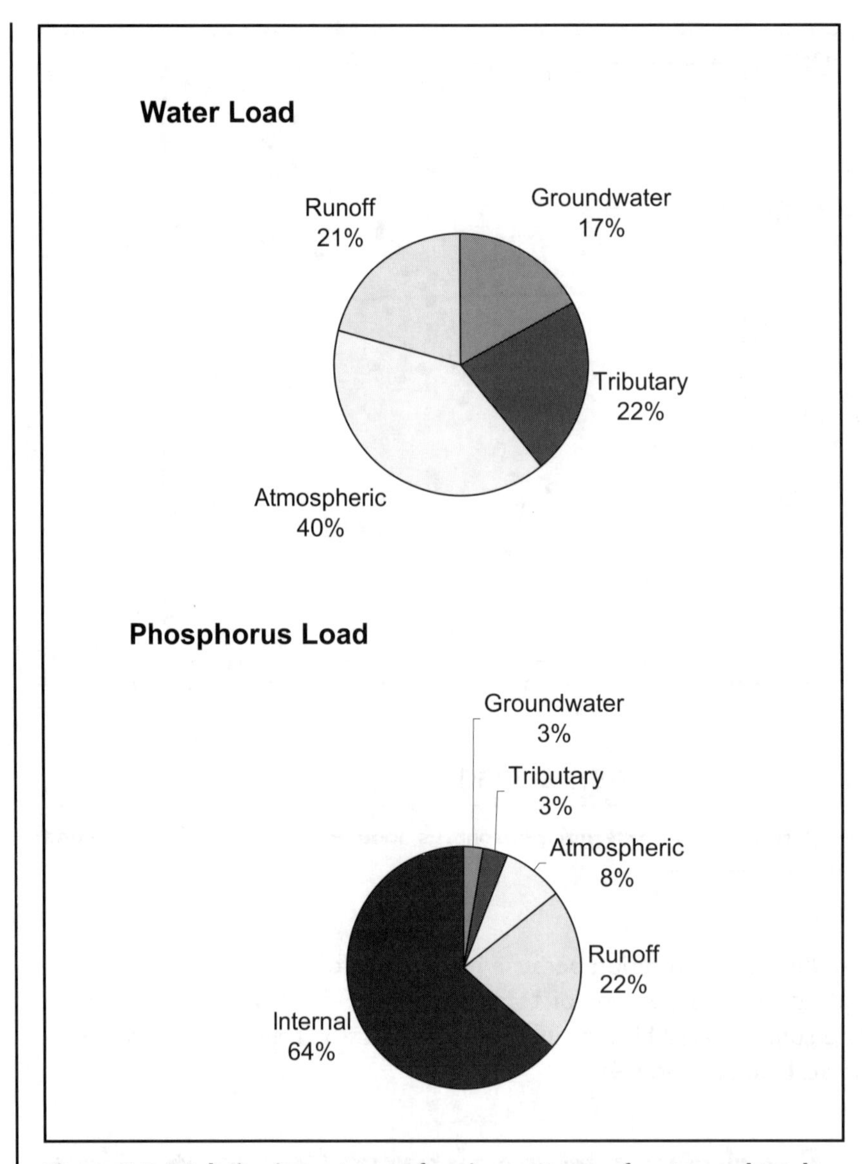

Figure 5-5.—Relative importance of various sources of water and total phosphorus for Lake Wilcox, Ontario.

2. The **average water residence time** (often called hydraulic retention time) (τ) is the average length of time water spends in a lake or reservoir before being discharged through the outlet.

   Average water residence time (τ) = lake volume/outflow

   Again, outflow is calculated annually to even out seasonal variation. If total inflow equals outflow, τ equals the time required for the lake to refill if it were completely drained. As residence time increases, interactions between the water column and bottom sediment have more influence on water quality. For a given inflow concentration, phosphorus sedimentation usually increases and lake phosphorus concentration decreases as residence time increases. At very short residence times (less than one to two weeks), algae may not have enough time to respond to the inflowing nutrient supply.

3. **Mean depth** ($\bar{z}$) is a basic morphometric characteristic of the lake.

   Mean depth ($\bar{z}$)= lake volume/lake surface area

   Other factors being equal, lakes and impoundments with shallower mean depths are generally more susceptible to eutrophication problems. In shallower lakes, light penetrates a larger proportion of total depth to support photosynthesis, and the greater sediment/water contact can encourage nutrient recycling.

4. **Phosphorus retention** defines the fraction of incoming phosphorus kept in lake sediments, as only part of a lake's incoming phosphorus leaves by the outflow. It is the most difficult variable to determine as it depends on sedimentation which in turn depends on morphometry and hydrology.

   Net retention (R_{net}) reflects the net result of all physical, chemical, and biological processes causing vertical transfer of phosphorus between the water column and lake sediments. Hence, it is the difference between annual phosphorus sedimentation and internal phosphorus load.

   R_{net}= (sedimentation – internal load)/external load

   When internal phosphorus sources exist, R_{net} decreases. If R_{net} is known, the steady state lake phosphorus concentration can be calculated as:

   Lake phosphorus concentration = average input phosphorus concentration x (1– R_{net})

   However, because sedimentation and internal load are difficult to measure, R_{net} is rarely available. When complete water and phosphorus budgets are available, you can estimate net P retention by the difference between external and export loads:

   R_{meas} = (external load – export load)/external load

   because (sedimentation – internal load) = (external load – export load). If R_{meas} is substituted for R_{net} in the lake phosphorus concentration equation, average lake phosphorus is estimated by the average outflow P concentration.

   In pre- or post-development scenarios, or when outflow loads are not available, even R_{meas} will not be available. For these cases, empirical retention models (R_{pred}) have been developed that predict retention from water budget variables and lake morphometry. One such model (Nürnberg, 1984) that incorporates sedimentation only (as it was developed using data from lakes without any internal load) is:

   R_{pred} = 15/(18+q_s)

   Where q_s (areal water load) = $\bar{z}/\tau$

   If internal load is significant, this R_{pred} will overestimate net retention and thus, when used in the lake phosphorus concentration equation, underestimate predicted lake phosphorus, as happened in Cedar Lake, Wisconsin (see Chapter 4).

Conversely, predicted phosphorus concentration will be overestimated in situations where physical or chemical processes enhance sedimentation more than predicted by the empirical retention model. This has been observed in some hard, calcium-rich lakes, in some soft, iron-rich lakes, and in reservoirs containing large amounts of silt. Until retention models are developed for these types of lakes, their retention must be measured directly (R_{meas}).

Typically, these equations are used to predict annual average lake phosphorus concentrations; however, you can use different combinations of variables (see Table 5-3) to predict seasonal phosphorus concentrations in stratified and frequently mixed lakes. These are especially useful in stratified lakes with large internal loads, where annual average and fall turnover phosphorus concentrations may be out of proportion to their far smaller epilimnetic concentration in early summer (Fig. 5-6). In these lakes, use the corresponding seasonal phosphorus average to predict other lake water quality variables. For example, if chlorophyll concentration was determined from annual phosphorus rather than epilimnetic summer average, it would be greatly overestimated in mesotrophic Lake Wilcox (observed: 9 μg/L; predicted from epilimnetic phosphorus: 7 μg/L; from annual average phosphorus: 30 μg/L).

After a change in loading, the lake will take a while to reach a new steady state. The duration of this lag time increases with the annual water residence time (τ) and the mean depth ($\bar{z}$) of the lake.

Table 5-3.—Models to predict phosphorus averages for different seasons in stratified lakes with and without anoxic hypolimnia (and hence internal load), and in polymictic lakes (Nürnberg, 1998).

Model predictions were compared to measured phosphorus averages in different lake groups. External load (L_{ext}, mg m^{-2} yr^{-1}) values are gross estimates before any settling, internal load (L_{int}, mg m^{-2} yr^{-1}) estimates are either "gross" or "net" values in the stratified lakes, depending on whether the estimates are based on a whole year budget ("net L_{int}", estimate 1 of Table 5-2) or estimated from sediment phosphorus release rates and anoxia ("gross L_{int}", estimate 4 of Table 5-2). If estimate 5 (Table 5-2) is used as "gross L_{int}" in model #4 of the table below, it becomes identical to model #1. In polymictic lakes, an "in-situ" estimate of L_{int} was used that represents a partially net estimate, since it incorporates some loss via sedimentation. Measured phosphorus concentration averages are: P_{ann}, annual; P_{epi}, summer epilimnetic; P_{fall}, fall turnover. Retention was either measured from a P budget (R_{meas}) or predicted as R_{pred}= 15/(18+q_s), where q_s, annual areal water load (m yr^{-1}). The expression of L_{ext}/q_s in these models is the same as average input P concentration as defined in the main text. n.a., not applicable

		LAKE TYPE		
		STRATIFIED		POLYMICTIC
#	MODEL	OXIC	ANOXIC	
(1)[1]	L_{ext}/q_s (1-R_{meas})	=P_{ann}	=P_{ann}	=P_{ann}, =P_{epi}
(2)	L_{ext}/q_s (1-R_{pred})	=P_{ann}	<P_{ann}	<< P_{ann}
		>P_{epi}	=P_{epi}	<< P_{epi}
(3)[2]	L_{ext}/q_s (1-R_{pred}) + net L_{int}/q_s	=P_{ann}	=P_{ann}	n.a.
(4)	(L_{ext} + gross L_{int})/q_s (1-R_{pred})	=P_{ann}	=P_{ann}	n.a.
(5)	L_{ext}/q_s (1-R_{pred}) + gross L_{int}/q_s	>P_{fall}	>P_{fall}	n.a.
(6)	L_{ext}/q_s (1-R_{pred}) + in-situ L_{int}/q_s	n.a.	n.a.	>P_{ann} >P_{epi}
(7)	(L_{ext} + in-situ L_{int})/q_s (1-R_{pred})	n.a.	n.a.	<P_{ann}, <P_{epi}

[1]Dillon and Rigler (1974) [2]Nürnberg (1984)

For example, in a stratified anoxic lake with sediment release, epilimnetic summer phosphorus is usually smaller than fall turnover and annual average phosphorus concentration. The modeling of annual phosphorus requires consideration of internal load (3 or 4), while prediction of epilimnetic early-summer phosphorus concentration may be based on external load only (2).

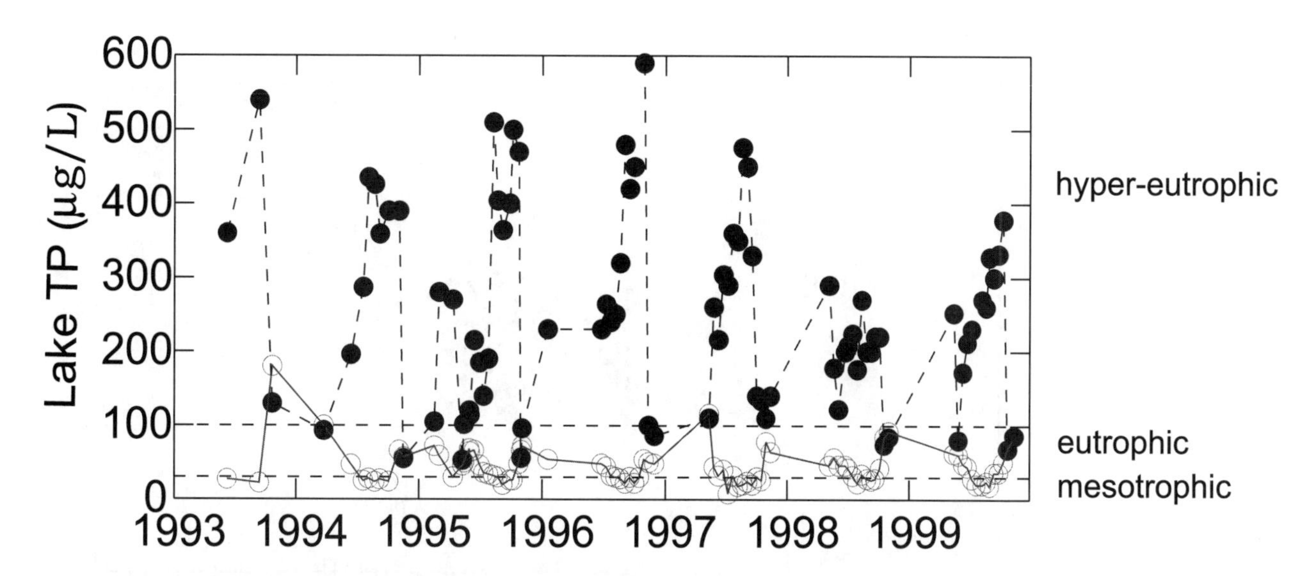

Figure 5-6.—Seasonality of phosphorus concentration in the surface (0–4 m or circa 0–12 feet) and bottom water (1 m or circa 3 feet above bottom) of a stratified mesotrophic lake (Lake Wilcox, Ontario [after Nürnberg, 1997]). Note that the surface phosphorus concentration (open circles) is only around 30 µg/L during summer, but can increase to over 100 µg/L at fall mixing events. Hypolimnetic phosphorus concentration (filled circles) is much higher during summer than surface phosphorus and ranges between 100 and 600 µg/L .

STEP 3: *Relationships Between Phosphorus and Other Water Quality Variables*

The value of determining lake phosphorus concentrations becomes obvious when comparing phosphorus levels to other water quality data. Phosphorus concentrations, especially epilimnetic summer averages, correlate highly with algal biomass indicators like summer averages of chlorophyll and Secchi disk transparency (Cooke et al. 1993). Use regression equations to predict algae abundance over a wide range when only phosphorus data are available (Fig. 5-7).

Other important water quality variables pertaining to oxygen levels in the bottom water (Fig. 5-8) correlate to phosphorus as well, especially when you consider the shape of the lake basin (Nürnberg, 1996). Several models use phosphorus or chlorophyll to predict variables related to fish (number of species, biomass, or yield) for certain geographic regions (Nürnberg, 1996). For example, the number of coldwater fish species (of the families Salmonidae, Coregonidae and Gadidae) is considerably higher in oligotrophic lakes as compared with eutrophic lakes (Fig. 5-9).

To simplify the assessment of water quality in specific lakes, Carlson (1977) developed the Trophic State Index (TSI, see Chapter 4). This system, used by many states to classify lakes, essentially relates phosphorus and chlorophyll concentrations to Secchi disk transparency in an index that is consistent with northern lake behavior (Table 5-4; Fig. 5-4). The actual equations to compute TSI (Table 5-4) were derived from a data set of 60 to 150 northern natural lakes (Carlson, 1977) calculated similar to those depicted in Figure 5-7. The TSI helps compare water quality variables; its scale is calibrated so that an increase of index units corresponds to a decrease of transparency.

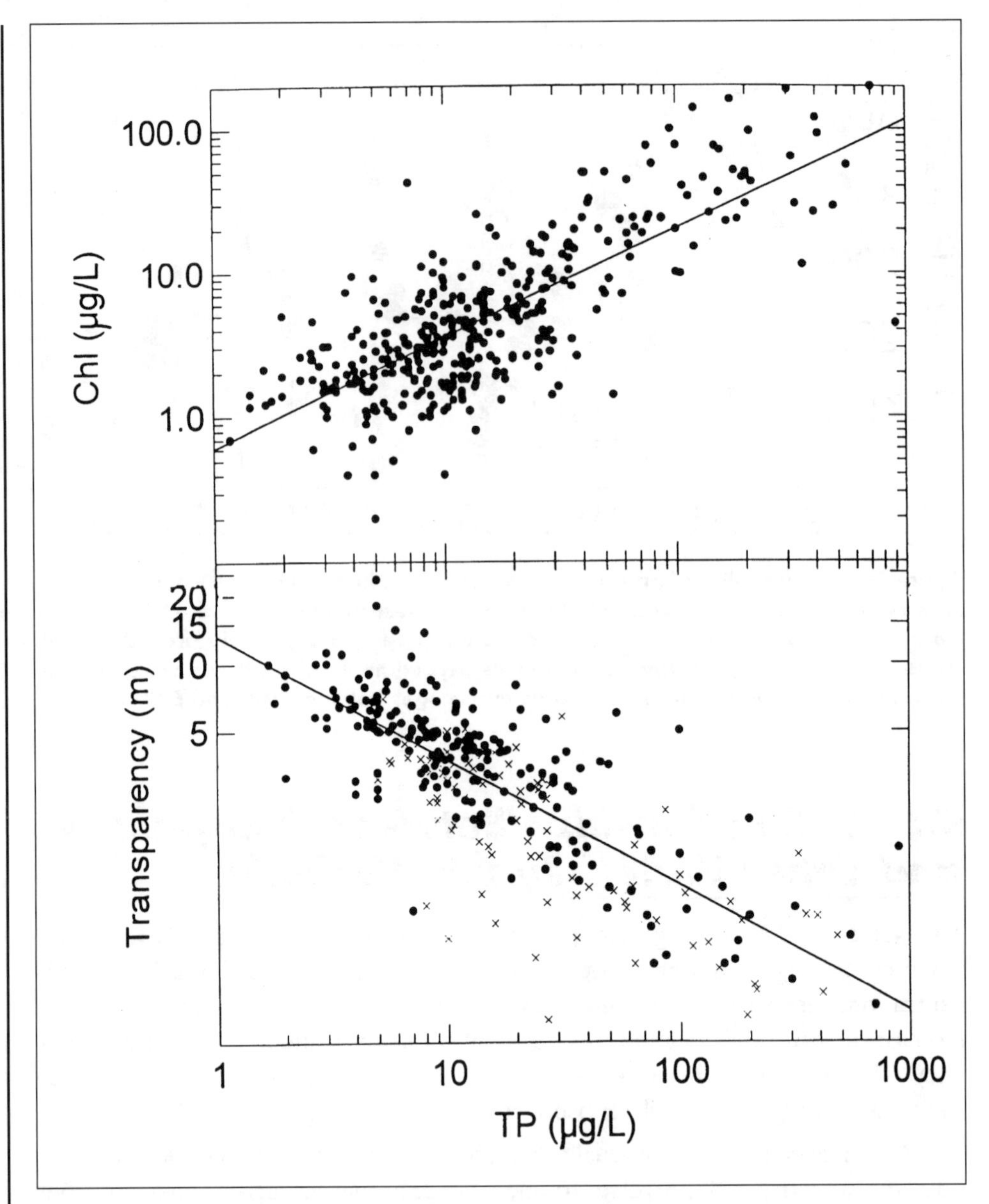

Figure 5-7.—Summer epilimnetic averages of algae biomass indicators versus total phosphorus summer averages in temperate freshwater global lakes. Regression lines are for equations in Nürnberg (1996): Chlorophyll, Secchi disk transparency. Some of the turbidity in colored lakes (x) is due to colored organic acids, not only algae. Therefore, it is useful to know the color value in brown water lakes.

Table 5-4.—Carlson's Trophic State Index.

Trophic State Index values can be computed from the various summer average water quality variables (Carlson, 1977; U.S. EPA, 1998c).

TSI of transparency = 60 – 14.41 ln(transparency)
TSI of chlorophyll = 30.6 + 9.81 ln(Chl)
TSI of phosphorus = 4.15 + 14.42 ln(P)
TSI of nitrogen = 4.45 + 14.43 ln(N)

Hypolimnetic anoxia can be considered in the TSI concept as well. The TSI of the anoxic factor (Table 5-1) can be approximately computed as:

TSI of anoxic factor = 10 + AF (Nürnberg, unpublished).

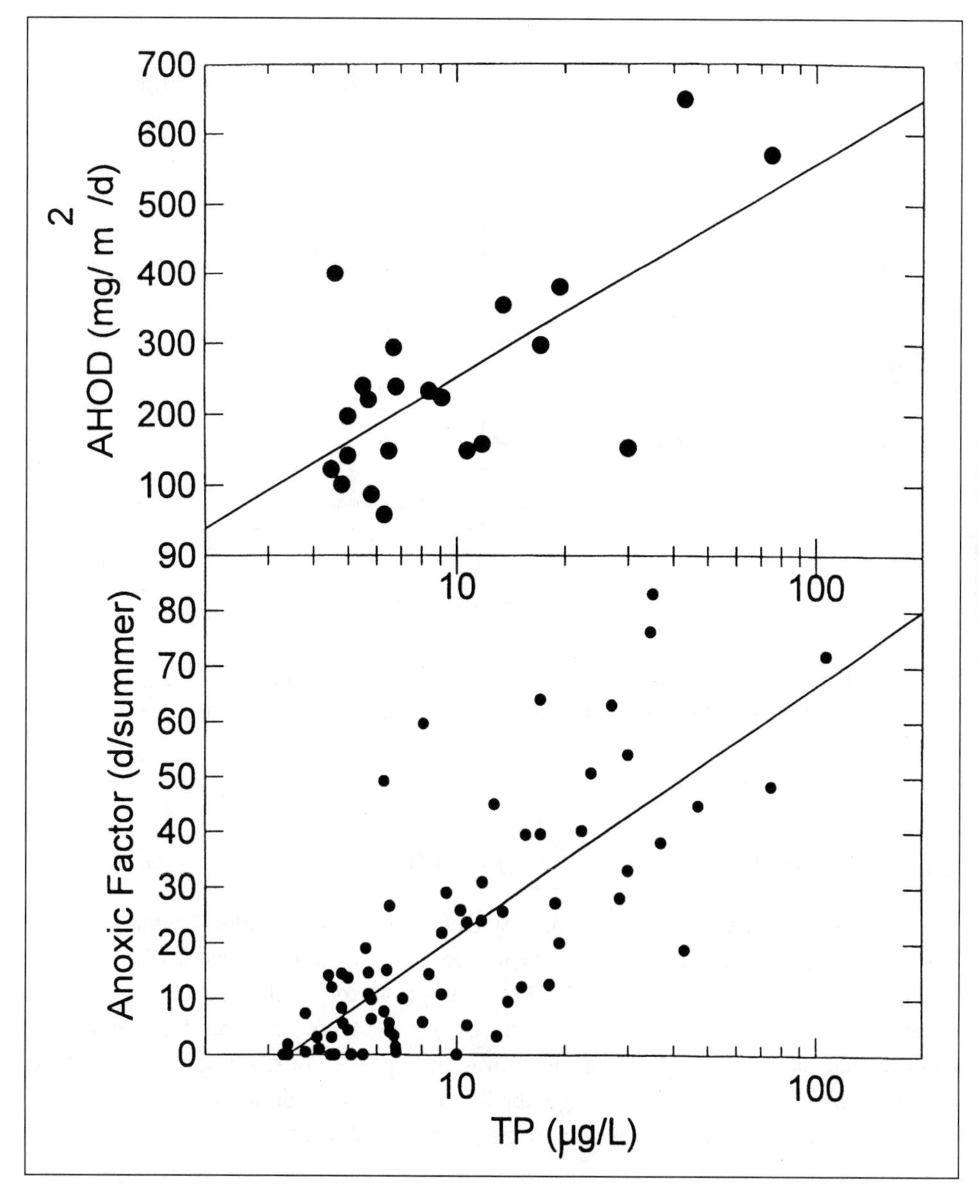

Figure 5-8.—Indicators of hypolimnetic oxygen versus total phosphorus concentration averages in temperate freshwater lakes. Regression lines are for equations in Nürnberg (1996): areal hypolimnetic oxygen depletion rates (AHOD); anoxic factor (Table 5-1).

AHOD: The areal hypolimnetic oxygen deficit describes the rate at which oxygen concentration declines in the hypolimnion (units of mg oxygen per square meter of hypolimnetic area per day). It does not indicate extent or duration of anoxia.

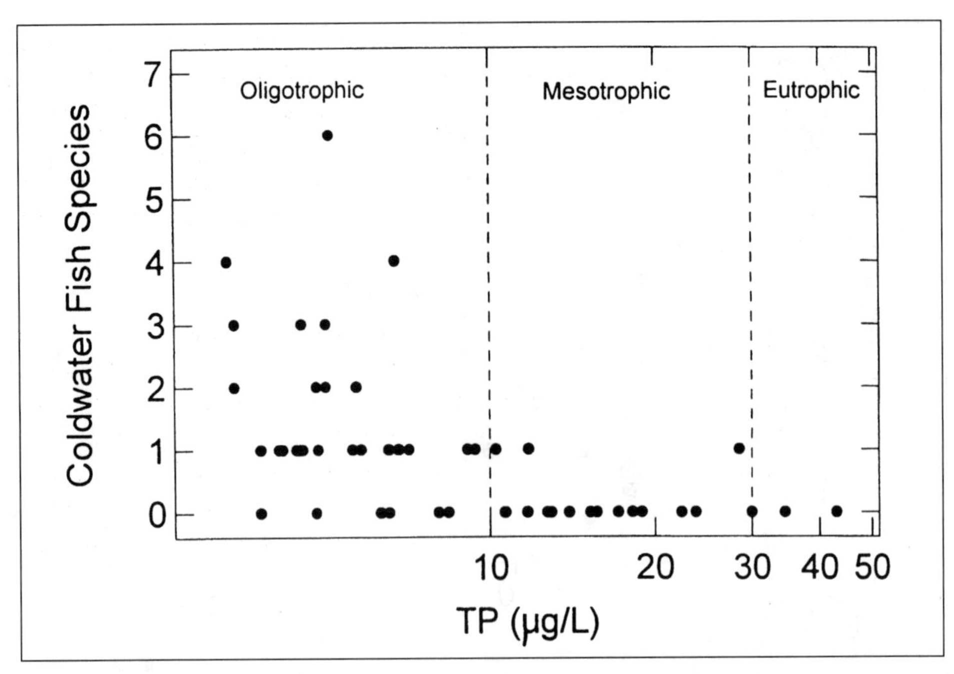

Figure 5-9.—Number of coldwater fish species versus total phosphorus concentration averages in Ontario lakes on the Precambrian Shield. Lines indicate trophic limits.

STEP 4: *Model Verification*

Water quality models must be tested with real field data under baseline conditions to ensure that they work as expected before they are used. Many of the simpler models are designed to predict average water quality conditions over a certain period of time (a growing season or year) and often over the whole lake, even though water quality varies over time and space. Thus, when collecting field data to calibrate a model, averaging usually covers several dimensions. Remember: actual observations can comprise a large range of values.

- ▼ ***Depth:*** The top, mixed layer (epilimnion) is the part of the water column that is generally averaged. Vertical variations within this layer are usually small. But for "whole-lake" phosphorus estimates, you must also account for accumulation in a potentially anoxic hypolimnion.

- ▼ ***Sampling Station:*** Although a small, round lake is so homogenous it should need only one station, generally, sampling stations should be located in different places in the lake. Water quality in most large lakes and reservoirs may differ significantly (from oligotrophic to hypereutrophic) from station to station. In such situations, a measurement for the "average water quality" may be meaningless; it would be more appropriate to divide the lake or reservoir into segments for modeling purposes where outflow from one segment serves as inflow to the next.

- ▼ ***Seasonal:*** Phosphorus and especially transparency and chlorophyll concentrations can vary significantly at a given station from one sampling date to the next during the growing season. It is not unusual, for example, for the maximum chlorophyll concentration to exceed two to three times the seasonal average.

In lakes with internal phosphorus sources (i.e., phosphorus release from anoxic bottom sediments), phosphorus levels vary as well, depending on the relative importance of the internal phosphorus source. In these lakes, surface water phosphorus concentrations are lowest in early spring and highest in the fall after the upwelling of bottom water (Fig. 5-6). Specialized models have been developed to account for internal load in these lakes (Table 5-4).

▼ ***Annual:*** Average water quality also varies year to year because of climate fluctuations, particularly stream flows and factors that control thermal stratification. For example, compared to a dry year, a wet summer can significantly increase runoff and external nutrient load and may trigger algae blooms. Monitoring programs extending over at least several years, depending on the flushing rate, are often recommended to characterize baseline conditions and provide an adequate basis for lake diagnosis and modeling.

Given that analytical error as well as natural variability will affect field measurements, it may be more realistic to consider measured water quality as a range of values rather than as a specific value. With care, you may arrive at an estimate of lake water quality with small "Confidence Intervals." A Confidence Interval will give the probability (usually 90 percent or 95 percent) that the "true" average is contained within that interval. Any slight improvement or deterioration in water quality within the Confidence Interval will be difficult to detect, but when it becomes comparable to or greater than the expected variation (as indicated by the Confidence Interval), the change will become "statistically significant" and detectable.

Similarly, model predictions of lake water quality may have a range or Confidence Interval associated with them because of the natural and analytical variability of data inputs to the model and from any empirical relationships used. Model predictions can be confirmed or rejected on the basis of whether the Confidence Interval of their prediction overlaps the Confidence Interval of the observed baseline parameter. If there is no overlap, then possible systematic errors in the model's input parameters or in model assumptions and structure must be evaluated. In some cases, it may become clear that model assumptions or structure are inappropriate; this conclusion in itself can be a useful diagnostic tool for assessing lake function.

It is possible that predicted or observed Confidence Intervals are too wide to provide a useful test of the model. Only an improved modeling program can resolve this. Another problem might be unknown systematic errors in the model leading to its rejection. These may be canceled out by predicting relative changes in lake conditions rather than absolute values when two scenarios are compared. Informed interpretation of model results based on an adequate lake and watershed monitoring program reduces the risk of errors that could lead to false conclusions and poor management decisions.

STEP 5. Forecasting and Tracking Changes in Water Quality

Step 5 is probably the most important reason for lake modeling because it is the only way to predict future conditions. In this step, models developed and verified in the previous steps are applied to future (or past) scenarios. You can use changes in phosphorus loading to predict changes in average phosphorus concentrations, chlorophyll, and transparency. These predictions can then help you set future water quality targets based on phosphorus.

Be sure to select an appropriate target for the problem at hand. For example, if the management goal in mesotrophic Lake Wilcox was to decrease annual average phosphorus concentration, in-lake restoration might decrease the internal load (more than 60 percent of total load; Fig. 5-5). However, if early summer algal biomass is targeted, reducing the internal load may not work, since, in this stratified lake, it won't affect the epilimnion until late summer.

You should formulate targets as concentrations rather than loads because it is the concentration that reflects a lake's trophic status and water quality. In particular, loading targets would not adequately protect water quality in situations (such as reservoirs) where water flow could be decreased. In these cases, lake concentrations could increase despite constant loading.

In a rare situation loading targets may be too restrictive. For example, 21 possible development scenarios were modeled for Lake Wilcox to find out which combination of additional developed areas would result in "an external phosphorus load no greater than existing conditions" (the legal management goal). External load values predicted for several scenarios are shown in Figure 5-10. Only scenarios 4, 5, and "Hyp" complied with the external load target; but phosphorus concentrations, algae biomass, and anoxia are predicted to decline for some of the less stringent scenarios as well (Fig. 5-10) because extreme measures in runoff treatments from developed areas will probably dilute the increased phosphorus loading.

Some typical restrictions include a certain percentage increase of phosphorus concentration above pre-development concentrations. The Ministry of the Environment in Ontario, Canada, proposes a factor of 1.5 increase above prehistoric annual average phosphorus concentrations as target values for its more than 100,000 lakes on the Canadian Shield (Hutchinson et al. 1991). The Swedish government endorses a maximum of twice the background levels of phosphorus (and nitrogen) as a national target (Swedish EPA, 1994). Compliance with these targets can be evaluated only by using phosphorus models in which current anthropogenic sources are first included and used to verify the model and then removed to arrive at a pre-development phosphorus concentration.

The TMDL (total maximum daily load) program is based on target values as well. States must identify and list water bodies where state water quality standards are not being met and establish TMDLs to restore them. A TMDL specifies the amount a pollutant needs to be reduced to meet water quality standards (which ideally should be based on concentration), allocates pollutant load reductions among pollutant sources in a watershed, and provides the basis for restoring a water body through point source and nonpoint source controls. To set TMDLs for eutrophication-related problems, use variations of Steps 1 to 5, including phosphorus modeling and predicting water quality variables (Table 5-5).

In another approach, the ecoregion concept helps assess the quality of individual lakes (Omernik, 1991). This concept realizes that the natural trophic status of lakes depends on its watershed's land surface form and use, natural vegetation, and soils. Typically, your watershed is assigned to a pre-defined ecoregion based on its location. Medians of trophic state variables based on lakes in EPA's Storet database are available for different ecoregions. Next, certain lake characteristics, like average phosphorus or summer chlorophyll concentration, are compiled for each region separately. Then the characteristic of each individual lake is compared with measures of the central tendency (median) for all lakes in that region. A lake should fall below the median, i.e., it should belong to the half of the better lakes.

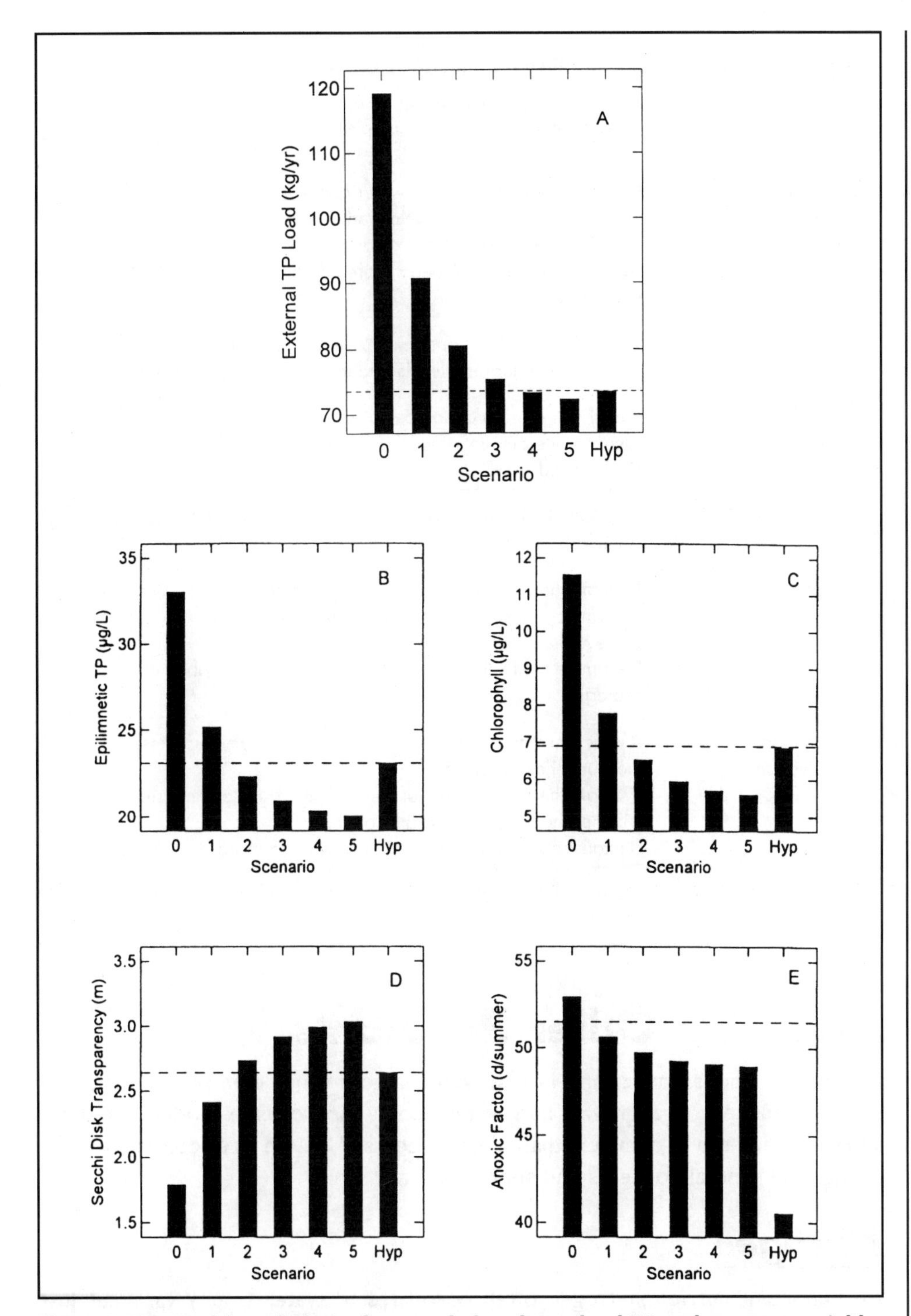

Figure 5-10.—Model predictions of external phosphorus load (A) and response variables (B, C, D, E) for several different hypothetical scenarios in Lake Wilcox, Ontario. Existing conditions are indicated by broken lines. Existing phosphorus load (A) was the legal target value and must not be exceeded in future development. Scenarios 0 to 5 include additional development in the watershed: "0", without any treatment, "1" includes several storm treatment ponds, "2" to "5" include storm treatment ponds and various levels of additional treatment. "Hyp" is the model prediction for existing watershed conditions with the in-lake treatment of hypolimnetic withdrawal (see Chapter 7).

Ideally, it should fall in the quarter of the best lakes or should be restored so that it does. To set limits based on the ecoregion concept, apply variations of Steps 1 to 4, including phosphorus modeling and the prediction of water quality variables.

Hypolimnetic withdrawal: Lake outflow is taken from the hypolimnion, not from the surface of a lake. The surface outlet is usually dammed.

Table 5-5.—Benefit of modeling for the TMDL process.

Several steps in TMDL development and implementation planning may require data gathering and the use of predictive water quality models. For example, the following information needs are often associated with the first six components (a–f) described below:

a. Target Identification	—Develop numeric target for water quality conditions (e.g., criterion) —Translate criterion to numeric loading capacity level (quantified pollution load from all sources, including background, necessary to meet criterion, e.g., through a predictive analysis of pollution in the waterbody)
b. Deviation from Target	—Quantify the amount and timeframe of deviation between current/future loading levels and the loading capacity level
c. Source Identification	—Identify all sources or source categories —Quantify the amount of load from sources, including natural background
d. Allocation of Pollution Loads	—Ensure that allocations will lead to attainment of water quality standards
e. Implementation Planning	—Estimate the effectiveness of controls/management measures —Determine that controls/management measures are sufficient to achieve the TMDL allocations —Determine the likelihood of actual implementation of control strategies
f. Monitoring/ evaluation	—Assess whether the implementation of controls/management measures has occurred —Evaluate the effectiveness of controls/management measures and whether they are meeting allocations —Demonstrate attainment of water quality standards

Chapter 5.2, Section 5, from Report of the Federal Advisory Committee on the TMDL Program, U.S. EPA, 1998.

Modeling Other Pollutants

In some situations, particularly in reservoirs, factors other than phosphorus may strongly influence algal growth and water quality. Appropriate models for these situations are more complex than those discussed in the previous section, although the general concepts and approaches are similar.

Nitrogen

An important nutrient, nitrogen often correlates to water quality variables just like phosphorus. Epilimnetic summer averages of total nitrogen and phosphorus positively correlate over a wide range of concentrations in large data sets of North American and worldwide lakes (Fig. 5-11, Nürnberg, 1996). Nonetheless, most lakes are phosphorus- rather than nitrogen-limited, or both, and attempts to reduce lake nitrogen levels in these lakes may have little effect on algal biomass.

In fact, a comparison of apparently nitrogen-limited lakes (as determined from their N:P ratios) with those that were phosphorus limited could find no difference in their phosphorus to chlorophyll relationships (Nürnberg, 1996). When nitrogen actually affects algae, it's usually because either the phosphorus level is high or nutrient inputs are very low. In the first case, it makes more sense for lake managers to target phosphorus rather than nitrogen and to establish phosphorus

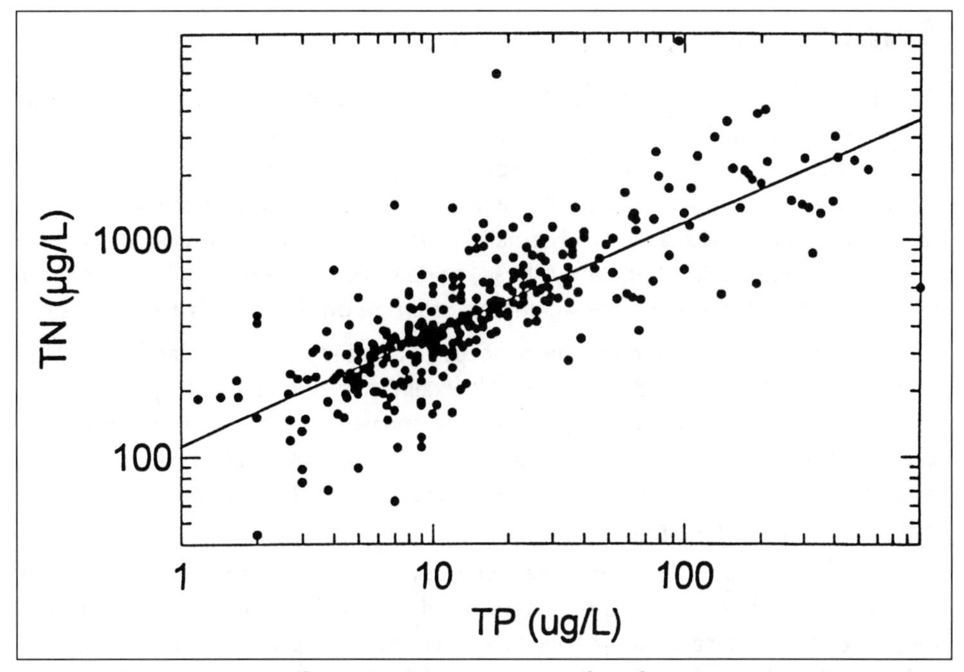

Figure 5-11.—Summer epilimnetic nitrogen versus phosphorus concentration averages in temperate freshwater lakes. Regression lines are for equations in Nürnberg (1996).

budgets instead of nitrogen budgets. In the second case, the lake probably does not have a eutrophication problem.

The ocean, however, is usually nitrogen-limited, and thus nitrogen pollution from coastal watersheds can cause eutrophication in estuaries and other coastal salt water. Northern European communities have developed strict nitrogen export limits to prevent the Baltic Sea from becoming more eutrophic (Swedish EPA, 1994), and similar controls exist in some American coastal states (U.S. EPA, 1998a). Lake nitrogen budgets can help predict and eventually control nitrogen exports from lakes in coastal watersheds to sensitive marine environments.

Nitrogen mass balance models are more difficult to construct than those for phosphorus (see Steps 1 and 2). Although nitrogen export rates are available for various land uses and regions, nitrogen retention can be measured in the field only with great difficulty, primarily because nitrogen is transformed so easily by blue-green algae and bacteria. Atmospheric nitrous oxides must also be taken into account. Some nitrogen models have been developed, however, based on empirical (observed) data (Bachmann, 1980; Windolf et al. 1996).

Suspended Sediments

Siltation has been identified as the third major problem after nutrients and metals; it severely affected 7 percent of all U.S. lakes surveyed in 1998 (U.S. EPA, 2000) and was the major problem in a quarter of all lakes rated with impaired water quality.

Suspended sediments can cause turbidity (thus limiting light), impair fish spawning and feeding habitat, and create taste and odor problems. A sediment budget can be established much like a nutrient budget:

External load = outflow load + sedimentation – resuspension + change in storage

As with phosphorus, net retention can be estimated as the difference between in- and outflowing mass over incoming mass. In addition, sediment loss equations are available that model the settling of suspended sediment.

Acidity

Many lakes throughout the world have been acidified by the deposition of acids from the atmosphere or acid drainage from mines. The majority of models addressing these problems emphasize soil conditions and soil-water interactions, because atmospheric acid deposition is neutralized primarily by the soil around lakes. Only in lakes with a relatively small watershed-to-lake-area ratio will in-lake processes dominate. Consequently, these models are not very useful in predicting the effects of acid lake liming or acid mine drainage on the chemistry of lakes.

Most models are based on the concept of alkalinity (or ANC, acid neutralizing capacity), rather than pH. Unlike pH, alkalinity is not affected by changes in weak acids such as carbon dioxide, making its measurement and prediction more reliable. The appropriate definition and measurement of alkalinity in a lake depends on the major weak acid buffering systems present (e.g., carbonic acid, weak organic acids, or aluminum hydroxides).

Detailed alkalinity budgets have been constructed for some acidified lakes and catchments, but they have only been used to calculate direct alkalinity retention or release. The retention of alkalinity has not been predicted in a more general way as has been done for phosphorus in lakes.

Summary and Conclusions

Mathematical models are useful both in diagnosing lake problems and in evaluating possible solutions.

Eutrophication modeling is a step-by-step process, including construction of hydrological and phosphorus budgets, calculation of average phosphorus concentration, prediction of water quality from lake phosphorus, model verification, and forecasting of alternative scenarios.

For in-depth coverage of this subject, consult Reckhow and Chapra (1983), Welch (1992), and Cooke et al. (1993). Case studies, including restoration efforts using different models and scenarios as well as historical data, have been documented by federal environmental protection agencies (e.g., U.S. EPA, 1998a; Swedish EPA, 1994).

References

Bachmann, R.W. 1980. Prediction of total nitrogen in lakes and reservoirs. *In* Restoration of Lakes and Inland Waters. EPA 440/5-81-010. U.S. Environmental Protection Agency, Washington, DC.

Carlson, R.E. 1977. A trophic state index for lakes. Limnol. Oceanogr. 22: 361-9.

Cooke, G.D, E.B. Welch, S.A. Peterson, and P.R. Newroth. 1993. Restoration and Management of Lakes and Reservoirs. Lewis Publishers, Ann Arbor, MI.

Chapra, S.C. and R.P. Canale. 1991. Long-term phenomenological model of phosphorus and oxygen for stratified lakes. Water Res. 25: 707-15.

Dillon, P.J. and F. H. Rigler. 1974. A test of a simple nutrient budget model predicting the phosphorus concentration in lake water. J. Fish. Res. Board Can. 31: 1771-8.

Duarte, C.M. and J. Kalff. 1990. Patterns in the submerged macrophyte biomass of lakes and the importance of the scale of analysis in the interpretation. Can. J. Fish. Aquat. Sci. 47: 357-63.

Hutchinson, N. J., B. P. Neary, and P. J. Dillon. 1991. Validation and use of Ontario's trophic status model for establishing lake development guidelines. Lake Reserv. Manage. 7: 13-23.

Nürnberg, G. K. 1984. The prediction of internal phosphorus load in lakes with anoxic hypolimnia. Limnol. Oceanogr. 29: 111-24.

———. 1995. Quantifying anoxia in lakes. Limnol. Oceanogr. 40: 1100-11.

______. 1996. Trophic state of clear and colored, soft- and hardwater lakes with special consideration of nutrients, anoxia, phytoplankton, and fish. Lake Reserv. Manage. 12: 432-47.

______. 1997. Coping with water quality problems due to hypolimnetic anoxia in Central Ontario Lakes. Water Qual. Res. J. Can. 32: 391-405.

______. 1998. Prediction of annual and seasonal phosphorus concentrations in stratified and polymictic lakes. Limnol. Oceanogr. 43: 1544-52.

Omernik, J. M., C. M. Rohm, R.A. Lillie, and N. Mesner. 1991. Usefulness of natural regions for lake management: Analysis of variation among lakes in Northwestern Wisconsin, USA. Environ. Manage. 15: 281-93.

Reckhow, K. H. and S. C. Chapra. 1983. Engineering Approaches for Lake Management. Vol. 1: Data analysis and empirical modeling. Butterworth, New York.

Swedish Environmental Protection Agency. 1994. Eutrophication of Soil, Fresh Water and the Sea. Solna.

U.S. Environmental Protection Agency. 1997. Compendium of Tools for Watershed Assessment and TMDL Development. EPA-841-B-97-006. Washington, DC.

———. 1998b. Report of the Federal Advisory Committee on the Total Maximum Daily Load (TMDL) Program, July 1998. EPA-100-R-98-006. Washington, DC.

———. 1998c. Lake and Reservoir Bioassessment and Biocriteria. EPA-841-98-007. Washington, DC.

———. 1999. Regional Guidance on Submittal Requirements for Lake and Reservoir Nutrient TMDLs. Office of Ecosystem Protection, New England Region. Boston, MA.

———. 2000a. National Water Quality Inventory. 1998 Report to Congress. EPA-841-R-00-001. Off. Water, Washington, DC.

——— 2000b. Nutrient Criteria Technical Guidance Manual: Lakes and Reservoirs. 1st ed. EPA-922-B00-001. Off. Science Tech., Washington, DC.

Walker, W.W. Jr. 1987. Empirical models for predicting eutrophication in impoundments; Report 4, Phase III: Applications manual. Tech. Rep. E-81-9. U.S. Army Corps Eng. Waterways Experiment Station, Vicksburg, MS.

Welch, E. B. 1992. Ecological Effects of Waste Water — Applied Limnology and Pollutant Effects. 2nd ed. Chapman & Hall, New York.

CHAPTER 6

Watershed Management

Watershed Management: Principles, Processes, and Practices

Why Watersheds?

A watershed is the land from which rain and surface water drain toward a central collector such as a stream, river, or lake (Chapter 2, *Ecological Concepts*). What happens as that water runs off the land and into the stream or lake in large part determines the quality of the lake water. Many lake problems — muddy waters, aquatic weeds, green scum, poor fishing, and more — reflect the land use and land cover in the watershed.

Lake management, then, cannot ignore watershed management. In fact, lake restoration and management should begin in your own backyard, in your community — in the watershed. To restore and manage a lake is to work with both the lake and its watershed.

This chapter defines the lake and its watershed as the management unit and introduces the concept of watershed planning (see Chapter 8), including a framework for identifying sources of pollutants and watershed management practices to reduce these sources and their transport into lakes.

A number of watershed management practices have developed over the years to protect and sustain both land uses and the bodies of water that receive runoff from the watershed. Known as best management practices, they reduce runoff, minimize erosion and sedimentation, reduce nutrient and contaminant loads, and provide better stream habitat for fish and other aquatic organisms. The last half of this chapter deals specifically with various types of best management practices.

The importance of the lake and watershed as the management unit cannot be overemphasized. This manual often uses the term *lake system* — always keep in mind that the watershed-lake is the true management unit. While this chapter emphasizes watershed management practices that are applicable to large watersheds with multiple owners and/or organizations controlling the land, these same principles and practices can also be used by lake homeowners, lake associations, and lake communities in smaller watersheds.

Best Management Practices (BMPs): Methods, measures, or practices selected by an agency to meet its nonpoint source control needs. BMPs include but are not limited to structural and nonstructural controls and operation and maintenance procedures. BMPs can be applied before, during, and after pollution-producing activities to reduce or eliminate the introduction of pollutants into receiving waters.

—*Federal Register* 40 CFR 130.2

The Lake/Watershed Relationship—The Management Unit

Pollution, as defined by the Clean Water Act and current regulations, is human-made or human-induced alteration of the chemical, physical, biological, and radiological integrity of a water body. A **pollutant**, as defined by the Clean Water Act and current regulations, is dredged spoil, solid waste, incinerator residue, sewage, garbage, sewage sludge, munitions, chemical wastes, biological materials, radioactive materials, heat, wrecked or discarded equipment, rock, sand, cellar dirt, and industrial, municipal, and agricultural waste discharged into water. For example, pollution would be the loss or destruction of a streambank or lakeshore habitat. Sediment or phosphorus discharged into a water body would be pollutants.

The lake and its watershed are inseparably linked — the lake or reservoir does not exist without its watershed (see Chapter 2). The management unit, then, is the lake and its watershed. It is not cost effective to manage the lake if the problem arises in the watershed or to manage watershed activities that have no effect on lake quality. You must manage the lake and its watershed.

A problem in a lake or reservoir is often the symptom of poor watershed management. As Chapter 2 points out, the watershed contributes both the water required to maintain a lake or stream and most of the pollutants that enter the lake. Obviously, addressing the symptoms of the problem without correcting the source and cause of the problem is not only shortsighted — it doesn't work for lakes!

Understanding the lake/watershed relationship requires some knowledge about the myriad of activities and land uses in the watershed. Such pursuits as farming, gardening and landscaping, logging, construction and development, and their resulting land cover — pastures, fields, forests, factories, subdivisions, and parking lots — can significantly affect water quantity and quality (see Fig. 6-1 for examples of watershed activities and uses that link the lake with its surroundings). While many of these activities occur in every watershed, large or small, the relative importance of each can vary from watershed to watershed.

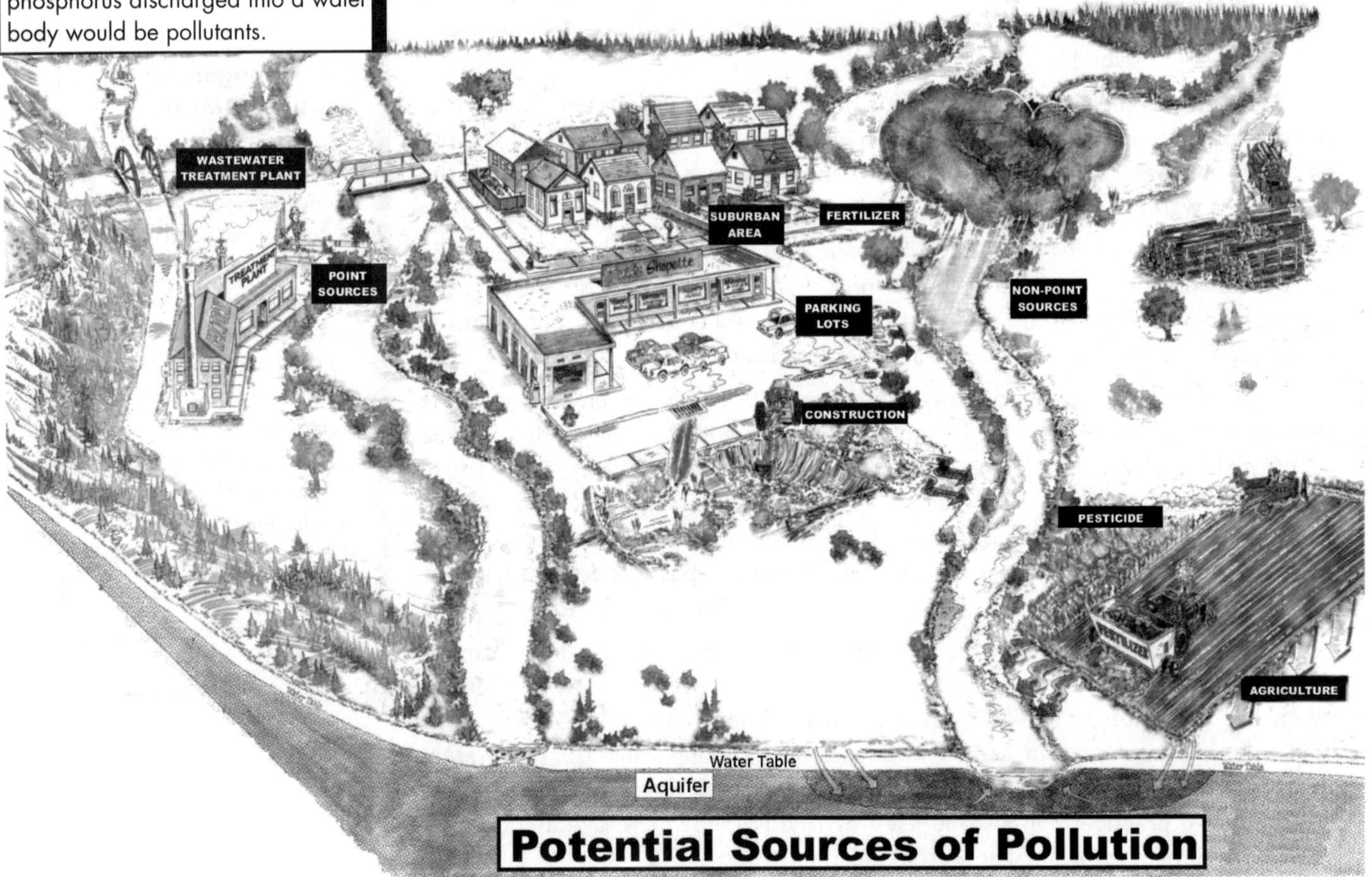

Figure 6-1.—Watershed activities and land uses that contribute to both point and nonpoint source pollution of lakes and reservoirs.

Another part of this management unit — groundwater — contributes flow and dissolved constituents such as nutrients, contaminants, and minerals. Groundwater can comprise the major part of the flow during dry periods (Fig. 6-1).

A primary objective of watershed management is to maintain the integrity of the major hydrological pathways (water flow) that people can affect. So you want to focus on how people have changed the landscape (e.g., urbanization) in a way that introduces materials and contaminants into the hydrological pathways and/or reduces the function and assimilative capacity of these pathways (e.g., loss of wetlands, streambank cover, trees, and shrubs). Altering these hydrological pathways harms the amount, timing, and quality of water that enters streams and lakes. Addressing these questions before beginning a watershed management process will lead to more realistic lake restoration or management goals.

Where Pollutants Come From

Pollutants, such as sediments, organic matter, and nutrients like nitrogen and phosphorus, enter a lake either from point or nonpoint sources in the watershed (Fig. 6-1).

- ***Point sources*** come from a distinct source such as a wastewater (sewage) treatment plant, industrial facility, or similar source that discharges through a pipe or similar outlet (Fig. 6-1). You can identify them by tracing the discharge back to its specific source.

  Point sources are usually controlled by state or federal permits such as the National Pollutant Discharge Elimination System (NPDES). The NPDES permit program has significantly reduced point source discharges of pollutants since 1972 (the year the Clean Water Act was passed) — and eliminated point sources as the major source of water pollutants in many watersheds.

  By the mid-1990s, stormwater had also been designated as a point source (the portion discharged to water bodies through storm drains) and is, in most instances, subject to federal or state permit requirements.

- ***Nonpoint sources,*** in contrast, do not originate from a pipe or single source. Nonpoint source pollution generally results from precipitation, land runoff, infiltration, drainage, seepage, hydrologic modification, or atmospheric deposition. As runoff from rainfall or snowmelt moves, it picks up and transports soil, nutrients, organic matter, toxins (herbicides, insecticides, metals), and other pollutants and carries them to lakes and streams (and sometimes, groundwater). From a regulatory standpoint, nonpoint sources are sources that are not defined by statute as point sources.

  Water running off a lawn, driveway, or road during a rain is a common sight — that's nonpoint source runoff (Fig. 6-1). It happens everyplace in the watershed, but some land uses such as agriculture, construction, and city streets contribute more nonpoint source pollutants than other land uses such as forests and land covered by vegetation.

  It is not always easy to distinguish a point source from a nonpoint source. In this chapter, point sources will be defined as factories, other industrial concerns, municipal wastewater treatment plants, and similar facilities that discharge wastewater through a pipe. In addition, runoff from construction sites greater than one acre is now regulated as a point source.

Point source: any discernible, confined, and discrete conveyance, including but not limited to any pipe, ditch, channel, tunnel, conduit, well, discrete fissure, container, rolling stock, concentrated animal feeding operation, or vessel or other floating craft from which pollutants are or may be discharged. This term does not include agricultural stormwater discharges and return flows from irrigated agriculture.

—as defined in Section 502(14) of the Water Quality Act of 1987

Nonpoint sources: sources not defined by statute as point sources; include return flow from irrigated agriculture, other agricultural and silvicultural runoff and infiltration, urban runoff from small or unsewered urban areas, flow from abandoned mines, and hydrologic modification.

—regulatory definition

Nonpoint sources will include all other types of pollutant loadings to the lake or stream, including lawns, driveways, subdivision roads, small construction sites, agricultural areas, forests, abandoned mine sites, and airborne or atmospheric contributions (Fig. 6-1).

The Growing Trend for Watershed Planning and Management

More and more, communities and agencies are emphasizing watershed planning and management, and finding assistance in various aspects from many federal and state agencies. EPA's website (Watershed Information Network at www.epa.gov/win/) offers information and links for most government programs. WIN users can surf their own watershed for information, and also access many other websites that focus on other watershed management programs throughout the country.

Watershed management starts with a plan — a road map defining where you are now, where you want to go, and how you are going to get there.

Watershed Management Plans

Without a plan, watershed management activities will be disorganized and ineffective (see Chapter 8 for more detail on formulating effective lake management plans). The number of local watershed planning organizations is increasing across the country as communities realize the value of a watershed action plan. Watershed planning helps people understand the relationship between the materials that enter the lake from the watershed and the water quality values that need to be protected (Fig. 6-2).

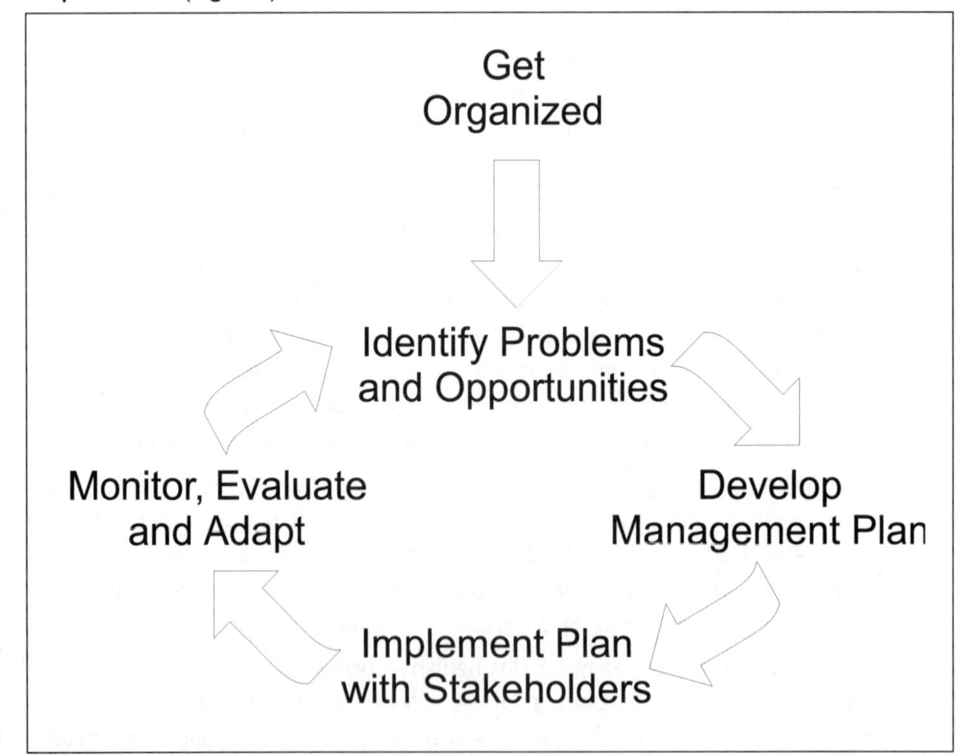

Figure 6-2.—Generalized watershed planning process.

The watershed planning process is the best approach for relating science, policy, and public participation to water resources management. As you begin building public support for your watershed planning process, include everyone who may have a significant impact on conditions in the lake — and those who might be affected by its restoration.

Keep in mind four important points about the process:

- The watershed plan does not need to be completed before activities can begin; rather, it guides the watershed group by mapping a strategy for improving or protecting the watershed.
- The planning process does not proceed in just one direction — you may have several activities going on simultaneously. What you learn in one step may cause the watershed group to revisit a decision made in a previous step.
- The watershed planning process and written plan are supporting, not prescriptive, tools. Be flexible.
- The watershed management plan is the beginning, not the end, of the management cycle (Fig. 6-2).

Some of the activities associated with each step in this management cycle follow.

Get Organized

- Recruit stakeholders and establish the partnership.
- Define the watershed.
- Create a mission statement.
- Establish points of contact, a decision process, and organizational structure.
- Facilitate information-sharing among participants.
- Document the process.

Identify Problems and Opportunities

- Collect information on the human and ecological features affecting water quality.
- Identify the predisturbance or reference conditions for the lake.
- Define objectives for water quality and other lake uses.
- Define the problems (to the extent that available data allow).
- Develop problem/opportunity statements for the watershed.

Develop Management Plan

- Define the future condition desired for the lake/reservoir.
- Define restoration goals and objectives.
- Identify restoration constraints and issues.
- Set priorities.
- Evaluate potential solutions for identified problems and objectives.

- Design restoration project selected to meet objectives.
- Identify measures of success for restoration project.
- Create an action plan, with schedules, task assignments, and a funding strategy.

Implement the Plan with Stakeholders

- Secure funding.
- Implement plan.

Monitor, Evaluate, and Adapt

- Monitor the restoration and evaluate progress.
- Revisit the management plan and make adjustments where needed.

Communication is essential because a watershed approach requires the informed participation of its stakeholders. To use your resources wisely you need to target objectives the stakeholders support. Stakeholders must participate at decision points; this ensures that final decisions will have sufficient support to succeed. Written watershed plans reflect the activities and decisions of the watershed planning group, so the planning process and its associated document should be designed to meet their goals.

Table 6-1 is a model outline of a watershed action plan included in *A Guide to Developing Local Watershed Action Plans in Ohio* (Ohio EPA, 1997). This format is a good starting point for designing a watershed plan, but you'll probably revise it to focus on your own lake's problems (the original has already been changed to add issues such as habitat and invasive species).

TMDLs — A Watershed Management Tool

Although not specifically designed for this purpose, the Total Maximum Daily Load (TMDL) Program provides an excellent framework for watershed management that can help you develop and implement a Watershed Management Plan (see www.epa.gov/owow/tmdl). Established by the original Clean Water Act in 1972, a TMDL calculates the maximum amount of a pollutant that a water body can receive and still meet water quality standards — a legal way of protecting that water body's desired uses. While the TMDL Program has become controversial and litigious, its overall goals are ultimately what we seek in watershed management:

- Identify the desired uses for the lake.
- Determine how much total loading of a pollutant(s) the lake can receive and still provide the desired uses.
- Assess how much of the pollutant load is coming from
  - → natural background,
  - → point sources, and
  - → nonpoint sources.

Table 6-1.—Ohio EPA template for a local watershed action plan (Ohio EPA, 1997).

1. Define the Watershed

- 1.1 Name, size, administrative boundaries of watershed
- 1.2 Geographic locators; USGS and state 305(b) identification numbers
- 1.3 Background/historical information on previous watershed protection and management activities, including previous planning documents
- 1.4 Purpose of the plan, a statement on the need for watershed action planning and why the plan was prepared
- 1.5 Scope and limitations of the plan
- 1.6 Who was involved in preparing the plan
- 1.7 Outline of the plan's content

2. Describe the Watershed

- 2.1 Natural features/characteristics of water source
  - 2.1.1 Special values: cultural, geologic, species
  - 2.1.2 Hydrology
  - 2.1.3 Land uses
    - point sources
    - nonpoint sources
- 2.2 Water quality
  - 2.2.1 Use designations/attainment
  - 2.2.2 Causes of non-attainment

3. Identify Problems

- 3.1 Identify sources of contaminants and quantify loads
- 3.2 Evaluate habitat conditions
- 3.3 Assess status of species of interest

4. Document Planned Activities

- 4.1 List goals
- 4.2 Describe specific management objectives (incorporate solutions) and actions
- 4.3 Link actions to individuals, committees, or organizations
- 4.4 Match actions with indicators or measures
- 4.5 Outline activities timeline
- 4.6 Describe adaptive management methods

- If the incoming load is greater than the maximum load the lake can handle, then reduce the load and allocate these reductions among the sources.
- If the incoming load is less than the maximum load the lake can handle, protect it to ensure the desired uses will continue.

Using the desired uses of the lake and the applicable water quality standards identified in the Watershed Management Plan as a base, you can review existing data to find out if they are being attained. In addition, you can determine the pollutant loading for constituents of concern. An important part of the TMDL is to learn the relative contributions of pollutants from natural sources, point sources, and nonpoint sources.

Lakes & TMDLs

Of a number of Clean Lakes Restoration projects Washington state has submitted to EPA as TMDLs, some have been approved and others have been labeled incomplete — either because:

- Eutrophication problems were not thoroughly documented; or
- The TMDL goal was established without numeric water quality criteria.

Two examples:

Lake Fenwick (approved) — EPA approved this Phase 1 diagnostic/feasibility study with a Phase 2 restoration proposal based on clear identification of the phosphorus load needed to achieve a TMDL goal of aesthetic enjoyment acceptable to the lake user community. Probable funding of the Phase 2 project and ordinances adopted by the City of Kent for stormwater runoff also provided reasonable assurance that the TMDL goals will be met eventually.

Lake Erie (incomplete) — This lake restoration had completed both Phase 1 and Phase 2, and was in Phase 3 (evaluation stage) when EPA determined it to be an incomplete TMDL — primarily because the Phase 1 study did not thoroughly document the eutrophication problems and associate them with a TMDL goal. To qualify as a TMDL, a quantitative analysis must demonstrate that the goals established for the TMDL will meet the narrative standard for support of a designated use (e.g., phosphorus levels needed for aesthetic enjoyment).

GIS: A Useful Watershed Evaluation Tool

Geographic Information Systems (GIS) are useful watershed evaluation tools because they can be used to display what the land uses are in the watershed, and to estimate loads from these land uses. GIS can show the location of different land uses around the lakes and receiving streams (Fig 6-3). The location of land use affects the loading to the system. Generally, forest, riparian zones, and grassed areas along the streambanks and lake shoreline result in lower loadings and better water quality than other land uses near the water body. Formulas or equations have been developed describing sediment, nutrient, and organic loading from different types of land uses. The GIS can use these formulas with the extent of the land use to estimate the loadings to the lake. The GIS can also be used to evaluate the reduction in loading that might occur if best management practices were implemented in the watershed.

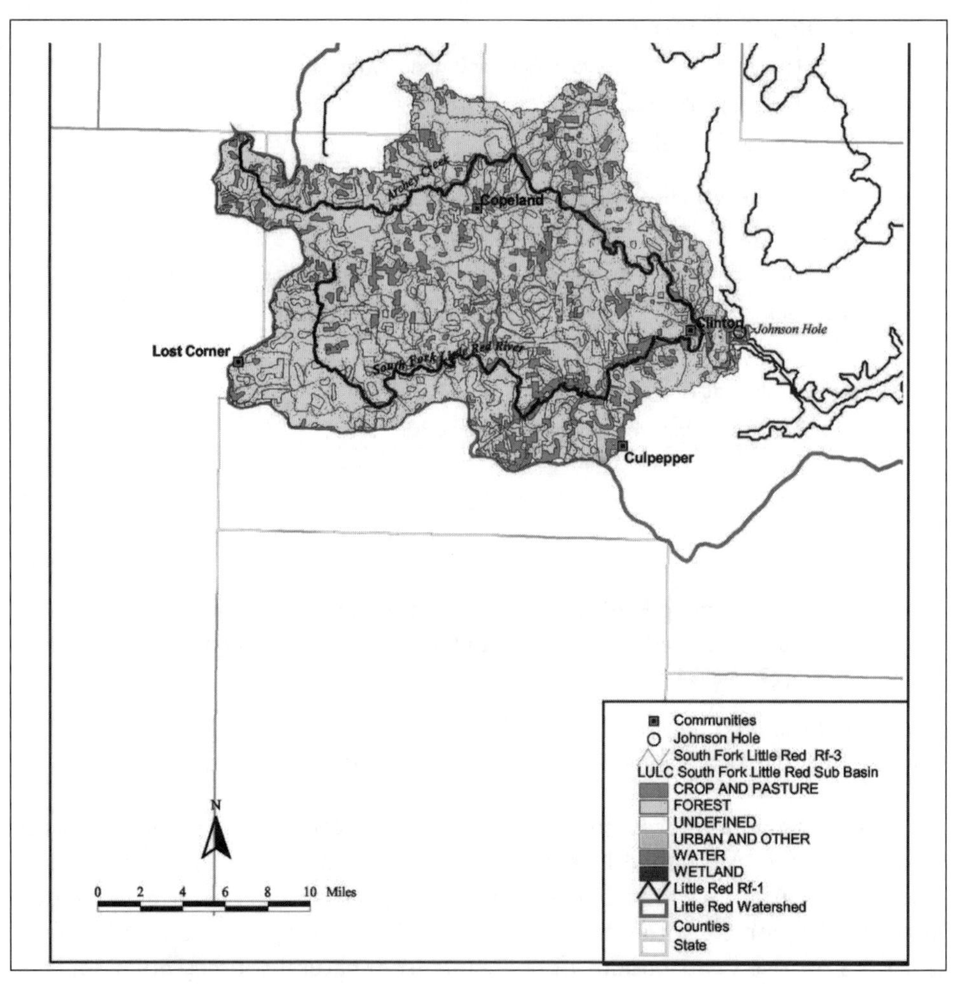

Figure 6-3.—Sample of a GIS.

Background Sources

For most constituents, including sediment, nitrogen, phosphorus, organic matter, and any naturally occurring metals (iron, manganese, zinc, copper, etc.), naturally occurring sources in the watershed contribute them to the receiving streams or lake (Fig. 6-1). For some lakes, the natural background loads may not permit some desired lake uses to be achieved. For example, organic loading from relatively undisturbed, forested watersheds result in zero dissolved oxygen in the bottoms of many southern lakes and reservoirs — a natural condition in this part of the country. This lack of oxygen in the bottom waters, however, means that these southern reservoirs won't support a coldwater fishery, which might be a desired use for the lake.

Some of the approaches for estimating background conditions include:

- Use lakes or streams with relatively undisturbed watersheds as a reference for what could reasonably be attained in the watershed.
- Use some of the models described in Chapter 5 to estimate constituent loads assuming no point or nonpoint sources.
- Use first principles (i.e., the fundamental relationships from which all others are derived) to estimate the erosion and transport of constituents based on the soils and geology in the watershed.

Point Sources

Because point sources are controlled under the National Pollutant Discharge Elimination System, information on the volume of discharge and constituent concentrations in the discharge can be obtained from EPA and the states. Point source loadings can be estimated by summing the point source discharges for both municipalities and industries in the watershed. Guidance manuals on how to estimate point source loads are listed on the EPA TMDL web site (www.epa.gov/owow/tmdl).

Nonpoint Sources

Nonpoint sources are both natural and human-influenced (Fig. 6-1). Estimating natural background loads was discussed previously. Human-induced loads can be estimated based on sampling or monitoring data or by using some of the modeling and estimation approaches discussed in Chapter 5. Once these estimates are obtained, the natural background load can be subtracted to determine how much of the loading comes from just human activities in the watershed, such as grazing cattle in pastures, adding fertilizers to crops, mining, timbering, building roads and highways, and similar human disturbances (Fig. 6-1). Again, guidance and manuals on how to estimate nonpoint source loads can be found on the EPA TMDL website, which also has links to other agency sites that focus on specific land-use types such as agriculture or silviculture.

Total Maximum Daily Loads

A total load for the pollutant(s) of concern can be estimated by adding:

Total Load = Natural Background Load + Point Source Load + Nonpoint Source Load

Compare this Total Load to the Total Maximum Daily Load that the lake can handle and still allow desired lake uses. If the Total Load exceeds the TMDL, then point and nonpoint source loads need to be allocated to the TMDL. Caution: you must allow a margin of safety in comparing the TMDL with the Total Load, just in case there were some unknowns in the estimates. This margin of safety should ensure that reducing pollutant loads will improve water quality.

If the Total Load is less than the TMDL, then watershed management practices and actions need to focus on protecting the watershed and lake from exceeding the TMDL.

In either case, the TMDL can guide watershed management plans and practices to achieve the lake uses desired by the community. See EPA's TMDL web site for examples of several approved TMDLs for pollutants such as bacteria, turbidity, sediment, nitrogen, and phosphorus.

Other useful tools for assessing watershed conditions include Ecosystem Analysis and Smart Growth. Ecosystem Analysis is used particularly in western watersheds, whereas Smart Growth is an approach to minimize the effects of urban sprawl on aquatic ecosystems.

Ecosystem Analysis is used particularly in western watersheds, whereas Smart Growth is an approach to minimize the effects of urban sprawl on aquatic ecosytems.

Ecosystem Analysis at the Watershed Scale

Ecosystem analysis is a procedure frequently used by many western U.S. resource management agencies such as Game and Fish agencies or the Bureau of Land Management (B.M.) to characterize the human, aquatic, riparian, and terrestrial features, conditions, processes, and interactions (collectively referred to as "ecosystem elements") within a watershed (Intergovernmental Advisory Committee, 1995).

Ecosystem: a system of interrelated organisms and their physical-chemical environment. In this manual, the ecosystem is usually defined to include the lake and its watershed.

Watershed analyses address:

- Erosion;
- Hydrology or water flow;
- Vegetation or land cover in the watershed;
- Stream channel habitat;
- Water quality;
- Biological organisms or species and habitats; and
- Human uses.

There are several forms of watershed analysis, but each is structured around a series of key questions which, if answered, provide a model of landscape and ecosystem function, disturbance history, and current and potential future conditions.

Watershed analysis supports decision-making priorities, because it generates the information required to make informed choices about how land uses will work in the watershed — within its ecoregion (Montgomery et al. 1995).

Watershed analysis is based on a six-step process that:

- Characterizes the watershed by identifying the dominant physical, biological, and human processes or features that affect ecosystem functions or conditions — and identifies primary ecosystem elements that require more detailed analysis;
- Identifies issues and key questions to focus the analysis on the key elements of the ecosystem most relevant to the management questions and objectives, human values, or resource conditions;
- Describes the current conditions of ecosystem elements;

- Describes reference conditions for ecosystem elements or how ecological conditions have changed over time;
- Synthesizes information by comparing existing and reference conditions and determining the capability of the system to achieve key management plan objectives; and
- Describes management plan recommendations that are responsive to watershed processes identified in the analysis.

This watershed analysis process has been the starting point for many watershed management forums for lakes in the western United States. Several land management agencies there have joined with local communities and Resource Conservation Districts to conduct watershed analyses as part of emerging watershed partnerships. The Northwest Forest Plan has encouraged watershed analysis, based on the experience of the Forest Service and BLM with the northern spotted owl (USDA, 1994).

For additional guidance on watershed analysis, see U.S. EPA (2000); www.epa.gov/owow/watershed/wacademy/wam/.

Smart Growth – An Approach to Urban Watershed Planning and Management

Smart Growth is a phrase to capture a new way of thinking about land development, urban renewal, and economic growth. Smart Growth is a holistic planning approach that factors in many considerations before making changes or developing solutions to problems. It is related to watershed management because it attempts to assess the cumulative effects of land uses and changes to the environment. And, like watershed management, it depends on information exchange and consensus among people, communities, stakeholders, and local, state, and federal agencies to make decisions that benefit the entire system in which a community lives.

Both Smart Growth and watershed-based approaches can be used to foster consensus, to develop objectives, and ultimately, to make positive impacts on the quality of the land on which we live — and the water we use.

A 1-acre parking lot generates 16 times more polluted runoff than a 1-acre meadow.

Development replaces natural vegetated land cover with roads, parking lots, driveways, sidewalks, and rooftops. These surfaces are impermeable to rainwater and tend to increase surface water runoff that then carries pollutants directly to streams, rivers, and lakes. A 1-acre parking lot generates 16 times more polluted runoff than a 1-acre meadow.

Examples of Smart Growth that parallel watershed planning practices include:

- Encouraging "best development practices" such as designing parking lots with natural buffers to capture runoff, and developments with common open spaces, parks, trails, and less impervious surfaces;
- Increasing incentives for revitalizing city centers and brownfields (areas degraded by past industrial use) for reuse rather than expanding into farmland and open spaces, thereby reducing suburban sprawl;
- Building communities that depend less on the automobile for getting around and more on public transit (which means increasing support for mass transit);

- Changing zoning to encourage conservation of natural areas; and
- Increasing use of building and landscaping practices that create riparian buffer zones.

Further information can be found at www.smartgrowth.org.

From Planning to Practice

With the Watershed Management Plan developed, information on point and nonpoint source loading, and an estimate of the TMDL, what management practices can you use either to reduce loading from point and nonpoint sources or to protect the lake from increases in these loadings?

The next section discusses watershed management practices that can be used to control point source loads; and the following section, best management practices (BMPs) to control nonpoint source loads. Neither section tries to describe all control practices, but rather to illustrate the types of control techniques and practices that are available.

Please see Appendix 6-A for an extensive list of links and references to information that can be used in watershed management. And contact your local and state agencies for more information.

Watershed Management Practices: Point Sources

Wastewater usually comes from a point source; it's discharged through pipes by industrial and municipal treatment plants. And even though it's treated to remove most pollutants, it still may contain organic matter, bacteria, nutrients, toxic and other substances — most of which can be extremely harmful to lake water quality.

For example, when incoming water carries a great deal of organic matter, the bacteria that decompose it may consume so much of the lake's dissolved oxygen in the process that the supply can't keep up with the demand. This is particularly dangerous in lakes whose bottom waters are already anoxic (see Chapter 2). The result: without oxygen, you have fishkills, odors, and noxious conditions. In addition, as organic matter decomposes, it can also contribute nutrients to the water.

Although stressful enough by itself, the combination of high, oxygen-demanding organic loads and low dissolved oxygen levels compounds when it coincides with the peak growing season for algae and aquatic plants. The incoming nutrients fertilize the algae and plants, which not only grow excessively but further deplete the oxygen as they die and decompose.

Most wastewater treatment plants discharge at low rates: over 75 percent of all publicly owned treatment plants discharge less than 1 million gallons per day (mgd). Sewage treatment ponds or lagoons — the most common type of wastewater treatment — typically have discharge rates less than 1 mgd. But low discharge rates do not translate into insignificant effects on lakes and streams.

At just 10 to 50 parts per billion (µg/L) total phosphorus concentration in the water, some lakes may develop algal blooms, murkiness, and other problems. The average total phosphorus concentration of wastewater treatment plant discharges is about 100 times greater than this.

In many streams, wastewater discharges may dominate streamflow during the dry summer period when the stream's total flow is low. Also, at the higher summertime water temperatures, water cannot hold as much dissolved oxygen as it does during the cooler periods of the year.

Contact your state water pollution control agency for information on permitted facilities discharging into your lake or into streams entering your lake.

Wastewater Treatment

If point sources are important contributors of organic matter, bacteria, nutrients, or toxic pollutants, good wastewater treatment will provide critical protection for your lake. The better the system is at removing pollutants, the fewer algal blooms, aquatic weeds, and odors will occur in the lake. Regardless of the treatment system, however, ***all treatment systems require proper design, operation, and maintenance.*** These requirements vary among treatment systems, but no system can be installed and then ignored. Systems must be maintained.

Several approaches for treating point source discharges are briefly described in this section, beginning with municipal treatment systems. In general, it is more efficient and cost effective to collect wastewater from homes and industries and treat it in one large facility than to have individual septic systems or treatment facilities. In some cases, however, smaller treatment systems are required. A site, soil, and TMDL assessment are necessary before the final decisions are made. In addition to the descriptions that follow, more information on any of these treatment systems can be found in Appendix 6-A or on the EPA Office of Wastewater Management website (www.epa.gov/owm).

Municipal Systems

Typical waste treatment systems for larger cities and municipalities are conventional sewer systems piped to treatment facilities. These large treatment plants include systems such as activated sludge, biofilters, contact stabilization, sequencing batch reactors, land treatment, and large-scale lagoons. Most municipal treatment systems have both primary and secondary treatment. Some treatment systems are even more advanced and also have tertiary treatment

- ***Primary wastewater treatment*** uses screens and sedimentation (settling) to remove the larger organic solids. But dissolved organic matter can still use considerable oxygen, so secondary treatment is used to reduce this oxygen demand before the wastewater is discharged into the lake or stream.

- ***Secondary treatment*** uses biological and chemical processes to remove 80 to 95 percent of the organic matter (Fig. 6-4). Primary and secondary treatment, however, do not significantly reduce dissolved phosphorus concentrations (Table 6-2).
  - Total phosphorus concentrations in untreated domestic wastewater are reduced about 5 percent by primary treatment and about 10 to 15 percent by secondary treatment.
  - Both primary and secondary treatment remove much more nitrogen: about 40 percent with primary treatment; 60 percent with secondary treatment.

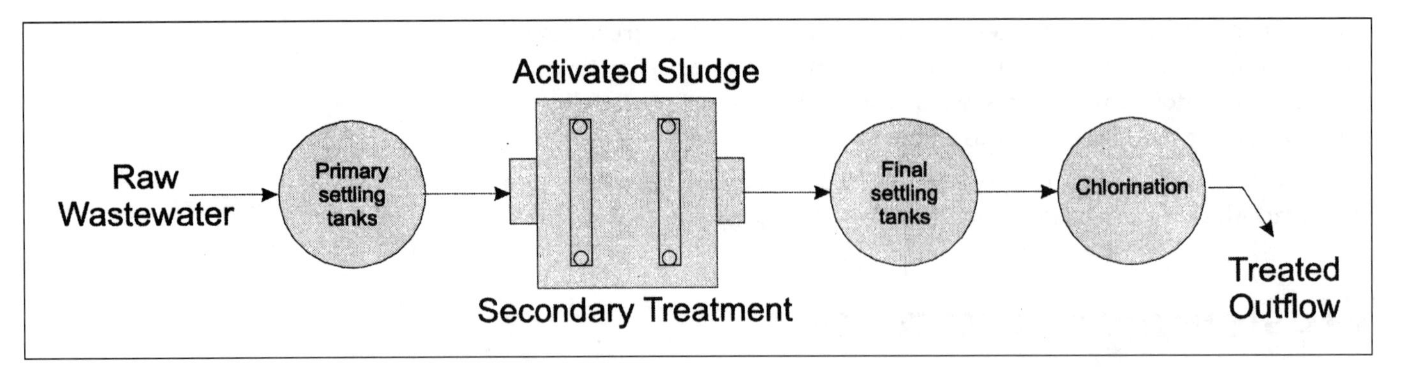

Figure 6-4.—Typical secondary treatment system with primary settling of water, chemical or biological treatment, and additional settling and disinfection before discharge.

Table 6-2.—Treatment removal efficiencies for wastewater treatment systems.

CONSTITUENT	TREATMENT CATEGORY		
	PRIMARY	SECONDARY	TERTIARY
Suspended Solids	50 – 70%	85 – 95%	–
Biological Oxygen Demand	20 – 40%	85 – 95%	–
Ammonia–Nitrogen	–	40 – 60%	~ 90%
Total Phosphorus	~ 5%	10 – 15%	75 – 90%

This means, however, that about half the total nitrogen and almost all the total phosphorus stay in the wastewater after it's treated at the second level.

▼ ***Another level of treatment — tertiary or advanced treatment —*** is required to significantly reduce nutrient concentrations in wastewater. This level of treatment used to be relatively expensive so it was not used to the same extent as secondary treatment. However, more cost-effective tertiary treatment systems, including constructed wetlands, ammonia stripping, multimedia filtration, and carbon adsorption are becoming available.

Normally, large municipal treatment systems are not suited to small communities. These complicated mechanical systems require skilled operators to run and maintain and typically use large amounts of energy — they're also costly for small communities to build. Fortunately, alternative treatments now exist for small communities and lake homeowners.

Small Community Systems

Several small-scale, simple, and reliable central treatment systems are suitable for the lake homeowner or lake association (Table 6-3; Fig. 6-5a,b). All of these well-established methods provide secondary or better levels of treatment at less cost to build and run than the larger municipal treatment plants. They also use less energy and are easier to operate and maintain.

If you're starting to plan a wastewater project, select an engineer who has experience with these small community technologies. Check with a local contractor who specializes in wastewater treatment, and with your health department, water pollution control agency, and EPA. You will need permits to discharge and the type of system and permit will vary by municipality, county, and state. The EPA Office of Wastewater Management has a specific program for small communities. The EPA "Small Communities Team" partners with community organizations to provide programs for technical assistance, financial assistance, and education and training (see www.epa.gov/own/smallc/) and Appendix 6-A.

Table 6-3.—Examples of small-scale treatment plants and designs.

EXAMPLE	DESCRIPTIVE NOTES
Septic Tank	A septic tank followed by a soil absorption bed is the traditional on-site system for the treatment and disposal of domestic wastewater from individual households or establishments. The system consists of a buried tank where wastewater is collected and scum, grease, and settleable solids are removed by gravity and a subsurface drainage system where wastewater percolates into the soil.
Septic Tank Mound System	Can be used as an alternative to the conventional septic tank–soil absorption system in areas where soil conditions preclude the use of subsurface trenches or seepage beds.
Septic Tank–Sand Filter	Surface discharge of septic tank effluent. Can be used as an alternative to the conventional soil absorption system in areas where subsurface disposal contains an intermediate layer of sand as filtering material and under drains for carrying off the filtered sewage.
Facultative Lagoon	An intermediate depth (3 to 8 feet) pond in which the wastewater is stratified into three zones. These zones consist of an anaerobic bottom layer, an aerobic surface layer, and an intermediate zone.
Oxidation Ditch	An activated sludge biological treatment process. Typical oxidation ditch treatment systems consist of a single or closed loop channel 4 to 6 feet deep, with 45° sloping sidewalls. Some form of preliminary treatment such as screening or removing normally precedes the process. After pretreatment, the wastewater is aerated in the ditch using mechanical aerators that are mounted across the channel.
Trickling Filter	The process consists of a fixed bed of rock media over which wastewater is applied for aerobic biological treatment (Fig. 6-5a). Slimes form on the rocks and treat the wastewater. The treated wastewater is collected by an underdrain system.
Overland Flow Treatment	Wastewater is applied by gravity flow to vegetated soils that are slow to moderate in permeability and is treated as it travels through the soil matrix by filtration, adsorption, ion exchange, precipitation, microbial action, and plant uptake (Fig. 6-5b). An underdrainage system recovers the effluent, controls groundwater, or minimizes trespass of wastewater onto adjoining property by horizontal subsurface flow.
Spray Irrigation	The wastewater is sprayed on crops or ground cover and the water is treated as it percolates through the soil. An under drainage system functions as with Overland Flow.
Treatment Wetlands	Wetlands are constructed specifically to function as wastewater treatment systems. In the wetland system, plants and soils remove nutrients for growth, provide a surface for micro-organisms and bacteria to break down waste, and promote settling of solids. Treatment wetlands serve as tertiary treatment for many communities.

Figure 6-5a.—Trickling filter works by trickling wastewater over layers of rocks. Bacteria growing on the rocks break the waste down.

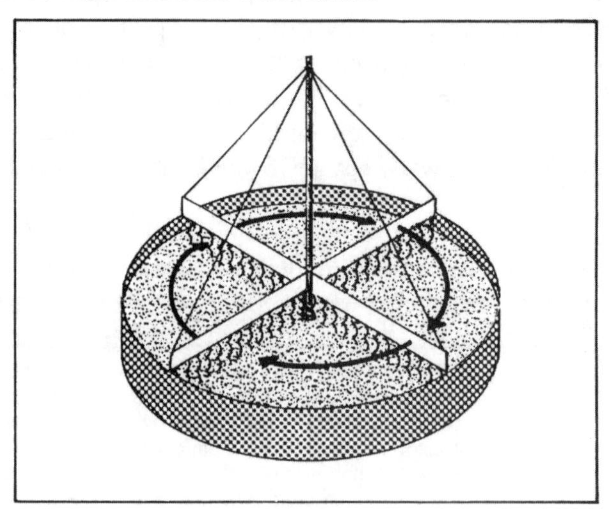

Figure 6-5b.—Overland flow of wastewater removes organic matter through soil bacteria, phosphorus by sorption to soils, and nitrogen through plant uptake.

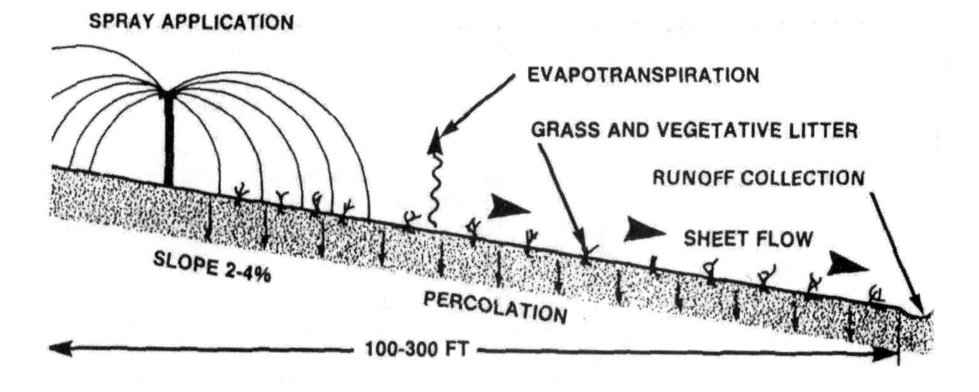

On-lot Septic Systems

The septic tank and drain field (Fig. 6-6) comprise the most common of these individual home sewage disposal systems. The septic tank traps solids, oil, and grease that could clog the drain field, storing sludge (solids that settle to the bottom), scum, grease, and floating solids until they can be removed during regular septic tank cleaning (every three to five years, depending on use). Specific recommendations for pumping out the tank can be obtained from county or state agencies. The wastewater that remains flows out of the septic tank and into the drain field where it seeps into the soil. The soil filters this partially treated sewage, and bacteria that began decomposing the waste in the tank continue to work. Characteristics of septic tanks are given in Table 6-4.

As the wastewater flows through the drain field, phosphorus may be absorbed by soil particles, and biological processes reduce the nitrogen. Bacterial decomposition in the drain field reduces the oxygen demand of this wastewater before it enters the lake or groundwater.

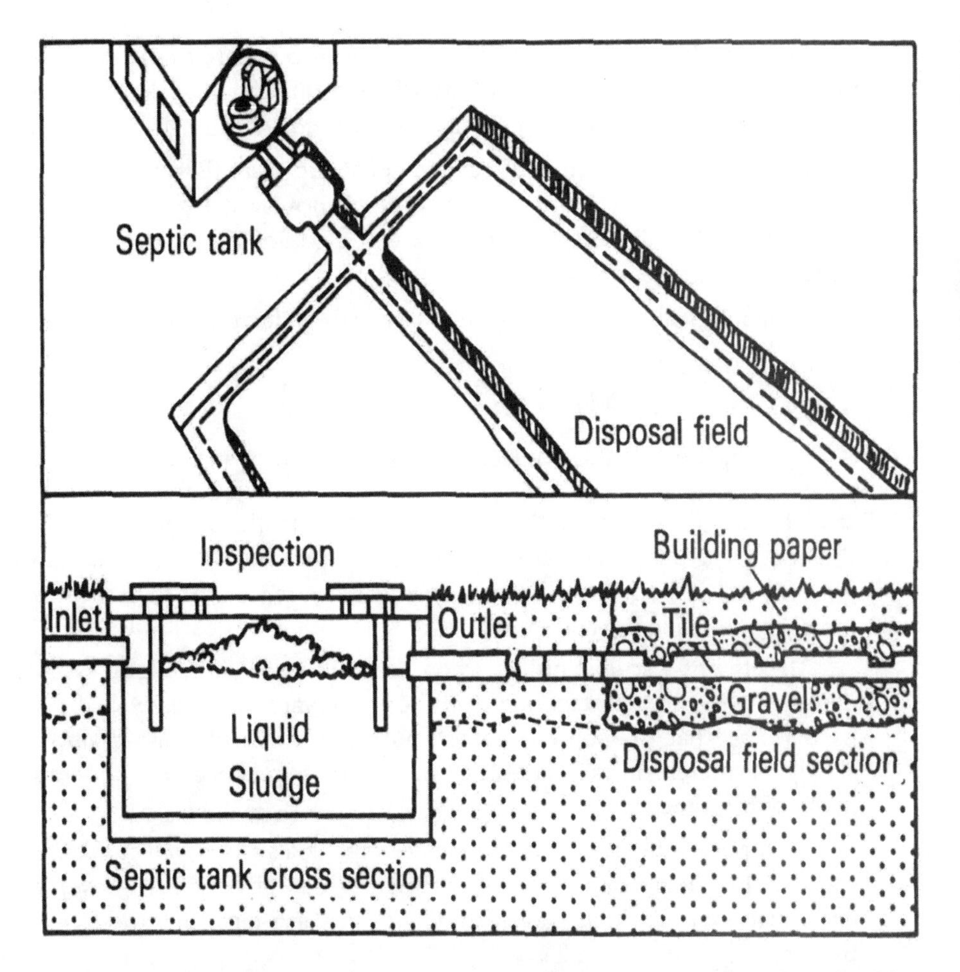

Figure 6-6.—Typical home septic system works well in many settings, but should not be used near lakes. Improper operations or failure will pollute the lake.

Septic systems are used by almost one-third of the U.S. population.

Table 6-4.—Characteristics of septic tanks.

CRITERIA	REMARKS
1. **Status**	Most widely used method of on-lot domestic waste disposal; used by almost one-third of the U.S. population.
2. **Applications**	Used primarily in rural and suburban areas. Properly designed and installed systems require a minimum of maintenance and can operate in all climates.
3. **Reliability**	Reliable, if properly designed and managed in appropriate soils.
4. **Limitations**	Properly designed, constructed, and operated septic tank systems are efficient and economical. System life may equal or exceed 20 years.
5. **Cleaning**	Sludge and scum in tank must be removed every 3 to 5 years, depending on soil and site conditions, the ability of the soil to absorb liquid, depth to groundwater, nature of and depth to bedrock, seasonal flooding, and distance to well or surface water.
6. **Treatment Side Effects**	May increase nitrates and other contaminants in groundwater when the soil does not remove them. Soil may also clog on the surface with potential health problems.

Some bacteria also convert nitrogen as ammonia to nitrate in the drain field, which tends to move with the water, eventually entering the lake. Ammonia and nitrate are fertilizers, and can encourage algal growth under the right conditions.

If they're properly designed and maintained, septic systems can remove organic matter, bacteria, and nutrients. They only work, however, in good site conditions — which don't exist on all lakeside lots. Conditions that interfere with proper septic system function:

- Unsuitable soils (you've heard the saying: they don't perc);
- High water tables;
- Steep slopes; and,
- The human element, as reflected in poor design or improper use.

▼ ***Soil*** plays a key role in the septic system. Tightly bound and poorly drained soil types (clays) are not effective filters. At the other extreme, gravel is also a poor filter because the wastewater drains through it too rapidly to be adequately treated.

Treatment also diminishes when the soil is too wet. Septic systems depend upon good contact between the wastewater and relatively dry soil particles so that the soil can absorb nutrients well. Soils that drain very slowly may be chronically saturated and the system, therefore, inoperative much of the time.

In a poorly drained soil, the wastewater is also likely to surface and run directly to the lake. A streak of bright green grass growing over the drain field indicates that wastewater nutrients are fertilizing the lawn on the way up.

▼ ***High groundwater tables*** can also prevent treatment by periodically flooding the drain system.

▼ ***Steep slopes*** cause either rapid flow-through or surfacing of wastewater.

Frequently, a septic problem can be traced to improper use, commonly arising from:

- Too small a tank;
- An inadequate drain field;
- Serving more people than the system was designed for;
- Using improper washing products;
- Following a poor septic tank maintenance schedule; or
- Using a garbage disposal, which overloads the system with fine solids.

Check with your health department or environmental agency for a reference on the functioning and design of septic systems. EPA also has a design manual for on-site wastewater treatment and disposal systems (U.S. EPA, 1980). In addition, many county extension offices have information on septic system installation and maintenance, including the Home*A*Syst software program.

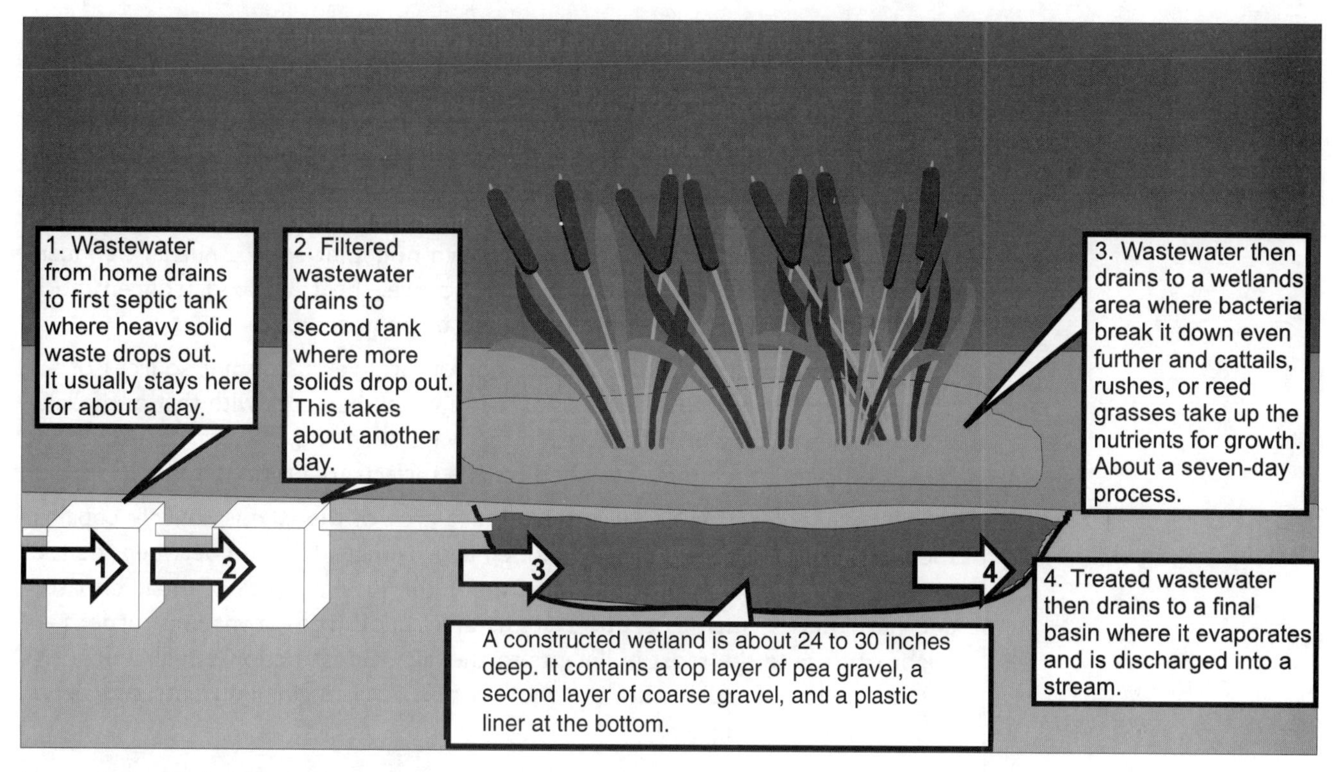

Figure 6-7.—Wetland treatment system.

Natural Treatments

Natural areas such as wetlands have occasionally been constructed around a lake to provide advanced wastewater treatment. Such treatment is typically used when conventional wastewater treatment cannot produce the lower nutrient concentrations needed. Wetlands can function as a biological filter to remove silt, organic matter, and nutrients from an inflowing stream, keep it out of the lake, and improve lake quality (Fig. 6-7). But, wetlands, under some conditions, can also contribute organic matter and nutrients to lakes. Nutrients released from wetlands can fertilize algal growth and contribute to lake problems.

Whether a wetland serves as a source or filter for nutrients and organic matter needs to be studied further. EPA has a *Design Manual for Constructed Wetlands for Wastewater Treatment Systems* (U.S. EPA, 1988). Additional information can be found on the EPA website (www.epa.gov/owm).

Watershed Management Practices: Nonpoint Sources

Nonpoint sources of pollution became apparent as municipal and industrial point sources were controlled. Point source controls have clearly made a difference in improving water quality since the passage of the Clean Water Act (1972), but water quality has not improved as much as expected.

Why? Point sources, long the major water polluter, had masked nonpoint source pollution problems. Once point source loadings declined, the impact of nonpoint sources became apparent.

Only by stepping away from the narrow viewpoint that point sources caused nearly all water quality problems were water quality managers able to see the lake and watershed as an integrated system affected by diverse sources of pollutants.

By approaching the management of lakes and streams from a broader perspective, water managers and scientists found that in many systems nonpoint sources contributed as much or more than point sources — particularly sediment, organic matter, and nutrients. Although nonpoint source nutrient concentrations were less than those in a point source, the total load (concentration times flow) can far exceed that contributed by point sources.

To determine the relative importance of local versus distant sources of nutrients and sediments to a lake, compare the watershed area with that of the lake. For example, if there are 100 acres in the watershed and the surface area of the lake is 100 acres, then the watershed to lake surface area ratio is 1 to 1 (1:1).

In small watersheds, where the surface area of the lake is roughly equal to the surface area of the watershed (i.e., an approximate 1:1 ratio between the lake surface area and the size of the watershed), the local sources of organic matter and nutrients — such as septic systems and runoff from lawns and gardens — might contribute the most pollutants to the lake. Construction can also be a significant source of sediment, with runoff from roads bringing nutrients, sediments, and heavy metals (Fig. 6-8).

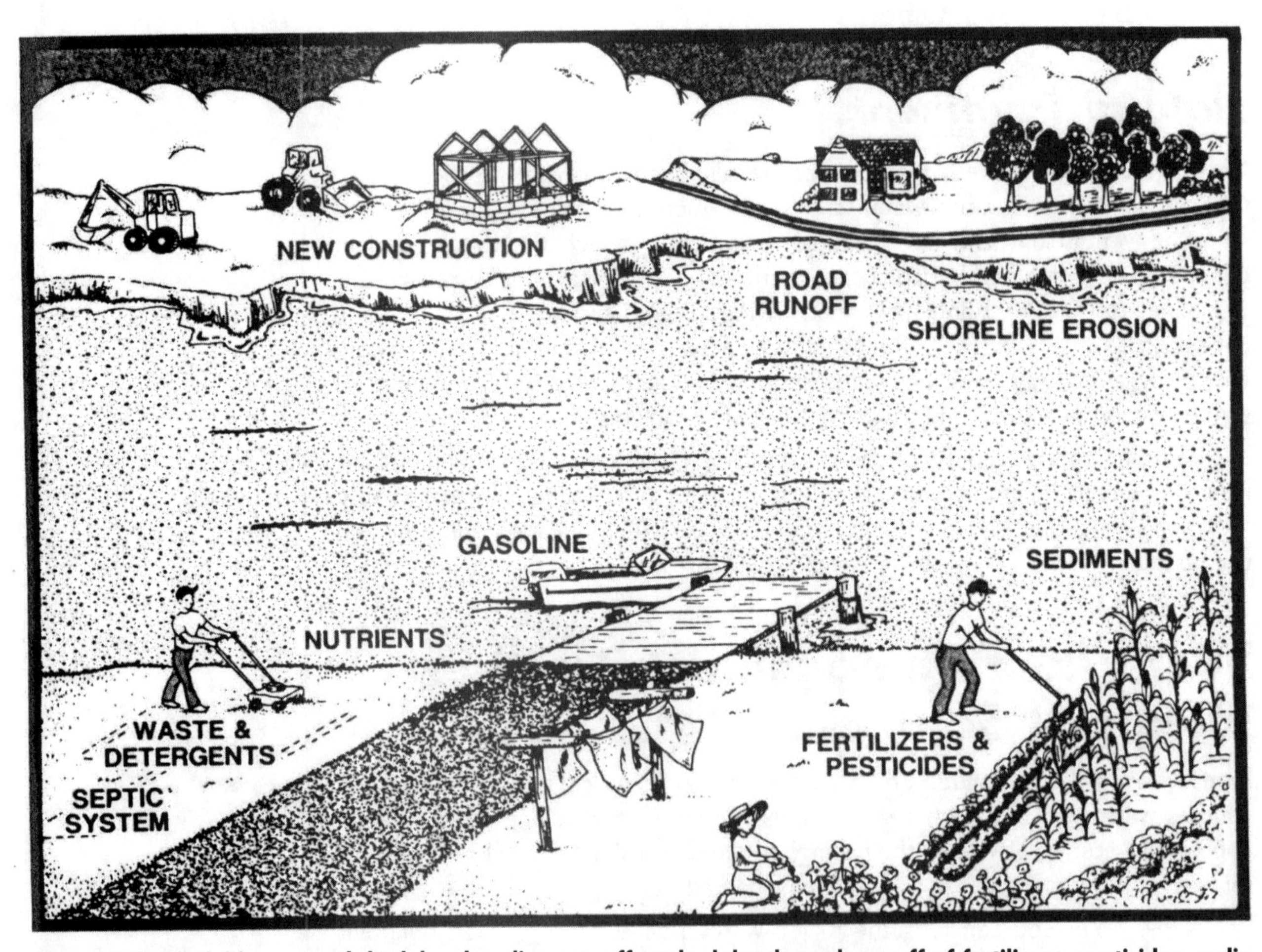

Figure 6-8.—Activities around the lake shoreline can affect the lake through runoff of fertilizers, pesticides, sediment, organic waste, and other contaminants.

As the watershed to lake surface area ratio grows, other sources of pollutants become increasingly important:

- Agricultural runoff carrying animal wastes, soil, and nutrients;
- Urban runoff from streets, yards, and rooftops carrying sediment, organics (oils and greases), nutrients, and heavy metals; and
- Forestry as a source of sediments.

In large watersheds, the contribution from urban, silvicultural, and agricultural areas is generally more significant than that from lakeshore homes.

What Are Best Management Practices?

So, how can we control these nonpoint sources of pollutants to protect and improve lake quality? If nonpoint sources don't discharge through a pipe, how do we reduce the amount coming from the watershed?

You have a number of options for improving the water quality of your lake — from picking up litter around the lake to implementing best management practices in the watershed. Best management practices have been developed for agricultural, forestry, urban, mining, construction, and similar land-use activities.

- **Agricultural BMPs**, for example, have been developed for cropland, pasture, barnyard and manure management, and fertilizer and pesticide control.
- **Forestry BMPs** manage activities such as road construction in timberlands, timber harvest techniques, forest lands cut or killed by disease or fire, and the use of pesticides.
- **Urban BMPs** have been designed to manage stormwater runoff in developing and established urban areas.
- **Construction BMPs** prevent erosion and runoff control.
- **Abandoned mine BMPs** protect lakes from excessive sediment runoff from tailings and leachate that frequently is very acidic and has high concentrations of metals. This book does not address abandoned mine mitigation and mining BMPs.

Best management practices were not initially designed to protect water quality, but to maintain productivity of farmland and reduce pesticide and fertilizer costs — or, in cities, to help protect homeowners from mud slides or flooding. Regardless of their original intent, many of these watershed best management practices are useful in lake management and restoration projects.

Managers of lakes and streams focus on best management practices to control four primary factors:

- Water, runoff, and soil moisture;
- Erosion;
- Nutrient loading; and
- Contaminant loading.

These factors are not independent, but highly interactive. Runoff control, for example, helps reduce sediments, nutrients, and pesticide runoff and contamination in lakes and streams. Figure 6-9 shows these factors for a typical construction site.

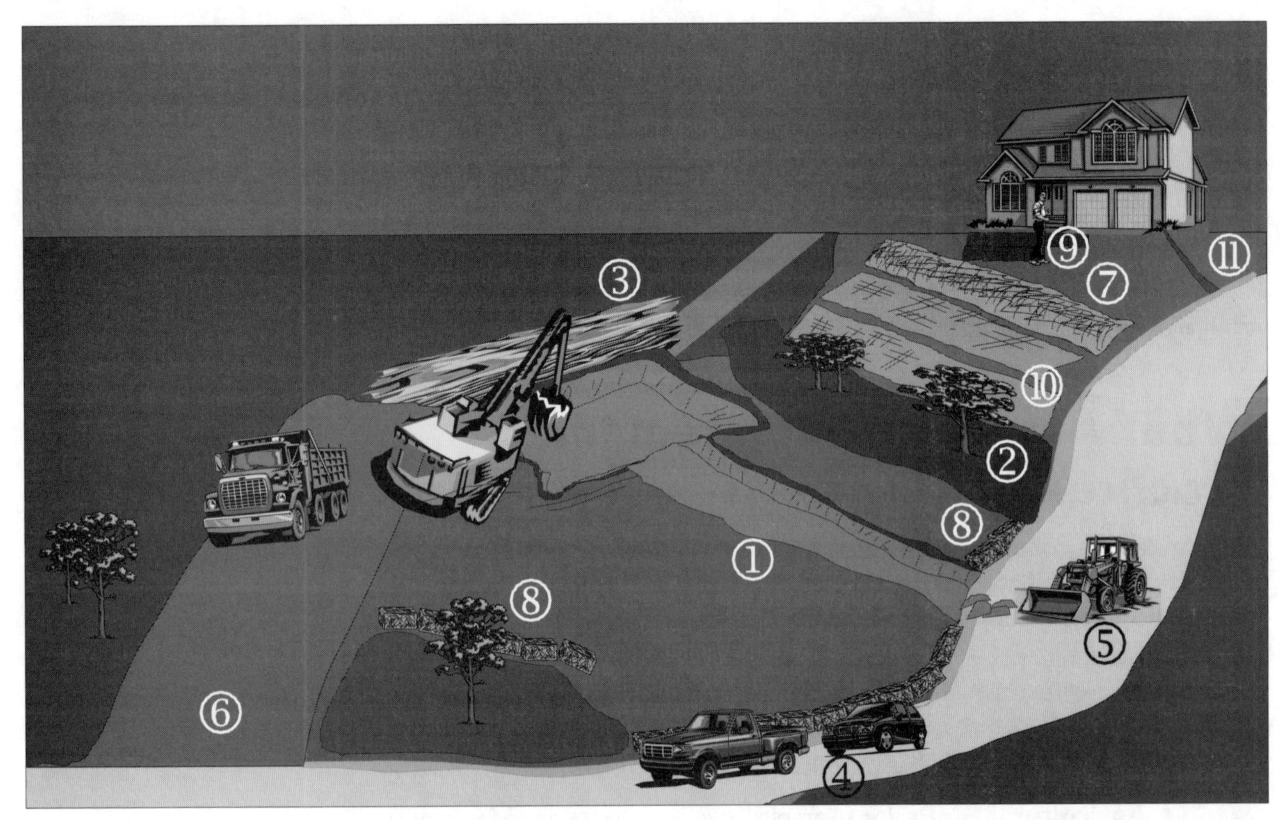

Figure 6-9.—No-cost or low-cost approaches to reducing erosion (from Wisconsin Dep. Natural Resources, Pub. 4-2300).

1 Plan your construction activities so that the soil is disturbed a minimal amount of time. For example, plan to install gas pipelines, sewer laterals, and other utilities at close time intervals.

2 Leave grass, trees, and shrubs in place wherever you can. The more vegetation, the less sediment-laden water leaves your site.

3 When you excavate the basement, pile the soil away from storm sewer drains — in the back- or side-yard area, for example. Once you backfill around the basement, remove any excess soil from the site.

4 Park cars and trucks on the street, not on the site. You will keep the soil less compacted and more water-absorbent, and you will keep mud from being tracked onto the street.

5 Arrange to have the street cleaned regularly while you are building to remove sediment that preventive measures failed to keep off the street.

6 Soon after you start construction, install a gravel driveway and encourage cars and trucks to use only this route on your site. Later, you can install the permanent driveway over the gravel.

7 Build a berm to divert rainwater away from steep slopes or other highly erodible areas.

8 Install straw bales or filter fences along curbs to filter rainwater before it reaches the gutter and storm sewer drains.

9 Seed and mulch, or sod your site as soon as you complete outside construction. You will control erosion, and — if you are building for a prospective buyer — you will increase the lot's salability by making it more attractive.

10 If you cannot seed and mulch the entire lot, cover any critical areas with a temporary protective material, such as filter fabric or netting. Later, you can remove the cover long enough to install utility lines.

11 Use roof downspout extenders and sump pump drain tubes to funnel water away from exposed soils and directly to the curb and storm sewer. After site is vegetated, downspout extenders and drain tubes should outlet to the vegetated area to maximize infiltration.

Generally, a best management practice (Table 6-5) involves some combination of three different approaches:

1. Reducing the generation of pollutants on-site by minimizing rainfall contact with the pollutant;
2. Restricting water runoff from on-site and up in the watershed to limit transport or movement of pollutants off the site into nearby waters; and/or
3. Capturing/trapping pollutants in the watershed and preventing them from entering the lake or groundwater.

Table 6-5.—Best Management Practices.

Best management practices used for various land-use activities. Although the names differ, the practices are all based on controlling runoff or water movement, erosion, and nutrient and contaminant loading.

AGRICULTURE	DESCRIPTIVE NOTES
Animal Waste Management	Reduces nutrient and organic matter loading by controlling timing, amount, and form of manure application to fields.
Conservation Tillage	Any tillage or planting system that maintains at least 30% of the soil surface covered by residue after planting to reduce soil erosion by water; examples of conservation tillage include no-till, ridge-till, or mulch-till.
Contour Farming	Conducting field operations, such as plowing, planting, cultivating, and harvesting, on the contours of the field.
Contour Stripcropping	Layout of crops in comparatively narrow strips in which the farming operations are performed approximately on the contour. Usually strips of grass or close-growing crops are alternated with those in cultivated crops or fallow.
Crop Rotation	Reduces soil erosion and nutrient applications by alternating with nitrogen-fixing legumes such as alfalfa.
Fertilizer Management	Reduces nutrient loading by controlling timing, amount, and type of fertilizer to crops.
Integrated Pest Management	Reduces pesticide applications, improves effectiveness of application, and uses more resistant cultivars.
Livestock Exclusion	Excluding livestock from highly erodible land and land near lakes and streams reduces erosion and nutrient loading.
Range and Pasture Management	Reduces runoff and erosion by maintaining vegetative cover. Reduces manure loadings to streams.
Terraces	Reduce erosion by shortening flow paths and improving drainage.
URBAN	**DESCRIPTIVE NOTES**
Flood Storage	Reduces runoff, sediment, and attached nutrient/contaminant loading by settling sediment particles out of the water.
Porous Pavement	Reduces runoff, erosion, and pollutant loading by rainfall soaking through the pavement into the underlying soil.
Street Cleaning	Reduces nutrient and contaminant loading by removing them from the pavement. Pollutants will not be washed into streams during storms.
FORESTRY	**DESCRIPTIVE NOTES**
Ground Cover Management	Reduces runoff and erosion by maintaining cover over soil so it is not exposed to raindrops or runoff.
Pesticide/Herbicide Management	Reduces contaminant loading by controlling the timing, amount, form, and location of pesticide applications.
Riparian Zone Management	Reduces runoff, erosion, and nutrient/contaminant loading by maintaining vegetation and ground cover along stream banks (riparian zone). **Buffer strips** and **streamside management** are other terms used for this BMP.
Road/Skid Trail Management	Reduces length of runoff flow path and reduces erosion. Erosion from roads and skid trails(i.e., paths where logs are dragged to the loading area)is the major source of sediments from forested watersheds.

Table 6-5.—Best Management Practices (cont.).

CONSTRUCTION	DESCRIPTIVE NOTES
Disturbed Area Limits	Reduces erosion by restricting the area of the construction site that is disturbed or has ground cover removal.
Nonvegetative Soil Stabilization	Reduces soil erosion by using matting, mulch, or similar ground cover over the soil to reduce rainfall eroding the soil surface.
Surface Roughening	Reduces the length of runoff flow paths to slow the water, creating pools or depressions and reducing the energy of water to dislodge and transport soil off the site.
MULTICATEGORY	**DESCRIPTIVE NOTES**
Detention/Sedimentation Basins	Reduces the flood peak, sediment, nutrient and contaminant loading by retaining runoff and letting soil particles and attached nutrients/contaminants settle out in the basin.
Grassed Waterways	Reduces erosion, nutrient, and contaminant loading by having runoff flow over a grassy area as it moves toward the stream. Soil is protected and grass helps trap nutrients and contaminants.
Interception or Diversion Practices	Reduces runoff erosion, nutrient, and contaminant transport by intercepting runoff before the flow path becomes too long or divesting the runoff away from the lake.
Maintenance of Natural Waterways	Natural stream banks, riparian zones, and wetlands trap sediment and nutrients and limit streamside erosion.
Riprap	A layer of broken rock, cobbles, boulders, aggregate, or fragments of sufficient size and thickness to resist the erosive forces of flowing water or wave action; such structures usually are used to protect channels with relatively high velocity flow, shores, slopes on dams, or outlets of structures.
Streamside Management Zones	Reduces runoff, erosion, nutrient and contaminant loading by maintaining vegetative and ground cover next to the streambank. Typically vegetative strips 30 to 100 feet wide.
Streambank Stabilization	Reduces erosion and in-stream sediment by protecting and maintaining the streambank so it does not erode or fall into the stream.
Vegetative Stabilization	Reduces runoff, erosion, nutrient, and contaminant loads by maintaining good vegetative cover at critical locations throughout the watershed such as highly erodible areas and streamsides and banks.
Zoning	Reduces runoff, erosion, nutrient, and contaminant loadings through legally enforceable regulations for permissible businesses and land uses and management needed to protect lakes and streams.

From: Resource Conservation Glossary. 1982. 3rd ed. Soil and Water Conservation Society of America. Ankeny, IA. For more detailed lists of BMPS, see U.S. EPA (1993).

Runoff and Erosion Control Practices

One of the most powerful forces on earth is rainfall. Even gentle rains can destroy slopes and create mud slides, gulleys, and flooding — making sediment (plus the contaminants it carries) the most widespread pollutant in North America. Controlling water and associated runoff from the watershed, then, is the first step in implementing best management practices.

One of the most powerful forces on earth is rainfall. Even gentle rains can destroy slopes and create mud slides, gulleys, and flooding – making sediment (plus the contaminants it carries) the most widespread pollutant in North America.

Erosion is a natural process, so watershed best management practices mimic the natural processes that control water and soil moisture; thus, they are very similar across the wide spectrum of land uses (e.g., agriculture, forestry, mining, construction). Features found in forested watersheds, for example, form the basis for most efforts to reduce runoff and erosion (Fig. 6-10).

The forest canopy or tree cover intercepts rain as it falls — shrubs and small trees below the tree tops intercept the rainfall moving through and dripping from the canopy. A dense cover of pine needles, leaves, and vegetation on the forest floor further keeps rain drops from hitting bare earth, permitting it to seep into the soil. And small depressions at all three levels store water to keep it from running off.

When rainfall does begin to run off a forest, it soon encounters a barrier such as a depression or log that slows it down or a channel that collects and conveys it to other channels. Woody debris, rocks, and gravel in these channels slow the water's flow and thus, reduce soil erosion.

Following rain, the trees and other vegetation take water out of the soil through their roots. Remaining soil water percolates downward to enter the groundwater system where it can be discharged into stream channels, thereby reducing the water in the soil — giving it the capacity to store water from the next rainfall.

These natural features and processes have become the key features of best management practices to control water and runoff:

- Intercepting or impeding rain drop impacts;
- Creating short flow paths and impeding flow; and
- Designing and protecting channels for collecting the flow.

▼ ***A multi-story vegetative cover*** over the site — leaves at several levels — is more effective in intercepting rainfall than leaves at just one level. Most best management practices establish or maintain ground cover.

- For example, in agriculture, conservation tillage (keep the crop residues on the soil; do not plow them under) both covers the soil and retains moisture.
- Forest timber practices maintain ground cover to protect forest soils from rainfall.
- Forest loggers may remove haul roads and seed the roadbed following harvest.

▼ ***Creating short flow paths and slowing the flow*** of water to reduce its velocity (and thus, erosion) is an integral part of almost all watershed BMPs.

- In agriculture, contour farming, contour stripcropping, conservation tillage, terraces, and grassed waterways are used to create barriers to slow runoff.

Figure 6-10.—Watersheds with good vegetative cover like those in the left-hand side of the figure have lower and slower runoff, less erosion, and lower nutrient and contaminant loading than watersheds where vegetative cover has been removed. BMPs try to slow runoff, reduce erosion, and trap or retain nutrients and contaminants and prevent them from entering streams and lakes.

- In forest management, water stops along forest roads (they look like little speed bumps), woody debris, and buffer strips near streams create short flow paths and slow the flow.
- Proper design of logging roads and drainage ditches is important to forestry management.
- Construction sites roughen the ground surface to create small depressions and barriers and use hay bales to impede runoff.
- In-stream features mimic the pools that store water in natural stream channels, and their S-shaped meander that slows the flow and reduces the size of flood waves.
- Stormwater detention/sedimentation basins, interception/diversion systems, and similar features are BMPs used in many construction and urban areas.

▼ ***Channels designed to collect and carry runoff are universal BMPs,*** regardless of the land use. They include:

- Grassed swales;
- Constructed channels lined with large rocks or riprap;
- Synthetically lined channels that, while expensive, may be better protection for critical areas; and
- Tree root wads, logs, and natural boulders.

For every land use, most watershed best management practices try to minimize impacts of rain on the soil.

Nutrient and Contaminant Control Practices

Eutrophication, like erosion, is a natural process, but people can increase nutrient loadings to lakes and streams. In general, the major sources of nutrients and contaminants come from agricultural and urban land use.

Nutrient and contaminant control practices build on runoff and erosion control practices. Less runoff means lower erosion and lower transport of both dissolved nutrients in the runoff and nutrients sorbed onto soil particles.

But two other principles are vital to controlling nutrients and contaminants:

- Reducing the amount of nutrients and contaminants applied; and,
- Using plant uptake to remove nutrients and contaminants from runoff and groundwater.

▼ ***Don't add more nitrogen and phosphorus than plants need.*** This is throwing money away: these nutrients are not used to increase crop production; instead, the excess washes from the watershed into lakes and streams and damages these ecosystems.

Agriculture uses soil nutrient testing to reduce the nutrients applied to crops, particularly nitrogen. Phosphorus and pesticides, however, generally adsorb onto soil particles so reducing soil erosion also helps control them.

▼ ***Install vegetative buffer strips*** next to streams and lakes to help prevent nutrients and contaminants from entering the water body by trapping those carried by eroded soil particles.

- Streamside vegetation uses nutrients from the water for growth.

▼ ***Use plants to remove*** certain types of heavy metals or organic contaminants.

- In a process known as phytoremediation, plants are being used to clean up certain types of heavily contaminated soils by absorbing the contaminants from the soil.

Reducing nutrient concentrations in surface runoff, however, can increase their concentrations in subsurface or groundwater. If nutrients such as nitrate, which are readily soluble in water (but do not adsorb to soils), are excessively applied to soils and not permitted to run off, they will seep into the soils and continue moving down until they encounter a barrier such as clay or enter the groundwater.

- Too much animal manure, liquid fertilizers containing nitrogen, and ammonium nitrate can significantly contaminate groundwater, particularly in watersheds with permeable soils or bedrock such as limestone. Rain leaches nutrients out of soils as it seeps into them and can cause high nitrate concentrations in groundwater and in domestic wells.

Herbicides such as atrazine have been found in the spring in relatively high concentrations in lakes, streams, and groundwater throughout the Midwest, illustrating the problems caused by excessive applications over large areas.

Although simply following manufacturers' directions for application can help prevent contamination of lakes and streams and groundwater, several programs can be useful:

- Manure management programs can help farmers contain the manure and apply it when needed by crops, eliminating excess.
- Well-head protection programs have been developed to help reduce soluble nutrient and herbicide concentrations in watershed areas that contribute to groundwater recharge.
- Integrated pest management programs for both agriculture and forestry reduce the use of pesticides by integrating weed- and insect-resistant crop varieties, crop rotation, stripcropping, and biological agents with pesticides to control pests.

Zoning and ordinances go hand in hand with other BMPs in managing watershed activities.

Zoning and Ordinances

Zoning requirements and ordinances that regulate various types of land uses and development can be effective, long-term approaches in the watershed management tool box. Example ordinances protecting stream buffer zones, open space development, greenways, stormwater drainage, and other activities can be found on the EPA website at www.epa.gov/owow/nps.ordinance and the Center for Watershed Protection website at www.cwp.org.

Zoning and ordinances go hand in hand with other BMPs in managing watershed activities.

Water Conservation Practices

With a lake nearby, conserving water might not seem critical. But reducing water usage also reduces wastewater discharges.

- Water-saving devices such as flow-reducing showerheads and water-saving toilets (or just using less water in the tank) can cut household wastewater flows by as much as 25 percent (U.S. EPA, 1981).

Most of these procedures are very simple, even obvious (see Table 6-6), but if many of you living around the lake follow them, the water you conserve may allow your community to use a smaller wastewater treatment facility — at the very least, using less water can lower day-to-day operating costs for treatment chemicals and utilities.

And look at these other options:

- If your treatment capacity is nearly maxed out, conserving water may be the best step you can take.
- If you're connected to a regional sewer system, conservation can reduce treatment charges, which are usually based on the volume of sewage treated — based on your water meter readings and prorated by household.

Water conservation, then, not only costs less in the long run but also protects your lake from excessive organic matter and nutrients, as your wastewater discharges decline. More careful usage may also lower nonpoint sources from activities such as watering lawns. So as you prepare your management plan, remember to include water conservation as an essential best management practice. The EPA website, www.epa/gov/owm, also has ideas and information on water efficiency and recycling and reuse.

So as you prepare your management plan, remember to include water conservation as an essential best management practice.

Integrated Watershed — and Lake — Management

Watershed and lake management starts with a plan — what are the lake problems, where are they occurring, why are they occurring, what do you want to achieve, how can you go about it?

As you have found in this chapter, many management practices can be used to reduce these problems and help achieve lake and watershed management objectives.

But there is no magic bullet: rather, it is the combination of multiple practices that best control point and nonpoint sources of pollution in the watershed and in your lake. Many approaches can be used and many local, state, and federal agencies, universities, and organizations can provide technical and financial assistance and/or cost-sharing for watershed management. Use them! It is the integration of these watershed management practices and the implementation of a watershed plan that build the foundation for effective lake management.

Table 6-6.— Conscientious use of water can prevent excess runoff and reduce the volume of wastewater treated, both of which help protect lake water quality.

WATER CONSERVATION TECHNIQUES
• Inspect the plumbing system for leaks.
• Install flow-control devices in showers.
• Turn off all water during vacations or long periods of absence.
• Check the frequency with which home water softening equipment regenerates and backaches. It can use as much as 100 gallons of water each time it does this.
• Insulate hot water pipes to avoid having to clear the "hot" line of cold water during use.
• Check all faucets, inside and out, for drips. Make repairs promptly. These problems get worse—never better.
• Reduce the volume of water in the toilet flush tanks with a quart plastic bottle filled with water (bricks lose particles, which can damage the valve).
• Never use the toilet as a trash basket for facial tissue, etc. Each flush uses 5 to 7gallons of water. Items carelessly thrown in could clog the sewage disposal system.
• Accumulate a full laundry load before washing, or use a lower water level setting.
• Take showers instead of baths.
• Turn off shower water while soaping body, lathering hair, and massaging scalp.
• Bottle and refrigerate water to avoid running excess water from the lines to get cold water for meals. Shake bottle before serving to incorporate air in the water so it doesn't taste flat.
• To get warm water, turn hot water on first; then add cold water as needed. This is quicker this way and saves water too.
• Wash only full loads of dishes. A dishwasher uses about 9 to 13 gallons of water per cycle.
• When washing dishes by hand, use one pan of soapy water for washing and a second pan of hot water for rinsing. Rinsing in a pan requires less water than rinsing under a running faucet.
• Use rinse water—"gray water"—saved from bathing or clothes washing to water indoor plants. Do not use soapy water on indoor plants. It could damage them.
• Vegetables requiring more water should be grouped together in the garden to make maximum use of water applications.
• Mulch shrubs and other plants to retain moisture in the soil longer. Spread leaves, lawn clippings, chopped bark or cobs, or plastic around the plants. Mulching also controls weeds that compete with garden plants for water. Mulches should permit water to soak into the soil.
• Try "trickle" or "drip" irrigation systems in outdoor gardens. These methods use 25 to 50 percent less water than hose or sprinkler methods. The tube for the trickle system has many tiny holes to water closely spaced plants. The drip system tubing contains holes or openings at strategic places for tomatoes and other plants that are more widely spaced.
• Less frequent but heavier lawn watering encourages a deeper root system to withstand dry weather better.
• Plan landscaping and gardening to minimize watering requirements.
• When building or remodeling, consider: —Installing smaller than standard bath tubs to save water. —Locating the water heater near where hottest water is needed—usually in the kitchen/laundry area.

Source: from a bulletin issued by the Arkansas Cooperative Extension Service (USDA,1984).

References

Arkansas Cooperative Extension Service. 1984. Bulletin on Water Conservation Techniques. U.S. Dep. Agriculture, Washington, DC.

Colorado Department of Public Health and Environment. 1997. The Total Maximum Daily Load Process. Water Quality Control Div., Denver.

Federal Interagency Stream Restoration Working Group. 1998. Stream Corridor Restoration: Principles, Processes, and Practices. Internet at www.usda.gov/stream_restoration.

Intergovernmental Advisory Committee. 1995. Ecosystem Analysis at the Watershed Scale: Federal Guide for Watershed Analysis–Version 2.2. Portland, OR.

Minnesota Department of Natural Resources. 1999. Landscaping for Wildlife and Water Quality. St. Paul.

Montgomery, D. G. Grant, and K. Sullivan. 1995. Watershed analysis as a framework for implementing ecosystem management. Water Retour. Bull. 31(3):369-86.

Ohio Environmental Protection Agency. 1997. A Guide to Developing Local Watershed Action Plans in Ohio. Columbus.

Soil and Water Conservation Society of America. 1982. Resource Conservation Glossary. 3rd ed. Ankeny, IA.

U.S. Environmental Protection Agency. 1993. Guidance Specifying Management Measures for Sources of Nonpoint Pollution in Coastal Waters. EPA-840-B-93-001c. www.epa.gov/owow/nps/MMGI/. Off. Wetlands Oceans and Watersheds, Assess. Watershed Prot. Div., Washington, DC.

———. 1997. Technical Guidance Manual for Developing Total Maximum Daily Loads: Book 2. Streams and Rivers–Part 1: Biochemical Oxygen Demand / Dissolved Oxygen and Nutrients/Eutrophication.EPA 823-B-97-002. Off. Sci. Tech., Stand. Appl. Sci. Div., Washington, DC.

———. 1998. Clean Water Action Plan: Restoring and Protecting America's Waters. USDA Natural Resource Conserv. Serv. and U.S. Environ. Prot. Agency. EPA-840-R-98-001. National Center for Environmental Publications and Information (800/490-9198).

U.S. Geological Survey. 1997. Poster Series. Water Quality. Grade school ed. USGS Branch Information Services, Denver, CO.

See Appendix 6-A for additional Guidance Manuals, Web sites, and Links.

APPENDIX 6-A

Best Nonpoint Source Resources

GENERAL NOTES:
* means the organization is likely to charge for the listed document
NSCEP means the document is free from the National Service Center for Environmental Publications and Information at (800) 490-9198 or (513) 489-8190.

AGRICULTURAL BMP MANUALS

Animal Facilities

Animal Waste Management Field Handbook: USDA NRCS National Engineering Handbook (NEH): Part 651. U.S. Department of Agriculture, Natural Resources Conservation Service. The Animal Waste Management Field Handbook is an excellent resource for animal waste system designers or anyone interested in how animal waste is typically handled and stored. The document covers all aspects of the design and management of animal facilities and has lots of very useful diagrams and figures. See: http://www.ftw.nrcs.usda.gov/awmfh.html.

Earthen and Manure Storage Design Considerations. Natural Resource, Agriculture, and Engineering Service, Cooperative Extension (1999). Publication number NRAES-109 covers environmental policy as well as manure storage and management. The text can be technical but also covers the basics of environmental issues and risk reduction. NRAES, Cooperative Extension, 152 Riley-Robb Hall, Ithaca, NY 14853-5701, (607) 255-7654. See: http://www.nraes.org/.

Liquid Manure Application Systems Design Manual. Natural Resource, Agriculture, and Engineering Service, Cooperative Extension (1998). Document number NRAES-89 focuses on the characteristics and land application of liquid manure. Evaluations of application sites for environmental risk, manure handling, and safety are key issues. NRAES, Cooperative Extension, 152 Riley-Robb Hall, Ithaca, NY 14853-5701, (607) 255-7654. See: http://www.nraes.org/.

On-farm Composting Handbook. Natural Resource, Agriculture, and Engineering Service, Cooperative Extension (1992). NRAES-54 describes the composting process in detail as well as discusses the benefits and drawbacks of using composting in an operation. Raw materials, various composting methods, how to use compost, and how to market compost are all covered in this 186-page manual. For copies, contact NRAES, Cooperative Extension, 152 Riley-Robb Hall, Ithaca, NY 14853-5701, (607) 255-7654. See: http://www.nraes.org/.

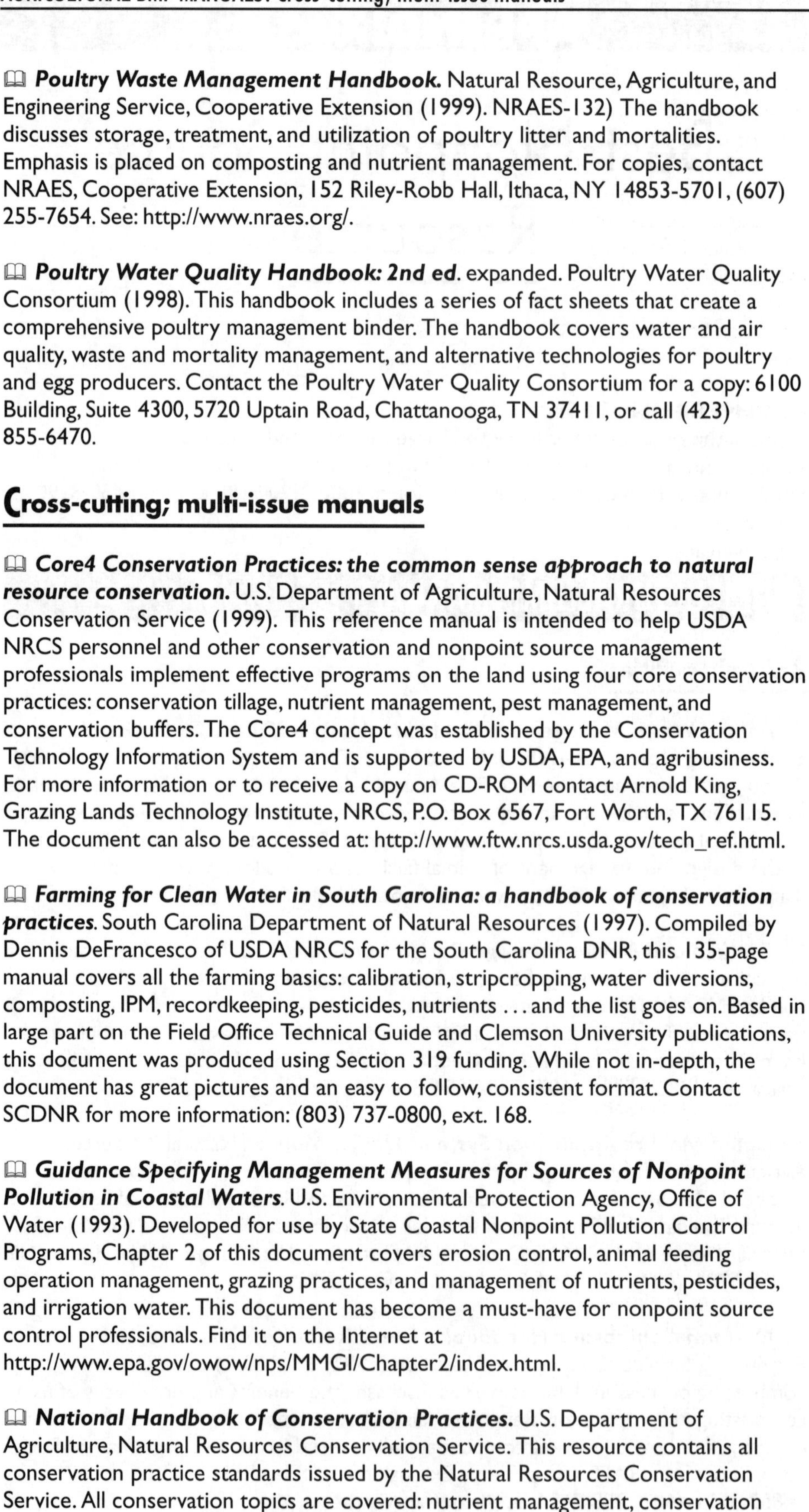

AGRICULTURAL BMP MANUALS: Cross-cutting; multi-issue manuals

Poultry Waste Management Handbook. Natural Resource, Agriculture, and Engineering Service, Cooperative Extension (1999). NRAES-132) The handbook discusses storage, treatment, and utilization of poultry litter and mortalities. Emphasis is placed on composting and nutrient management. For copies, contact NRAES, Cooperative Extension, 152 Riley-Robb Hall, Ithaca, NY 14853-5701, (607) 255-7654. See: http://www.nraes.org/.

Poultry Water Quality Handbook: 2nd ed. expanded. Poultry Water Quality Consortium (1998). This handbook includes a series of fact sheets that create a comprehensive poultry management binder. The handbook covers water and air quality, waste and mortality management, and alternative technologies for poultry and egg producers. Contact the Poultry Water Quality Consortium for a copy: 6100 Building, Suite 4300, 5720 Uptain Road, Chattanooga, TN 37411, or call (423) 855-6470.

Cross-cutting; multi-issue manuals

Core4 Conservation Practices: the common sense approach to natural resource conservation. U.S. Department of Agriculture, Natural Resources Conservation Service (1999). This reference manual is intended to help USDA NRCS personnel and other conservation and nonpoint source management professionals implement effective programs on the land using four core conservation practices: conservation tillage, nutrient management, pest management, and conservation buffers. The Core4 concept was established by the Conservation Technology Information System and is supported by USDA, EPA, and agribusiness. For more information or to receive a copy on CD-ROM contact Arnold King, Grazing Lands Technology Institute, NRCS, P.O. Box 6567, Fort Worth, TX 76115. The document can also be accessed at: http://www.ftw.nrcs.usda.gov/tech_ref.html.

Farming for Clean Water in South Carolina: a handbook of conservation practices. South Carolina Department of Natural Resources (1997). Compiled by Dennis DeFrancesco of USDA NRCS for the South Carolina DNR, this 135-page manual covers all the farming basics: calibration, stripcropping, water diversions, composting, IPM, recordkeeping, pesticides, nutrients . . . and the list goes on. Based in large part on the Field Office Technical Guide and Clemson University publications, this document was produced using Section 319 funding. While not in-depth, the document has great pictures and an easy to follow, consistent format. Contact SCDNR for more information: (803) 737-0800, ext. 168.

Guidance Specifying Management Measures for Sources of Nonpoint Pollution in Coastal Waters. U.S. Environmental Protection Agency, Office of Water (1993). Developed for use by State Coastal Nonpoint Pollution Control Programs, Chapter 2 of this document covers erosion control, animal feeding operation management, grazing practices, and management of nutrients, pesticides, and irrigation water. This document has become a must-have for nonpoint source control professionals. Find it on the Internet at http://www.epa.gov/owow/nps/MMGI/Chapter2/index.html.

National Handbook of Conservation Practices. U.S. Department of Agriculture, Natural Resources Conservation Service. This resource contains all conservation practice standards issued by the Natural Resources Conservation Service. All conservation topics are covered: nutrient management, conservation tillage, erosion control, irrigation, grazing, etc. This handbook is available on-line at http://www.ftw.nrcs.usda.gov/nhcp_2.html.

AGRICULTURAL BMP MANUALS: Cross-cutting; multi-issue manuals

🕮 ***Soybean Management and the Land: A Best Management Practices Handbook for Growers.*** American Soybean Association (2000). This manual is a two-for-one bonus. The "Resource Book" presents information on BMPs for the farmstead, cropland, pastureland, and other areas. All types of BMPs are covered: erosion, pest management, nutrients, well protection, buffers, etc. The BMP discussion includes real world examples of how these practices work through testimonials from real farmers. The "Workbook" allows soybean growers to assess the conditions on their farms and determine their environmental risk level. The "Workbook" also helps the producer make a plan for improvement. Contact the American Soybean Association, (800) 688-7692, ext. 210; ASA, 12125 Woodcrest Executive Drive, Suite 100, St. Louis, MO 63141, bmphandbook@soy.org (cost $36).

🕮 ***50 Ways Farmers Can Protect Their Groundwater.*** University of Illinois, College of Agriculture, Cooperative Extension Service (1993). The title says it all. While focusing on the management of fertilizers and pesticides, this 190-page book briefly covers livestock waste, wells, hazardous chemicals, and water testing. Contact Information Services, (217) 333-2007, for a copy or look at it on-line at http://web.aces.uiuc.edu/watershed/training edu.

🕮 ***60 Ways Farmers Can Protect Surface Water.*** University of Illinois, College of Agriculture, Cooperative Extension Service (1993). Topics include residue management, water flow control, nutrient management, livestock waste handling, and pesticide management. Contact Information Services, (217) 333-2007, or find it on-line at http://web.aces.uiuc.edu/watershed/training.edu.

Erosion

🕮 **(See list for Cross-cutting Manuals)**

Grazing

🕮 ***Best Management Practices for Grazing.*** Montana Department of Natural Resources and Conservation (1999). This manual describes the BMPs developed as part of Montana's Prescribed Grazing Standard NRCS Conservation Practice Standard. The manual covers grazing management plans, riparian areas, forestlands, and winter feeding areas. For copies, contact the Conservation Districts Bureau, Department of Natural Resources and Conservation, P.O. Box 201601, Helena, MT 59620-1601, or call (406) 444 6667.

🕮 ***Managing Change: livestock grazing on western riparian areas.*** U.S. Environmental Protection Agency Region 8 (1993). Written for "the men and women who move the livestock," this 31-page booklet encourages ranchers to look at the water quality and habitat impacts of their grazing practices. Excellent photographs illustrate how streambanks and water quality are degraded by improper grazing and how improved management can restore the health of the streams. Quick-fix structural components for stream training are cautioned against as an inadequate substitute for long-term responsible herd management. U.S. Environmental Protection Agency Region 8, 999 18th Street, Suite 500, Denver CO 80202-2466, and Northwest Resource Information Center, Inc., P.O. Box 427, Eagle, ID 83616.

AGRICULTURAL BMP MANUALS: Grazing

📖 ***National Range and Pasture Handbook.*** U.S. Department of Agriculture, Natural Resources Conservation Service, Grazing Lands Technology Institute (1997). Part of the NRCS Field Office Technical Guide, this manual covers inventorying, monitoring, and managing grazing lands as well as livestock nutrition, behavior, and husbandry. Special sections deal with the economics of grazing, wildlife management, and hydrology. See: http://www.ftw.nrcs.usda.govtech_ref.html.

Irrigation

📖 ***Irrigation Guide: USDA NRCS National Engineering Handbook: Part 652.*** U.S. Department of Agriculture, Natural Resources Conservation Service (1997). This manual describes NRCS-recommended processes for planning, designing, evaluating, and managing irrigation systems. GSA National Forms and Publications Center, 7CAF, 501 W. Felix Street, Warehouse 4 Dock 1, Fort Worth, TX 76115.

📖 ***Irrigation Management Practices to Protect Ground Water and Surface Water Quality State of Washington.*** Washington State Department of Ecology and Washington State University Cooperative Extension (1995). The handbook emphasizes a systems approach to irrigation management and water quality protection. Introductory material covers water quality issues and the basics of soil-water-plant relationships and irrigation processes. Contact State of Washington, Department of Ecology, P.O. Box 47600, Olympia, WA 98504-7600.

Nutrient Management

📖 (See list for Cross-cutting Manuals)

📖 ***Agricultural Phosphorus and Eutrophication.*** U.S. Department of Agriculture, Agricultural Research Service, ARS-149 (1999). A small booklet co-authored by USDA ARS and U.S. Environmental Protection Agency staff, this resource details the relationship between phosphorus application to agricultural fields and eutrophication of our nation' s waterways. A good primer on the phosphorus cycle and actions that can help control phosphorus. For copies contact USDA ARS, Pasture Systems & Watershed Management Research Laboratory, Curtin Road, University Park, PA 16802-3702 (while supplies last) or purchase copies from the National Technical Information Service (703) 605-6000.

Pesticides

📖 ***Best Management Practices for Agrichemical Handling and Farm Equipment Maintenance.*** Florida Department of Agriculture and Consumer Services and Florida Department of Environmental Protection (May 1998). This 42-page booklet covers pesticides, fertilizers, solvents, and degreasers. Emphasis is placed on storage, mixing, loading, spill management, and disposal. Emergency reporting is also stressed. See: http://www.dep.state.fl.us/water/slerp/nonpoint_stormwater/documents/ pubinfo.htm#agriculturalpollutionprevention.

📖 ***National Integrated Pest Management Network.*** U.S. Department of Agriculture, Cooperative State Research, Education, and Extension Service (2000). The National Integrated Pest Management Network (NIPMN) is the result of a federal-state extension partnership dedicated to making the latest and most accurate pest management information available on the World Wide Web. Participating institutions have agreed to a set of standards which ensure science-based, unbiased pest management information. See: http://ipmworld.umn.edu.

FORESTRY BMP MANUALS

BMP Implementation and Effectiveness

📖 ***Evaluating the Effectiveness of Forestry BMPs in Meeting Water Quality Goals or Standards***. U.S. Department of Agriculture, Forest Service (Publication 1520) (July 1994). The purpose of this document is to help forest managers and their staff develop water quality monitoring plans to evaluate the effectiveness of forestry BMPs in meeting water quality goals or standards. It deals with the design of monitoring projects and the selection of variables and methods for monitoring them. Contact: U.S. Department of Agriculture, Forest Service, Southern Region, 1720 Peachtree Road NW, Atlanta, GA 30367.

📖 ***Techniques for Tracking, Evaluating, and Reporting the Implementation of Nonpoint Source Control Measures for Forestry***. U.S. Environmental Protection Agency (July 1997). This guidance is intended to assist state, regional, and local environmental professionals in tracking the implementation of BMPs used to control nonpoint source pollution generated by forestry practices. Information is provided on methods for sample site selection, sample size estimation, sampling, and result evaluation and presentation. The focus of the guidance is on the statistical approaches needed to properly collect and analyze data that are accurate and defensible. See: http://www.epa.gov/owow/nps/forestry/index.html.

📖 ***Wisconsin's Forestry BMPs for Water Quality: The 1997 BMP Monitoring Report.*** Wisconsin Department of Natural Resources, Bureau of Forestry (PUB-FR-145-99). This document describes the monitoring results from forestry BMP monitoring on timber sales. The monitoring was done to determine the extent to which BMPs were being applied throughout the state, the effectiveness of properly applied BMPs in protecting water quality, and the effects of not applying BMPs where needed. In addition to the findings, the report discusses the general framework and methods used to design the audit, as well as conclusions and recommendations. Phone requests for publication: (608) 267-7494.

BMP Manuals

📖 ***Georgia's Best Management Practices for Forestry.*** Georgia Forestry Commission (January 1999). The purpose of this manual is to inform landowners, foresters, timber buyers, loggers, site preparation and reforestation contractors, and others involved with silvicultural operations about common sense, economical, and effective practices to minimize soil erosion, stream sedimentation, and thermal pollution. See: http://www..forestry.uga.edu/efr/docs/bmp-contents.html.

📖 ***Montana BMPs for Forestry.*** Montana Department of State Lands, Missoula. Montana's water quality protection program for forestry involves a combination of regulatory and non-regulatory approaches. Since the i 970s, these non-regulatory Forestry Best Management Practices have provided guidance as minimum water quality protection standards for forestry operations. Phone requests for publication: (406) 542-4200. See: http://www.dnrc.state.mt.us/forestry/sfburea.htm.

📖 ***Montana Guide to the Streamside Management Zone Law and Rules***. Montana Department of State Lands, Missoula. This booklet explains and illustrates the SMZ law and rules as they apply to forest practices. Phone requests for publication: (406) 542-4200. See: http://www.dnrc.state.mt.us/forestry/sfburea.htm.

📖 ***Sustaining Minnesota Forest Resources: Voluntary Site-level Forest Management Guidelines for Landowners, Loggers, and Resource Managers.*** Minnesota Forest Resources Council (February 1999). This guidebook was developed as a collaborative statewide effort involving diverse forestry and water quality stakeholders. It provides a set of integrated guidelines that serve as a menu of options which address impacts on forest and water resources. See: http://www.frc.state.mn.us.

📖 ***Wisconsin's Forestry BMPs for Water Quality: A Field Manual for Loggers, Landowners, and Land Managers***. Wisconsin Department of Natural Resources, Bureau of Forestry (1997). This field manual lists over 119 BMPs for forestry activities including road building, timber harvesting, prescribed burning, and the application of chemicals. To request a publication call: (608) 267-7494.

Managing Private Non-industrial Forests

📖 ***Forest*A*Syst: A Self-assessment Guide for Managing Your Forest.*** The objective of this publication is to encourage owners of forests — large or small — to manage that forest for recreation, wildlife, and timber, while protecting water quality resources. It is a national model, intended for states to tailor to their own needs and purposes. See: www.forestasyst.net.

Regulations

📖 ***Regulation of Private Forestry Practices by State Governments.*** Minnesota Agricultural Experiment Station (Bulletin 605-1995)(1995). A comprehensive overview of state forestry programs, with background information on history and evolution of the legal system relating to forestry practices. The document also describes current program effectiveness and constraints, as well as emerging policy and management issues. Contact: Minnesota Agricultural Experiment Station, 1420 Eckles Avenue, St. Paul, MN 55108.

📖 ***Water Quality and BMPs for Loggers***. An Internet resource providing detailed state-by-state information on BMPs, public agencies, laws, ordinances, maps, related links, and training and education opportunities. See: hffn://www.forestry.uga.edu/bmp.

Riparian, Wetland and Bottomland Forests

📖 ***Chesapeake Bay Riparian Handbook: A Guide for Establishing and Maintaining Riparian Forest Buffers***. U.S. Department of Agriculture, Forest Service, Northeastern Area (NA-TP-02-97) (1997). The purpose of this handbook is to provide professional land managers and planners with the latest information on the functions, design, establishment, and management of riparian forest buffers. See: www.chesapeakebay.net/pubs/subcommittee/nsc/forest/handbook

Roads

📖 ***Fish Passage Through Culverts.*** U.S. Department of Transportation, Federal Highway Administration (FHWA-FL-90-006), and U.S. Forest Service, San Dimas Technology and Development Center (November 1990). This report is intended to review, summarize, and update current information on fish passage through culverts. It is geared primarily toward fish biologists, hydrologists, and engineers who will be designing projects that pass fish. Phone requests for publication: (909) 599-1267, ext. 246.

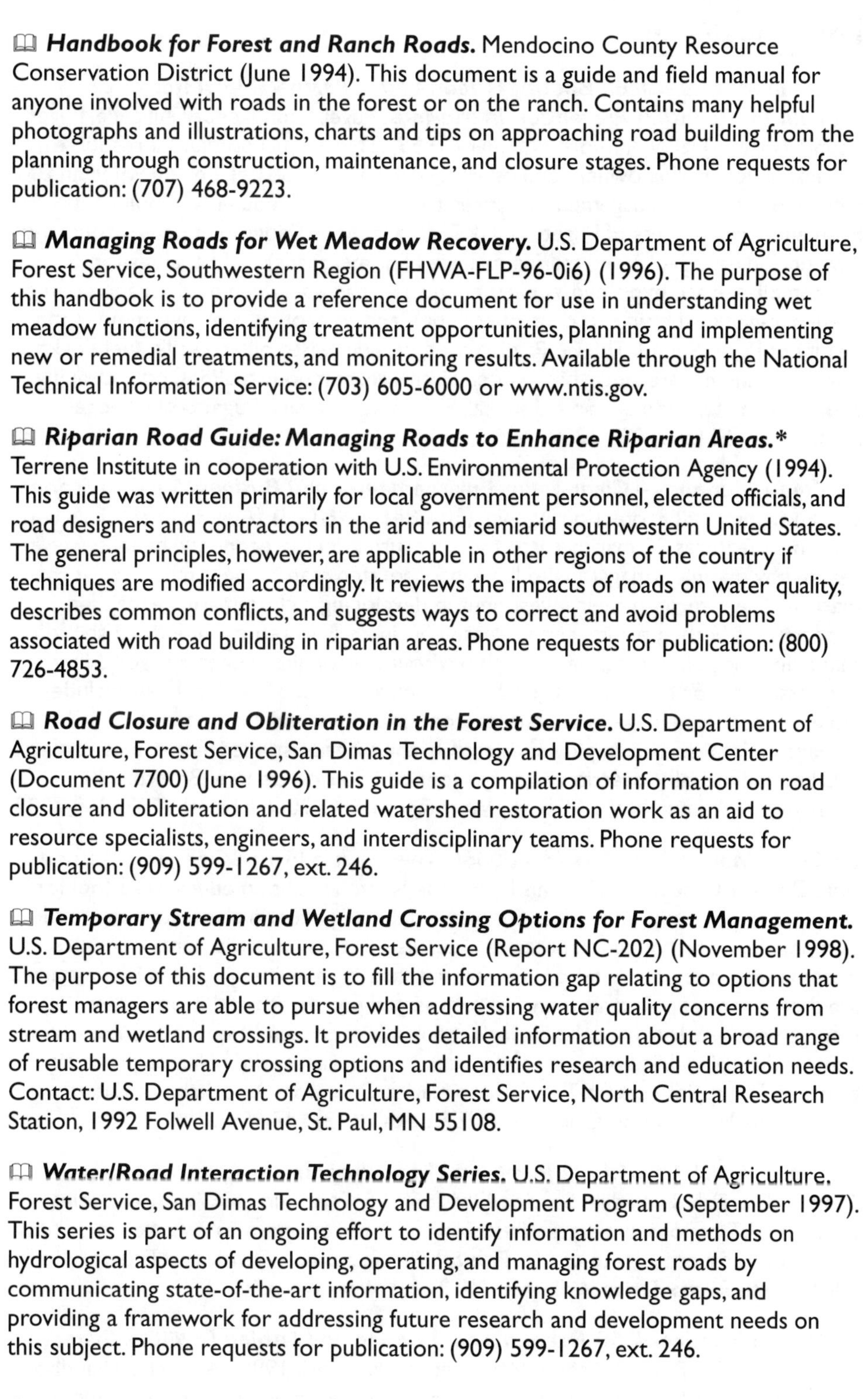

FORESTRY BMP MANUALS: Roads

Handbook for Forest and Ranch Roads. Mendocino County Resource Conservation District (June 1994). This document is a guide and field manual for anyone involved with roads in the forest or on the ranch. Contains many helpful photographs and illustrations, charts and tips on approaching road building from the planning through construction, maintenance, and closure stages. Phone requests for publication: (707) 468-9223.

Managing Roads for Wet Meadow Recovery. U.S. Department of Agriculture, Forest Service, Southwestern Region (FHWA-FLP-96-0i6) (1996). The purpose of this handbook is to provide a reference document for use in understanding wet meadow functions, identifying treatment opportunities, planning and implementing new or remedial treatments, and monitoring results. Available through the National Technical Information Service: (703) 605-6000 or www.ntis.gov.

Riparian Road Guide: Managing Roads to Enhance Riparian Areas.* Terrene Institute in cooperation with U.S. Environmental Protection Agency (1994). This guide was written primarily for local government personnel, elected officials, and road designers and contractors in the arid and semiarid southwestern United States. The general principles, however, are applicable in other regions of the country if techniques are modified accordingly. It reviews the impacts of roads on water quality, describes common conflicts, and suggests ways to correct and avoid problems associated with road building in riparian areas. Phone requests for publication: (800) 726-4853.

Road Closure and Obliteration in the Forest Service. U.S. Department of Agriculture, Forest Service, San Dimas Technology and Development Center (Document 7700) (June 1996). This guide is a compilation of information on road closure and obliteration and related watershed restoration work as an aid to resource specialists, engineers, and interdisciplinary teams. Phone requests for publication: (909) 599-1267, ext. 246.

Temporary Stream and Wetland Crossing Options for Forest Management. U.S. Department of Agriculture, Forest Service (Report NC-202) (November 1998). The purpose of this document is to fill the information gap relating to options that forest managers are able to pursue when addressing water quality concerns from stream and wetland crossings. It provides detailed information about a broad range of reusable temporary crossing options and identifies research and education needs. Contact: U.S. Department of Agriculture, Forest Service, North Central Research Station, 1992 Folwell Avenue, St. Paul, MN 55108.

Water/Road Interaction Technology Series. U.S. Department of Agriculture, Forest Service, San Dimas Technology and Development Program (September 1997). This series is part of an ongoing effort to identify information and methods on hydrological aspects of developing, operating, and managing forest roads by communicating state-of-the-art information, identifying knowledge gaps, and providing a framework for addressing future research and development needs on this subject. Phone requests for publication: (909) 599-1267, ext. 246.

Annotated Bibliography

Water Quality Effects and Nonpoint Source Control for Forestry: An Annotated Bibliography. EPA-841/B-93-005 (August 1993). Covers in-stream studies, roads, timber harvest, streamside management areas, wetlands, water quality monitoring, and modeling. Available through the National Service Center for Environmental Publications (NSCEP): (800) 490-9198.

MARINA BMP DOCUMENTS

Programs/Manuals

Clean Marina+Clean Boating+Clean Water Partnership: A Resource Manual for Pollution Prevention in Marinas. Puget Soundkeeper Alliance, Puget Sound Water Quality Authority (January 1995). This manual provides a resource workbook for marina owners and operators. It contains boater tips on when and where to use best management practices to prevent and reduce NPS pollutants from entry into waters of Puget Sound. Federal and Washington State environmental laws (including CZARA and marina management measures) and penalties for noncompliance are explained. A partnership approach between marinas and boaters is encouraged as the way to control pollution and to protect the environment and beauty of Puget Sound. The BMPs discussed include waste oil and spills, fuel dock operation and maintenance, bilge water discharge, pumpout facilities, boat cleaning and, solid and hazardous waste disposal. For copies contact: Puget Soundkeeper Alliance, or call (260) 286-1309.

Clean Marinas — Clear Value: Environmental and Business Success Stories. U.S. Environmental Protection Agency (EPA-841-R-96-003) (August 1996). This document features 25 marina case studies focusing on the economic benefits realized by marina managers who have incorporated management measures at their marinas. The return on their investment in developing a clean marina exceeded expectations both in profits and increased business. This report intends to show that managing polluted runoff through environmental enhancements is good for business, good for boating, and good for the environment. The document includes tables on cost/benefit, general benefits from environmental changes, and management measures employed at each of the marinas studied. For copies contact: U.S. Environmental Protection Agency, Nonpoint Source Control Branch, 1200 Pennsylvania Avenue N.W., (4503-F) Washington, D.C. 20460; or call (202) 260-7009.

Clean Marina Practices Handbook. The Ontario Marina Operators Association, Ontario, Canada (1997). This handbook is intended as an educational tool for owners and operators of marinas and yacht clubs on pollution prevention and reduction practices. The handbook presents a detailed discussion of measures to control water, air, and land pollution from recreational boating activities. Included in the manual are descriptions of pollutants found at marinas, pumpout needs, and stormwater management; excerpts of applicable legislation; and recommendations to promote clean marinas through public awareness and voluntary compliance with pollution prevention practices. For copies contact: The Ontario Marina Operators Association, 4 Cataraqui Street, Suite 211, Kingston, Ontario K7KIZ7; or call (613) 547-6662.

Clean Vessel Act of 1992 Pumpout Grant Program: American Success Stories. U.S. Fish & Wildlife Service, Division of Federal Aid (1997). Case studies from USFWS regions and states illustrating how the program has worked in its first five years. For a copy, write to U.S. Fish & Wildlife Service, Division of Federal Aid, 4401 North Fairfax Drive, Arlington, VA 22203; www.fws.gov.

Design Handbook for Recreational Boating and Fishing Facilities. States Organization for Boating Access, Washington, DC (April 1996). A practical handbook describing the techniques and best practices for shoreline protection, boat ramp design, location of restroom facilities, and operations and maintenance of waterside components. For copies contact: SOBA, P.O. Box 25655, Washington, DC 20007.

National Management Measures Guidance to Control NPS Pollution from Marinas and Recreational Boating. U.S. Environmental Protection Agency (EPA 841-D-01-001). The latest in EPA series of "management measures guidance"; e.g., Coastal guidance. Contact NSCEP: (800) 490-9198.

State Manuals

📖 ***Best Management Practices for Coastal Marinas.*** Connecticut Department of Environmental Protection (August 1992). This report overviews potential impacts from marina facilities, identifies common pollutants targeted for control, and describes operational BMPs for addressing potential impacts. The report includes selected state policies and mechanisms for implementing management practices. Six existing state marinas were surveyed to review a variety of harbor settings and functions to determine impacts attributable to pollutants found at these marinas. The document is generally based on the CZARA management measures guidance with similar marina and recreational boating BMPs. For copies contact: Connecticut DEP, 79 Elm Street, Hartford, CT 06106-5127.

📖 ***Best Management Practices for Marinas and Boatyards: Controlling Nonpoint Pollution in Maine.*** Maine Department of Environmental Protection (December 1995). This practical manual discusses common pollutants and their potential impacts in marina waters. Lists BMPS as guidance to manage runoff from stormwater, solid waste, fuel and hydrocarbons, hazardous materials, liquid waste, and sewage discharge from boats at both recreational and commercial marinas and boatyards. The manual includes CZARA and CWA requirements and descriptive material from EPA's management measures guidance, but stresses that it is not a regulatory document. Included in the manual are useful fact sheets, a model oil spill response plan, and an operations and maintenance plan based on the Rhode Island model. For copies contact: Maine DEP, Augusta, ME 04333; or call (207) 287-7688.

📖 ***Environmental Guide for Marinas: Controlling Nonpoint Source and Storm Water Pollution in Rhode Island.*** Rhode Island Sea Grant, University of Rhode Island (September 1996). This environmental guide provides information on pollutants and potential impacts at marinas, state operations and maintenance plan requirements, oil spill response plans, and BMP worksheets that marina operators can use to develop the required O&M Plan for controlling polluted runoff at marinas. The guide is designed to meet the state's Coastal Nonpoint Source Control Plan requirements for marinas and recreational boating under CZARA. For copies contact: Rhode Island Sea Grant, Communications Office, University of Rhode Island, Narragansett, RI 02882-1197

📖 ***Marina Pollution Prevention Manual.*** University of California Sea Grant Extension Program (1995). A useful manual in fact sheet format outlining BMPs for voluntary compliance by marina operators and boaters and calling attention to potential problems caused by improper operation and maintenance. Topics cover boat cleaning, engine and hull maintenance, oil and hazardous waste disposal, fuel handling, and marina staff training. For copies contact: California Sea Grant Program, University of California, La Jolla, CA 92093-0232; or call (858) 534-4440.

📖 ***Maryland Clean Marina Guidebook.*** Maryland Department of Natural Resources (1998). This Guidebook was developed in response to the CZARA program and the state's need to strengthen its nonpoint source pollution controls at marinas and recreational boating facilities. The material presented covers key CZARA management measures with supporting BMPs organized under environmental concerns, legal setting, applicable BMPs, and information sources. Also presented are state and federal laws and regulations as enforceable policies and mechanisms affecting marinas. For copies contact: Maryland DNR, Waterway Resources Division, Annapolis, MD 21401; or call (410) 260-8770.

NONPOINT SOURCE MONITORING

🕮 ***Monitoring Guidance for Determining the Effectiveness of Nonpoint Source Controls.*** U.S. Environmental Protection Agency, Office of Water (EPA 841-B-96-004) (1997). The manual gives an overview of nonpoint source pollution and covers the development of a monitoring plan, data analysis, quality assurance/quality control, and biological monitoring. Contact the National Service Center for Environmental Publications at (800) 490-9198.

🕮 ***National Handbook of Water Quality Monitoring. Part 600, National Water Quality Handbook (1996).*** Covers statistical design, variable selection, sampling location, sample analysis, and much more. Contact U.S. Department of Agriculture, Natural Resources Conservation Service for a copy. See: http://www.wcc.nrcs.usda.gov/water/quality/frame/wqam/Guidance_Documents/guidance_documents.

🕮 ***Rapid Bioassessment Protocols for Use in Wadeable Streams and Rivers: periphyton benthic macroinvertebrates and fish. 2nd ed.*** U.S. Environmental Protection Agency (1999). This document describes refined and revised methods for conducting cost-effective biological assessments of streams and small rivers. It focuses on periphyton, benthic macroinvertebrates, and fish assemblages, and on assessing the quality of the physical habitat. See: http://www.epa.gov/owow/monitoring/rbp/.

🕮 ***Techniques for Tracking, Evaluating, and Reporting the Implementation of Nonpoint Source Control Measures — Agriculture.*** U.S. Environmental Protection Agency (EPA 841-B-97-010) (1997). Focusing specifically on monitoring agricultural BMPs, this manual covers site selection, sample size estimation, sampling, and results evaluation and presentation. Contact the National Service Center for Environmental Publications (800) 490-9198; or see: http://www.epa.gov/owow/nps/agfinal.html.

🕮 ***Volunteer Stream Monitoring: a methods manual.*** U.S. Environmental Protection Agency (1997). This document covers the basic elements of stream monitoring, how to conduct a watershed survey, how to measure various water quality components, and how to manage and present monitoring data. See: http://www.epa.gov/owow/monitoring/volunteer/stream/.

URBAN NPS DOCUMENTS

Best General Web Addresses

💻 ***American Society of Civil Engineers (ASCE) National Stormwater BMP Database.*** Contains information on the effectiveness of urban BMPs in removing pollutants from urban runoff. Only studies which conform to established protocols are entered into the database. See: **www.bmpdatabase.org**.

💻 ***Center for Watershed Protection.*** Contains model ordinances, BMP effectiveness information, and other information on urban pollution and controls, with a concentration on watershed-based approaches. See: **www.cwp.org**.

💻 ***City of Fort Worth. Texas: Municipal and County Storm Water Programs.*** This web site contains hot links to municipal and county government stormwater programs around the United States. There is also a description of Fort Worth's NPDES Stormwater Phase I permit requirements, including construction and new development. See: **http://ci.fort-worth.tx.us/dem/stormcontacts.htm.**

URBAN NPS DOCUMENTS: Best General Web Addresses

Low Impact Development Institute Web Site. This web site contains general information on Low Impact Development BMPs and case studies as well as hot links to related organizations and projects. See: **www.lowimpactdevelopment.org**.

Nonpoint Education for Municipal Officials (NEMO). NEMO provides municipal officials with technical assistance and training in linking land use to water quality. See: **www.lib.uconn.edu/CANR/ces/nemo/**.

Terrene Institute. This organization has produced a number of very good general publications for use by the public as well as by watershed and local managers in addressing issues and problems concerning urban and other sources of nonpoint source pollution. See catalog on **www.terrene.org.** To order call (800) 726-4853 or e-mail terrinst@aol.com.

Citizens/Homeowners Guides

Baybook: A Guide to Reducing Water Pollution at Home.* Chesapeake Bay, Inc. and Alliance for the Chesapeake Bay (March 1993). A lay publication to create awareness and actions to reduce water pollution. Homeowners are the target audience for this document. Call: (410) 377-6270.

Bayscapes: Environmentally Sound Landscapes for the Chesapeake Bay. Alliance for the Chesapeake Bay (1994). A set of comprehensive fact sheets which contain detailed actions homeowners can take to implement environmentally friendly landscaping. Addresses pesticides, nutrients, erosion control, pesticides, and habitat diversity. Contains checklists for each fact sheet. Call: (804) 775-0951 or (717) 236-8825.

Clean Water in Your Watershed: A Citizens Guide to Watershed Protection.* Terrene Institute in cooperation with U.S. Environmental Protection Agency Region 6 (October 1993). An easy to understand and well illustrated guide to help citizens work with local, state, and federal government agencies to design and implement successful watershed protection and restoration projects. Step-by-step recommendations are provided. Call (800) 726-4853; e-mail terrinst@aol.com; see www.terrene.org..

HANDLE WITH CARE: Your Guide to Preventing Water Pollution.* Terrene Institute (1991). A simple but effective citizens' guide to problems due to rainfall and runoff in urban areas and what citizens and homeowners can do to reduce the harmful effects of runoff on water quality. Call (800) 726-4853; e-mail terrinst@aol.com; see www.terrene.org

Turning the Tide A Citizen's Guide to Reducing Nonpoint Source Pollution.* Harborwatch, Inc. and South Carolina Department of Health and Environmental Control. A concise brochure that describes urban nonpoint pollution and what actions citizens can take to reduce pollution in urban areas. Includes checklists. Call: (803) 734-5300.

Construction Controls

Storm Water Management for Construction Activities: Developing Pollution Prevention Plans and Best Management Practices. U.S. Environmental Protection Agency (EPA 833-R-92-001) (October 1992). EPA's guidance on how to prepare a stormwater pollution prevention plan for NPDES Storm Water Phase I construction activities. Includes erosion and sediment control BMPs and other control require- ments for construction sites, from site evaluation to final stabilization. Contact NSCEP: (800) 490-9198.

Funding

Catalog of Federal Funding Sources for Watershed Protection (2nd edition). U.S. Environmental Protection Agency (EPA 841-B-99-003) (December 1999). Provides a comprehensive summary of federal grant and loan programs that be used at the local level to support watershed projects. Also contains references to other publications as well as web sites on funding assistance. Contact NSCEP: (800) 490-9198.

Clean Water State Revolving Loan Web Site. A one-stop-shopping site for learning about the State Revolving Loan Fund (SRF). Includes information on eligibility, repayment, and restrictions. See: **www.epa.gov/OWM/finan.htm**.

A State and Local Government Guide to Environmental Program Funding Alternatives. U.S. Environmental Protection Agency (EPA 841-K-94-001) (January 1994). Provides an overview of traditional (nongovernment) funding mechanisms and innovative approaches to fund environmental programs. Contact NSCEP: (800) 490-9198.

General NPS Controls for Urban and Urbanizing Areas

Fundamentals of Urban Runoff Management: Technical and Institutional Issues*. R.R. Homer, J.J. Skupien, E.H. Livingston, and H.E. Shaver (August 1994). Terrene Institute in cooperation with U.S. Environmental Protection Agency. Part I contains clear and concise runoff and pollutant impact assessment and technical information on structural runoff controls in a logical sequence. Nonstructural alternatives are cleverly imbedded in Part II, which addresses institutional structures and frameworks which will help ensure implementation and continuance of control programs. Call (800) 726-4853.

Guidance Specifying Management Measures for Sources of Nonpoint Pollution in Coastal Waters. U.S. Environmental Protection Agency (EPA 840-B-92-002) (January 1993). Chapter 4 contains a description of water quality problems caused by urban nonpoint source. Documents pollution as well as management measures that represent performance expectations for urban controls to be implemented in states with approved coastal zone management programs. Management practices (referred to in other documents as "best management practices (BMPs)" are described that can be used to economically achieve the performance expectations. Contact NSCEP: (800) 490-9198; and see: http://www.epa.gov/owow/nps/MMGI.

Stormwater Strategies: Community Responses to Runoff Pollution.* Natural Resources Defense Council (May 1999). This study highlights some of the most effective and efficient watershed and municipal examples of nonpoint source and stormwater control programs and activities in the country. By example, communities can use these case studies in developing and implementing their own runoff control programs. See: www.nrdc.org.

Watershed Protection Techniques.* Center for Watershed Protection, Silver Spring, MD. A periodic bulletin on urban watershed restoration and protection tools including runoff management practices or BMPs. Contains often-cited technical notes that describe, compare, and evaluate urban controls as well as the effects of runoff both with and without controls. See: www.cwp.org.

Institutional/Program Development (Includes Utility Management Systems)

📖 ***Institutional Aspects of Urban Runoff Management: A Guide for Program Development and Implementation.**** E.H. Livingston, H.E. Shaver, R.R. Horner, and J.J. Skupien (May 1997). The Watershed Management Institute, Inc. in cooperation with U.S. Environmental Protection Agency. A comprehensive review of the institutional framework of successful urban runoff management programs at city, county, regional, and state levels of government. Recommendations are provided (based on surveys) that can help in all aspects of urban runoff program development and management. Contact WMI at: (850) 926-5310.

Monitoring

📖 ***Monitoring Guidance for Determining Effectiveness of Nonpoint Source Controls.*** U.S. Environmental Protection Agency (EPA 841-B-96-004) (September 1997). This guidance addresses design of monitoring programs to assess water quality to determine impacts of nonpoint sources and effectiveness of best management practices used as controls. Contact NSCEP: (800) 490-9198.

📖 ***Techniques for Tracking, Evaluating, and Reporting the Implementation of Nonpoint Source Control Measures — Urban Field Test Version.*** U.S. Environmental Protection Agency (EPA 841-B-937-01 1) (July 1998 – Update to Final in progress). Helps local officials focus limited resources by establishing statistical sampling to assess, inspect, or evaluate a representative set of BMPs, erosion and sediment controls, and on-site wastewater treatment systems. For more information, contact Rod Frederick at (202) 260-7054.

New Development Controls

📖 ***Better Site Design: A Handbook for Changing Development Rules in Your Community.**** The Center for Watershed Protection (August 1998). (See list for Site-Level Planning.) See: www.cwp.org.

📖 ***Caltran's Storm Water Quality Handbook.*** This handbook is intended to provide background information on Caltrans' (California Department of Transportation) program to control water pollution and to standardize the process for preparing and implementing the Water Pollution Control Program (WPCP) and Storm Water Pollution Prevention Program (SWPPP). Caltrans requires contractors to prepare and implement a program to control water pollution during the construction of all projects. See: http://www.dot.ca.gov/hq/construc/stormwater.html.

📖 ***Economic Benefits of Runoff Controls.*** U.S. Environmental Protection Agency (EPA 841-S-95- 002) (September 1995). This document contains a description of studies that document increases in property values and rental prices when properly designed runoff control facilities are used as visual amenities. Contact NSCEP: (800) 490-9198.

📖 ***Environmental Land Planning Series: Site Planning for Urban Stream Protection.*** T. Schueler (December 1995). Prepared by the Center for Watershed Protection for the Metropolitan Washington Council of Governments, Washington, DC. (See list for Site-Level Planning.) See: www.cwp.org

URBAN NPS DOCUMENTS: New Development Controls

🕮 ***Stormwater BMP Design Supplement for Cold Climates***.* Center for Watershed Protection in cooperation with U.S. Environmental Protection Agency (December 1997). This manual addresses some of the unique challenges of design and maintenance of runoff BMPs in cold climates and recommends strategies to make BMPs in these regions more effective. See: www.cwp.org.

🕮 ***Texas Nonpoint Source Book — BMPs.*** Contains stormwater pollution management information for public works professionals and other interested parties in the State of Texas. Also contains information on watershed assessment and design and implementation of nonstructural and structural BMPs. See: http://www.txnpsbook.org.

🕮 ***Washington State Department of Ecology Website.*** This website contains information on Washington State's Nonpoint Source program and NPDES stormwater programs. Refer to the publications list. The draft BMP manual for Western Washington State is worth reviewing to learn about Washington's approach to protecting the hydrological regime of streams. Also noteworthy are "Planning as Process: A Community Guide to Watershed Planning" (Ecology Publication #99-01-WQ); and "Watershed Urbanization and the Decline of Salmon in Puget Sound Streams" by Dr. Chris May. The Community Guide is a useful reference for community involvement in the planning process. The paper by Chris May draws a link between urbanization and salmonid impacts. See: http://www.wa.gov/ecology/biblio/9911.html.

🕮 ***2000 Maryland Storm Water Design Manual. Volumes I and II.*** Volume I contains information on BMP siting and design on new development sites to comply with the state's 14 stormwater performance standards. Stormwater credits for innovative design are a significant addition. Volume 2 contains detailed technical information including step-by-step design examples. See: www.mde.state.md.us.

Notable State Manuals and Guidance Manuals

🕮 ***Erosion and Sediment Control Planning and Design Manual. Field Manual and Inspector's Guide***.* North Carolina Department of Environment, Health, and Natural Resources, Division of Land Resources. Addresses planning design, implementation, and inspection of erosion and sediment control BMPs. Call: (919) 733-4574.

🕮 ***Sediment and Stormwater Management Certified Construction Reviewer Course and Associated Delaware State and DOT Standards/Specifications.**** State of Delaware, Delaware Department of Natural Resources and Environmental Control, Division of Soil and Water Conservation. Contains descriptions of Delaware's erosion and sediment control and runoff control BMPs as well as their certification requirements for contractors. Call: (302) 739-4411.

🕮 ***Tennessee On-line BMP manual. City of Knoxville*** *(under development)*. This draft on-line erosion and sediment control manual provides information on the design, inspection, and maintenance of structural and nonstructural BMPs that are used in Knoxville. The manual is similar to stormwater guidance prepared for the California Stormwater Quality Task Force and Caltrans. See: www.ci.knoxville.tn.us/reports/bmp manual/index.htm.

Operation and Maintenance

📖 ***Operation Maintenance and Management of Stormwater Management Systems.**** E.H. Livingston, H.E. Shaver, J.J. Skupien, and R.R. Horner (August 1997). Watershed Management Institute in cooperation with U.S. Environmental Protection Agency. Note: Includes Stormwater Management Inspection Forms as a separate supplement. The manual contains a comprehensive review of the technical, educational, and institutional elements needed to assure that stormwater management systems are designed, built, maintained, and operated properly during and after construction. Fact sheets on 13 commonly used BMPs are included. Call: (850) 926-5310.

Ordinances

📖 ***Model Ordinances to Protect Local Resources***. U.S. Environmental Protection Agency (November 1999). EPA has developed a web site that contains printable and create-your-own ordinances as well as links to other web sites. Ordinances include aquatic buffers, erosion and sediment control, open space development, stormwater control operation and maintenance, illicit discharges, post-construction runoff, source water protection, and miscellaneous ordinances (golf courses, etc.). See: http://www.epa.gov/owow/nps/ordinance/.

Restoration

📖 ***Izaak Walton League Save Our Streams Program*** (January, 1995). Includes ***A Citizen's Streambank Restoration Handbook,*** a video, and personal assistance.* A primer to help citizens, government planners, and decisionmakers understand channelization and streambank restoration techniques. Includes case studies, an annotated bibliography, and restoration contacts. Call: (800) BUG-IWLA.

📖 ***Restoring Streams in Cities: A Guide for Planners, Policymakers, and Citizens***. Ann Riley (1998). This book contains a logical sequence of land-use planning, site design, and watershed restoration measures along with stream channel modifications and floodproofing strategies that can be used in place of destructive and expensive public works projects. Contact Island Press' distribution center at P.O. Box 7, Covelo, CA 95428, phone: (800) 828-1302, fax: (707) 983-6414, ipwest@islandpress.org.

📖 ***Stream Corridor Restoration Principles, Processes, and Practices.*** Federal Interagency Stream Restoration Working Group (EPA 841-R-98-900) (October 1998). A cooperative effort of 17 federal agencies resulted in this compendium of stream corridor restoration expert advice and field-tested methods. A CD-ROM is also available. Contact NSCEP: (800) 490-9198.

📼 **Urban Stream Restoration: A Video Tour of Ecological Restoration Techniques.** * Includes information on six urban stream restoration sites with detailed instructions and graphic illustrations. Includes bioengineering stablilization techniques, recreating channel shapes and meanders, daylighting of buried creeks, and vegetated flood controls. Led by Ann Riley, Executive Director of the Waterways Restoration Institute in Berkeley, CA. See: www.noltemedia.com.

Site – Level Planning

🕮 ***Better Site Design: A Handbook for Changing Development Rules in Your Community.*** The Center for Watershed Protection (August 1998). This handbook was prepared for local planners, engineers, developers, and officials to help them understand development principles that can be used to create environmentally sensitive, economically viable, and locally appropriate development. See: www.cwp.org.

🕮 ***Conservation Design for Stormwater Management.**** The Environmental Management Center of the Brandywine Conservancy, Delaware Department of Natural Resources and Environmental Control (September 1997). This document describes methodologies for configuring new developments to increase water quality protection and pervious surfaces by reducing lot sizes and thereby increasing green space and common use areas. Call: (302) 739-4411.

🕮 ***Environmental Land Planning Series: Site Planning for Urban Stream Protection.*** T. Schueler (December 1995). Prepared by the Center for Watershed Protection for the Metropolitan Washington Council of Governments, Washington, DC. This series is for all audiences, presenting a clear and understandable description of the significance of imperviousness in a watershed. Also describes planning strategies to protect urban streams by reducing imperviousness and increasing green space. See www.cwp.org.

🕮 ***Low-Impact Development Design Strategies.*** Prince Georges County, Maryland (EPA 841-B-00-003) (January 2000). ***Low-Impact Development Hydrologic Analysis.*** Prince Georges County, Maryland (EPA 841-B-00-002) (January 2000). These two documents describe LID principles, programmatic considerations, and design strategies, and give an example of an analytic and computational procedure to use in designing appropriate runoff treatment systems. The strategies document (003) was prepared for local planners, engineers, developers, and officials to describe how to develop and implement LID methods from an integrated design perspective. The hydrologic analysis document (002) is a companion technical document that contains a methodology that can be used to estimate changes in site hydrology due to new development and also to design appropriate treatment systems to maintain the predevelopment hydrology of the site. Contact NSCEP: (800) 490-9198.

Watershed Planning

🕮 ***Rapid Watershed Planning Handbook.*** The Center for Watershed Protection (October, 1998). This handbook was written to assist watershed associations and local governments in developing effective and low-cost watershed protection plans. Eight steps are described in detail including how to identify and classify subwatersheds, protect and restore water resources, and evaluate progress. The document emphasizes resource identification, evaluation, and planning. See: www.cwp.org.

🕮 ***Technical Assistance and Team Training in Linking Land Use to Water Quality.*** Nonpoint Education for Municipal Officials (NEMO). NEMO provides municipal officials with technical assistance and training in linking land use to water quality. See: www.lib.uconn.edu/CANR/ces/nemo/

URBAN NPS DOCUMENTS: Watershed Planning

📖 ***Urbanization and Streams: Studies of Hydrologic Impacts***. U.S. Environmental Protection Agency (EPA-R-97-009) (December 1997). This report includes references and case studies that document the impacts of urbanization on water quality, habitat, and aquatic biota. Contact NSCEP.

📖 ***A Watershed Approach to Urban Runoff: Handbook for Decisionmakers***.* Terrene Institute in cooperation with U.S. Environmental Protection Agency Region 5, March 1996. An informative primer for local decisionmakers and watershed organizations on assessing the water quality of watersheds, identifying contributing sources, and prioritizing watershed resources to implement effective nonstructural and structural BMPs. Summarizes BMPs and lists resources to obtain additional information. Call (800) 726-4853; e-mail terrinst@aol.com; or see www.terrene.org.

CHAPTER 7

Management Techniques Within the Lake or Reservoir

Introduction

Like medical prescriptions, lake and reservoir management techniques have benefits, side effects, and limitations. None is suitable for every lake, for all problems, or even for a specific problem under varying circumstances. With that warning delivered, what can the reader expect to gain from this chapter?

- Insight into the ecological and management principles involved in crafting a management plan;
- Understanding of the range of available lake and reservoir restoration and management methods;
- An appreciation of the many factors involved in choosing the most appropriate technique(s); and
- Knowledge of the ecological basis for specific methods, including their applicability to specific problems, how they work, their advantages and disadvantages, and approximate costs.

Principles of Management

Restoration vs. Management vs. Protection

Although the terms restoration, management, and protection all have different meanings, they are often used interchangeably in dealing with lakes. The distinctions and overlap among them are important, however, and should be emphasized when planning a program to address lake problems.

Lake **restoration** is the use of ecologically sound principles (Chapter 2) to attempt to return a lake or reservoir to as close to its original condition as possible. Restoration suggests that some previously existing condition is to be regained. Sometimes the lake can be made even better than the original condition, which is less restoration than management.

Management is defined as improving the lake or reservoir to enhance stated uses, such as water supply, swimming, fishing, or wildlife habitat. Management may or may not involve restoration of past conditions, and might better be described as rehabilitation (Cooke, 1999).

Restoration: Use of ecologically sound principles to attempt to return a lake or reservoir as close to its original condition as possiblo.

Management: Improving the lake or reservoir to enhance stated uses, such as water supply, swimming, fishing, or wildlife habitat.

Protection: The prevention of adverse impacts.

Once restored or managed to the desired condition, a lake will usually require continued management to stay in that condition. This may involve repeated application of the technique that achieved the desired lake condition in the first place, or using different techniques to prevent undesirable changes. Here management grades into **protection**, defined as the prevention of adverse impacts (see Chapter 9).

We often think of protecting pristine lakes, but damaged lakes can also be protected by developing regulations that prohibit practices believed to threaten the lake. Hand pulling non-native plants can also be protective, just like pulling weeds from a garden. While removing vegetation can be a management technique, in this case it is protective: preventing the spread and dominance of an undesired species.

The difference between restoration, management, and protection is therefore a function of past and present lake conditions (Chapter 4) set in the context of goals for the lake's desired condition (Chapter 3). Although applicable techniques are basically the same, restoration and protection differ according to specific objectives. Management is the approach most used on U.S. lakes.

Selection of Goals

Successful management must be guided by a clear statement of goals and priorities (refer to Chapter 3). What are the intended uses of the lake; are any more important than others? Optimal conditions for swimming, boating, fishing, and water supply rarely exist in the same lake, although all may be accommodated to some degree. Balancing conditions and uses may be an appropriate goal, but not all lake management uses are completely compatible (Wagner and Oglesby, 1984), and goals must be set accordingly.

The choice of lake management techniques depends on water uses and management goals. Some techniques enhance some uses but not others. For example:

- Many herbicides cannot be applied to drinking water supplies, even though they control nuisance vegetation.
- Aeration that destratifies a lake may harm coldwater fisheries, even though oxygen concentrations increase lakewide and algae decline.

Definition of Problems

Normally, management goals are based on supporting specific water uses. This typically involves combating a variety of problems (see Chapter 4 for a full discussion of defining problems). All 11 of the general issues listed in Table 7-1 are common, but this chapter will emphasize controlling nuisance algae and rooted plants, the two most common management objectives.

▼ ***Algal blooms or rooted plant infestations:*** Many complex processes interact to cause these problems. To effectively control algae, you must understand the quality of water entering a lake and the cycling of nutrients within the lake. Rooted plants flourish in many lakes without significant inputs from the watershed, just because of the nature of the sediments present when the lake was formed.

Some of the other problems referred to in this chapter can be addressed by the same techniques applied to algae and vascular plant control.

Table 7-1.—Lake and reservoir problem definition.

GENERAL ISSUE	POSSIBLE SPECIFIC PROBLEMS
1. Nuisance algae	Loss of water clarity, taste/odor, algal toxicity, oxygen fluctuations and depression, disinfection byproduct formation upon chlorination, habitat impairment, human health risks
2. Nuisance vascular plants	Dense rooted or floating growths, dominance by aggressive introduced species, organic sediment accumulation, oxygen fluctuations and depression, habitat alteration
3. Sediment buildup	Loss of depth, undesirable sediment quality, sediment-water interactions (link to nutrient loading), habitat impairment
4. Non-algal color and turbidity	Observable color from humic substances or other agents, high levels of particulates in suspension, loss of water clarity, aesthetic impairment
5. Anoxia and related issues	Lack of oxygen, buildup of ammonia, hydrogen sulfide, carbonaceous gases, iron, manganese and phosphorus through anaerobic reactions, habitat impairment (usually a product of above problems)
6. Acidification	Reduced pH, fluctuating pH, pH-mediated water quality changes, habitat impairment
7. Toxic substances	Excessive levels of metals, pesticides, or organic metabolites, habitat impairment, human health risks
8. Pathogens	Excessive levels of bacteria, viruses, or other pathogens, aquatic fauna and human health risk
9. Undesirable fisheries	Small population size, undesirable size distribution, predator-prey imbalances, invasive or disruptive species, poor fish condition (e.g., weight vs. length, diseases), resuspension of bottom sediments
10. Animal nuisances	Excessive numbers of mosquitos, midges, other insects, ducks, geese, other waterfowl, leeches, zebra mussels, other undesirable fauna, swimmer's itch, human health risks
11. User conflicts	Interference among uses and users, including human vs. non-human uses, passive vs. active human uses, overcrowding, noise, and water-level conflicts

- ***Excessive sedimentation*** usually comes from either erosion in the watershed or organic matter (algae and vascular plants) produced in the lake. Once sediment accumulates, it may interact with the water column and accelerate eutrophication.

- ***Nonalgal color and turbidity*** can be used to control algae and rooted plants, but high levels of either generally have undesirable effects on water supply and recreational uses, and may also impair wildlife habitat.

- ***Anoxia*** is the absence of oxygen. Decaying algae, vascular plants, and other organic matter may use more oxygen than the atmosphere can supply, especially in bottom waters far removed from the atmosphere. Anoxic conditions often release phosphorus, iron, manganese, and sulfides from sediments; this may eventually establish a self-supporting cycle of excess production and decay. At that point, watershed inputs may no longer control in-lake conditions.

Anoxia: The absence of oxygen.

- ***Problems with acidification, toxic substances, and pathogens*** are largely watershed issues, and should be dealt with on that level (see Chapter 6). However, if controlling the source and trapping pollutants in the watershed do not succeed, techniques may be applied within the lake or reservoir to minimize their impact.

- ***Poor fishing*** may be rooted in several causes, ranging from habitat impairment within the lake to overfishing and poor stocking practices. Managing for an optimal fishery involves many choices and requires a thorough understanding of the physical, chemical, and biological features of the lake — plus the ability to deal effectively with people.

- ***Nuisance fauna*** such as geese and zebra mussels may threaten human health, annoy lake users, or disrupt ecological conditions for wildlife. Users and agencies may differ about when to declare nuisance conditions, but when they decide to address this problem, they must take care not to create a greater hazard. Nuisance fauna and fishing problems involve a human element that can cause user conflicts.

- ***User conflicts*** may arise over desired lake conditions, but more often, conflicts arise over access to the lake, allocation of space, or mutually exclusive uses occurring at the same time.
  - The conflict of motorized (especially personal) watercraft with other water-based recreation is reaching the same status as excessive algae or rooted plants in surveys of lake management needs.
  - The need for water in developing areas (for residential and other uses) is also competing more intensely with the preservation of aquatic habitat.

The techniques for dealing with these problems are as much social science as aquatic science, but lake managers frequently have to make judgments that affect how people use the lake, so this issue is most pertinent to lake management.

Understanding Existing and Potential Future Conditions

Not all lakes are created equal. Differences in origin, climate, and watershed features predispose lakes to certain conditions. When planning a lake management program and selecting management techniques, it is essential to completely understand existing conditions and limits to improvement.

Chapter 2 provides insight into the ecological principles upon which lake management is based, while Chapter 5 describes tools for predicting results. Chapter 4 addresses the information needed for characterizing problems and evaluating causes and effects — a key step between understanding lake ecology and making specific predictions.

While sustained management can accomplish major feats, it has both theoretical and practical limits. For a variety of reasons, some lakes cannot be adequately protected, or users' expectations are not realistic.

- A survey of Lake Champlain users (New York and Vermont) revealed that they expect clearer water than has ever been recorded for that lake (Garrison and Smeltzer, 1987).
- In the Upper Midwest, people expect much clearer water for more northern lakes (Heiskary, 1989), thus affecting management goals.

The origin of a lake must be considered in managing it. Lakes are created by natural or human forces, or by both combined. People-built lakes range from shallow ponds and impoundments predisposed to infestation by rooted plants to large, deep reservoirs that detain river water long enough to foster algal blooms when enough nutrients are present. Many impoundments were never deep, clear, infertile lakes, and must be managed just to maintain them, as both natural and cultural forces cause them to fill in.

Morphometry: Measurement of the physical shape and depth of a lake.

The depth of a lake, its morphometry, its division into multiple layers, and how often these layers mix are all extremely important to a lake's ecology, and often dictate the selection of management techniques. Although shallow (< 15 ft) lakes can stratify if light does not penetrate very far, usually a lake must be at least 20 feet deep before stratification becomes strong and prolonged. Lake depth influences the area that can support rooted plants, the internal cycling of phosphorus, flushing characteristics, and the nature of the fish community.

Some lakes are naturally fertile, largely because of the geology and soils around them (Rohm et al. 1995). Some ecoregions have richer, more erodible soils, and higher annual precipitation; these factors increase sediment and nutrients in those lakes, even without considering human uses of the land.

High fertility isn't always undesirable; management for fish production requires a substantial food base. Some lakes and reservoirs are so infertile they may not support many fish. Management might include adding nutrients to stimulate algae growth and the ensuing development of game fish populations, if that is a primary goal and regulatory agencies will permit it.

Ecosystem energetics: The flow of energy through trophic levels, including inputs to lakes from the watershed (such as leaves or insects) and the transformation of that energy within the lake.

Again, goals and priorities must be clearly stated if management techniques are to be selected correctly. Targeting balanced conditions and uses is usually the best course of action.

To understand and solve many lake problems, we must move beyond nutrients into ecosystem energetics (Kortmann and Rich, 1994). The flow of nutrients is but one aspect of energy and its transformations. Energetics plays a major role in nitrogen transformations and recycling of phosphorus, which in turn influence the amount and types of algae present. The position of the lake along a gradient from heterotrophy (dependence on the addition of organic matter) to autotrophy (dependence upon internally generated organic matter) significantly influences how the lake will react to management actions.

With an understanding of current and potential lake conditions, lake managers can evaluate management goals in terms of ecology, economics, and ethics.

- Is a proposed option consistent with the ecological function of the aquatic system?
- Is the proposed approach affordable, for both the short and long term?
- Does an action treat all lake users (human and otherwise) fairly?
- Can it be considered ethical from all rational viewpoints?

Developing a Watershed Management Program

Watershed management is the subject of Chapter 6; from a long-term perspective, the condition of a water body is dictated primarily by the quality and quantity of water entering it. Although biological interactions, sediment release of nutrients, and basin morphometry may all affect the lake, it is clear that nearly all attempts at lake management will be overwhelmed by continued high loading of silt, organic matter, and nutrients from the watershed. Watershed management is therefore a prerequisite for lake management or protection.

But watershed management may not be the sole answer. Once rooted plants infest a lake, no amount of watershed management is likely to eliminate them. Even if the flow of phosphorus to a lake is drastically reduced, internal recycling may maintain excessive algal production for many years.

In-lake techniques then become the next step in supporting lake uses, even if on a maintenance basis. An in-lake management program is meant to complement watershed management, however, not replace it.

In-lake Management: Matching Options to Problems

Lake management techniques can be grouped several ways. Table 7-2 simply lists them alphabetically, but alternative listings (described further in this chapter) include characterization as a physical, chemical, or biological technique, and grouping by problems to which they can be applied.

The list of available options in Table 7-2 is not very long (although subcategories would multiply them fivefold). Do not be deceived by the brevity of the list, however; properly applying these techniques will provide enough power to address almost all known lake problems.

Seldom can a single technique handle all problems faced by a single lake, and in many cases more than one technique must be used to successfully address even a single problem such as excessive algae or rooted plant growths.

Seldom can a single technique handle all problems faced by a lake, and in many cases more than one technique must be used to successfully address even a single problem such as excessive algae or rooted plant growths. These combinations and the scale and intensity of their application offer a wide range of possible management programs that can be tailored to a lake's unique properties.

Perhaps the greatest value of using professional help with lake problems is the professional's expertise in evaluating if a technique is appropriate, given the nature of the problem and the known (and potential) conditions and constraints. Many factors must be considered (Table 7-3); failure to address them can affect the success and/or cost of the management program.

Questions to ask when evaluating each technique in relation to a given problem:

- How effective is this technique likely to be?
- How rapidly will it achieve results and how long will those results last?
- What desirable or undesirable side effects might be expected?
- How much will it cost over the duration of the management period?
- Will a balance of appropriate uses be achieved by the proposed action(s)?

Success is a matter of matching the right technique(s) to the problem; this requires background information best obtained from a thorough diagnostic/feasibility study. It's always desirable to rapidly achieve the targeted results, but their longevity is usually more important. Longevity plays a large role in determining long-term costs, and since very few justifiable management programs are developed with only the short term in mind, this is an important consideration.

Table 7-2.—General management options within lakes and reservoirs.

TECHNIQUE	DESCRIPTIVE NOTES
1. Aeration or oxygen addition	Mechanical maintenance of oxygen levels.
2. Artificial or augmented circulation	Water movement to enhance mixing and/or prevent stratification.
3. Biocidal chemical treatment	Addition of inhibitory substances intended to eliminate target species.
4. Biomanipulation	Facilitation of biological interactions to alter ecosystem processes.
5. Bottom sealing	Physical obstruction of rooted plant growths and/or sediment-water interaction.
6. Chemical sediment treatment	Addition of compounds that alter sediment features to limit plant growths or control chemical exchange reactions.
7. Dilution and/or flushing	Increased flow to dilute or minimize retention of undesirable materials.
8. Dredging	Removal of sediments under wet or dry conditions.
9. Dye addition	Introduction of suspended pigments to create light inhibition of plant growth.
10. Hydroraking or rotovation	Disturbance of sediments, often with removal of rooted plants, to disrupt growth.
11. Harvesting, pulling, or cutting	Reduction of plant growths by mechanical means, with or without removal from the lake.
12. Nutrient inactivation	Chemical complexing and usually precipitation of nutrients, normally phosphorus.
13. Nutrient supplementation	Addition of nutrients to enhance productivity or alter nutrient ratios to affect algal composition.
14. Other chemical treatments	Addition of chemicals to adjust pH, oxidize compounds, flocculate and settle solids, or affect chemical habitat features.
15. Partitioning for pollutant capture	Creation of in-lake areas, such as forebays and created wetlands, to capture incoming pollutants.
16. Rules and regulations	Restrictions on human actions directed at minimizing impacts on lakes and lake users.
17. Selective withdrawal	Removal of targeted waters for discharge (undesirable features such as high nutrients) or intake (desirable features such as low algae).
18. Water level control	Drying or flooding of target areas to aid or eliminate target species.

Table 7-3.—Factors in selecting management techniques.

TECHNIQUE	APPLICABLE USES OF THE TECHNIQUE	PRIMARY INFORMATION NEEDS	KEY WATER ISSUES	KEY SEDIMENT ISSUES	KEY BIOLOGICAL ISSUES	KEY USE ISSUES	KEY COST FACTORS
1. Aeration or oxygen addition	◆ Algal control by reduced phosphorus release from sediment	◆ Hypolimnetic oxygen demand	◆ Restrictions on destratification	◆ Longevity of effects	◆ Potential for enhanced habitat	◆ Potential for interference with recreation	◆ Equipment and oxygen costs
	◆ Lowered dissolved Mn and Fe levels	◆ Sediment oxygen demand	◆ Longevity of effects	◆ Potential for resuspension	◆ Potential for gas bubble disease in which deadly gas bubbles form in a fish's blood stream		◆ Power supply
	◆ Improved fish habitat	◆ Relative resistance to thermal mixing			◆ Potential for thermal impacts through mixing		◆ Installation costs
	◆ Creation of deeper zooplankton refugia (places where predation is limited by light or temperature)	◆ Area and depth of epilimnion and hypolimnion					◆ Long-term operational costs
		◆ Annual phosphorus release rate for sediment					◆ Monitoring program
		◆ Concentrations of forms of P, Fe, Mn					
2. Artificial or augmented circulation	◆ Avoidance of stagnation	◆ Area and volume of water to be moved	◆ Possible water quality effects of mixing	◆ Potential for resuspension	◆ Potential to disseminate undesirable species	◆ Potential for interference with recreation	◆ Equipment costs
	◆ Destratification	◆ Rate of desired movement/oxygen demand			◆ Potential to disrupt life cycles of desirable species	◆ Effects on water supply quality	◆ Power supply
	◆ Increased mixing	◆ Relative resistance to thermal mixing					◆ Installation costs
	◆ Enhanced surface aeration	◆ Physical impediments to mixing					◆ Long-term operational costs
	◆ Disruption of algal mats and scums						◆ Monitoring program

Table 7-3.—Factors in selecting management techniques (continued).

TECHNIQUE	APPLICABLE USES OF THE TECHNIQUE	PRIMARY INFORMATION NEEDS	KEY WATER ISSUES	KEY SEDIMENT ISSUES	KEY BIOLOGICAL ISSUES	KEY USE ISSUES	KEY COST FACTORS
3. Biocidal Chemical Treatment	◆Algal control	◆Pond bathymetry and volume	◆Persistence of chemical and degradation products	◆Accumulation of contaminants	◆Anticipated impacts to target and non- target plant species	◆Downstream flow restrictions	◆Cost of chemical
	◆Vascular plant control	◆Flushing rate (by month or season)	◆Effects on oxygen levels	◆Accumulation of organic matter	◆Potential migration of chemical into hydraulically con-nected wetland areas	◆Use restrictions following treatment (severity and duration)	◆Application method
	◆Insect pest control	◆Choice of chemical	◆Potential for nutrient releases		◆Association of fauna with area to be treated	◆Alternative water supplies	◆Application labor
	◆Fish reclamation	◆Form of chemical to be used (pellets vs. liquid)	◆Other indirect water quality impacts		◆Presence of protected species	◆Public perception of risk of exposure to chemical	◆Monitoring program
		◆Chemical concentra-tion needed			◆Distance downstream at which chemical can be detected	◆Potential for much dead and decaying vegetation	
		◆Amount of chemical to be used			◆Possible impacts to downstream biotic assemblages		
		◆Duration of exposure to chemical needed					
		◆Timing and frequency of treatments					
		◆Outlet control features					
4. Bio-manipulation	◆Algal control	◆Biological inventory of lake	◆Possible effects of introduced species on water quality	◆Interactions of introduced species and sediments	◆Uncertainty of biological interactions	◆Interaction of introduced species with lake users	◆Interaction of introduced species with lake users
	◆Vascular plant control	◆Habitat evaluation for existing and proposed conditions			◆Longevity of effects		◆Cost of removal of individuals or species
	◆Fish control	◆Necessary stocking/planting or removal rate			◆Migration to unintended habitats/lakes		◆Cost of habitat alteration
	◆Nuisance fauna abatement	◆Ability to control inlets and outlets					◆Monitoring program
	◆Habitat enhancement						

Table 7-3.—Factors in selecting management techniques (continued).

TECHNIQUE	APPLICABLE USES OF THE TECHNIQUE	PRIMARY INFORMATION NEEDS	KEY WATER ISSUES	KEY SEDIMENT ISSUES	KEY BIOLOGICAL ISSUES	KEY USE ISSUES	KEY COST FACTORS
5. Bottom sealing	◆Reduced sediment-water interactions ◆Algal control ◆Rooted plant control ◆Improved recreational appeal	◆Physical sediment features ◆Chemical sediment features ◆Area to be treated ◆Water depth ◆Material to be used for sealing	◆Interactions of treatment with water column	◆Longevity of effects	◆Impacts on benthic organisms ◆Facilitation of colonization by new organisms ◆Impacts on biota of water column	◆Use restrictions to protect barrier ◆Safety concerns for contact recreation in barrier area	◆Cost of materials ◆Cost of application ◆Cost of maintenance ◆Monitoring program
6. Chemical sediment treatment	◆Reduced sediment-water interactions ◆Reduced phosphorus release from sediment ◆Algal control	◆Physical sediment features ◆Chemical sediment features ◆Area to be treated ◆Water depth ◆Chemical to be applied/dose	◆Interactions of treatment with water column	◆Longevity of effects	◆Impacts on benthic organisms ◆Facilitation of colonization by new organisms ◆Potential impacts on biota of water column	◆Water use restrictions during treatment ◆Use restrictions to minimize resuspension/burial of treated sediment	◆Cost of chemicals ◆Cost of application ◆Monitoring program
7. Dilution and/or flushing	◆Reduced detention time ◆Pollutant concentration reduction ◆Algal control	◆Sources of water ◆Quality of source water ◆Target concentration or detention time ◆Flow necessary to achieve dilution or flushing rate	◆Variability in water supply quantity or quality ◆Downstream flow restrictions	◆Potential for resuspension	◆Possible washout of zooplankton and fish larvae ◆Reduced fish production with lower nutrient base ◆Importance of attached algae in lake	◆Safety concerns associated with increased inflow/outflow ◆Possible use impairment with poor quality flushing source water	◆Cost of source water ◆Cost of any piping, control structures, and/or pumping ◆Monitoring program

Table 7-3.—Factors in selecting management techniques (continued).

TECHNIQUE	APPLICABLE USES OF THE TECHNIQUE	PRIMARY INFORMATION NEEDS	KEY WATER ISSUES	KEY SEDIMENT ISSUES	KEY BIOLOGICAL ISSUES	KEY USE ISSUES	KEY COST FACTORS
8. Dredging	◆Increased depth/access	◆Existing and proposed bathymetry	◆Possible peak flows	◆Classification of dredged material for disposal purposes	◆Preservation vs. restoration vs. restructuring of biotic communities	◆Possible loss of water supply and recreational use during project	◆Engineering and permitting costs
	◆Alteration of bottom composition	◆Volume of material to be removed	◆Expected mean flows	◆Physical handling limitations	◆Presence of any protected species	◆Potential short- and long-term habitat impacts	◆Construction of containment area
	◆Removal of nutrient reserves	◆Physical nature of material to be removed	◆Downstream flow needs	◆Drying time	◆Possible impacts on wetlands	◆Access and safety concerns	◆Equipment purchases
	◆Reduction in oxygen demand	◆Chemical nature of material to be removed	◆Turbidity generation/control	◆Bulking factor	◆Other habitats of special concern		◆Operational costs
	◆Algal control	◆Nature of underlying material to be exposed	◆Possible contaminants in discharge from containment area	◆Possible uses of dredged material	◆Potential for plants to impede dredging		◆Contract dredging costs
	◆Fe and Mn control	◆Protected resource areas					◆Ultimate disposal costs
	◆Rooted plant control	◆Dewatering capacity of sediments					◆Containment area restoration
	◆Habitat enhancement	◆Provisions for controlling water level					◆Possible sale of dredged material
		◆Equipment access					◆Monitoring program
		◆Pipeline route					
		◆Potential disposal sites					
		◆Dredging methodology restrictions					
9. Dye addition	◆Reduced light penetration	◆Water depth and volume to be treated	◆Increased surface temperature due to light absorption by dye	◆Adsorption of dye and reduced longevity of effects	◆Ecological impact of reduced light and increased surface temperature	◆Interference with use as a potable supply	◆Cost of dye
	◆Algal control	◆Flushing rate	◆Possible stratification of shallow water			◆Acceptability of water coloration	◆Cost of application
	◆Rooted plant control	◆Thermal regime	◆Downstream transport of dye				◆Monitoring program

Table 7-3.—Factors in selecting management techniques (continued).

TECHNIQUE	APPLICABLE USES OF THE TECHNIQUE	PRIMARY INFORMATION NEEDS	KEY WATER ISSUES	KEY SEDIMENT ISSUES	KEY BIOLOGICAL ISSUES	KEY USE ISSUES	KEY COST FACTORS
10. Hydroraking or rotovation	◆ Rooted plant control	◆ Characterization of plant assemblage	◆ Increased turbidity	◆ Post-treatment sediment resuspension susceptibility	◆ Recolonization of exposed sediment surfaces	◆ Treatment effects on water supply quality	◆ Cost of equipment and labor, usually as contracted project
	◆ Removal of physical obstructions (stumps, root masses)	◆ Area to be treated	◆ Water chemistry impacts of interaction with sediments	◆ Sediment-water interactions	◆ Impacts to benthic organisms	◆ Use restrictions during treatment	◆ Monitoring program
		◆ Sediment features					
		◆ Water depth					
		◆ Potential disposal areas					
11. Harvesting, pulling, or cutting	◆ Rooted plant control	◆ Characterization of plant assemblage	◆ Potential for increased turbidity and floating plant fragments	◆ Post-treatment sediment resuspension susceptibility	◆ Regrowth of plants	◆ Safety concerns during harvesting	◆ Cost of equipment and labor
	◆ Dense algal mat control	◆ Area to be treated		◆ Sediment-water interactions	◆ Recolonization of exposed sediment surfaces		◆ Contract vs. local staff vs. volunteer approach
		◆ Water depth			◆ Impact on organisms associated with plants		◆ Disposal location and restrictions
		◆ Physical obstructions			◆ Potential to spread plant species by fragmentation and drift		◆ Monitoring program
		◆ Potential disposal areas					
12. Nutrient inactivation	◆ Reduced phosphorus levels/algal control	◆ Phosphorus concentration and forms	◆ Flushing rate variability	◆ Interaction with settled floc	◆ Toxicity through aluminum or pH effects	◆ Possible use restrictions during/after treatment	◆ Cost of chemical
		◆ Water volume to be treated	◆ Circulation pattern	◆ Potential for reduced sediment release of phosphorus	◆ Impact on benthic organisms and zooplankton	◆ Restrictions to minimize resuspension of floc	◆ Need for buffering
		◆ Alkalinity and pH of water	◆ Stability of alkalinity and pH		◆ Possible reduced fish production with reduced fertility		◆ Cost of application
		◆ Expected reaction efficiency					◆ Monitoring program
		◆ Flushing rate					
		◆ Phosphorus load to lake from external and internal sources					

Table 7-3.—Factors in selecting management techniques (continued).

TECHNIQUE	APPLICABLE USES OF THE TECHNIQUE	PRIMARY INFORMATION NEEDS	KEY WATER ISSUES	KEY SEDIMENT ISSUES	KEY BIOLOGICAL ISSUES	KEY USE ISSUES	KEY COST FACTORS
13. Nutrient supplementation	◆Enhanced fish production ◆Control of algal composition	◆Detailed water chemistry ◆Loads of major nutrients ◆Flushing rate ◆Algal assemblage features	◆Possible water quality effects of additions	◆Possible increase in sediment accumulation rate	◆Changes in food web structure ◆Potential for shifts in algal species to undesirable forms	◆Potential for increased water supply treatment needs ◆Use impairment by decreased transparency	◆Cost of chemical ◆Cost of application ◆Monitoring program
14. Other chemical treatments	◆Particulate settling/algal control ◆Nuisance fauna control ◆pH adjustment ◆Oxidation/disinfection	◆Detailed water chemistry ◆Flushing rate ◆Algal assemblage features	◆Possible water quality effects of additions ◆Effect on pollutant release by sediments	◆Possible increase in sediment accumulation rate	◆Impacts on specific biota or community as a whole	◆Use restrictions associated with treatment	◆Cost of chemical ◆Cost of application ◆Monitoring program
15. Pollutant capture (basins within a lake)	◆Reduction in nutrient levels/algal control ◆Reduced loading of multiple pollutants	◆Target load reduction ◆Physical limits on detention ◆Treatment processes needed to achieve target reduction ◆Inflow rate ◆Inflow quality	◆Variability in flow and quality of incoming water ◆Flooding adjacent to detention areas	◆Need for removal of accumulated sediment from detention area	◆Habitat value of newly created detention area(s) ◆Loss of open, contiguous water area through partitioning	◆Access and safety concerns	◆Cost of engineering design ◆Cost of materials and construction ◆Maintenance costs ◆Monitoring program
16. Rules and regulations	◆User conflict resolution ◆Minimization of pollutant loading ◆Minimization of nuisance species introductions	◆Potential threats and control needs ◆User population demographics ◆Modeled or predicted results of control strategy	◆Potential for unanticipated impacts	◆Potential for unanticipated impacts	◆Potential for unanticipated impacts	◆Altered use patterns and user satisfaction	◆Cost of rule development ◆Cost of enforcement

Table 7-3.—Factors in selecting management techniques (continued).

TECHNIQUE	APPLICABLE USES OF THE TECHNIQUE	PRIMARY INFORMATION NEEDS	KEY WATER ISSUES	KEY SEDIMENT ISSUES	KEY BIOLOGICAL ISSUES	KEY USE ISSUES	KEY COST FACTORS
17. Selective withdrawal	◆Discharge of poor quality water and possible algal control	◆Vertical variation in water quality	◆Potential for withdrawal to exceed inflow and cause drawdown	◆Potential for resuspension and capture in discharge/intake	◆Potential for impingement or entrainment of organisms	◆Impairment of contact recreation by drawdown	◆Cost of pipe installation
	◆Prevention of hypolimnetic anoxia/algal control	◆Depth and volume of epilimnion and hypolimnion	◆Potential loss of hypolimnion and thermal stratification		◆Impact on thermally sensitive species	◆Impairment of fishing success by thermal alteration	◆Cost of control structure
	◆Intake of best quality water	◆Inflow rate/flushing rate	◆Water quality effects of removal of best quality water				◆Cost of any pumping
		◆Hypolimnetic/sediment oxygen demand	◆Possible need to treat discharge				◆Monitoring program
		◆Downstream constraints					
18. Water level control	◆Access to structures for maintenance or construction	◆Target level of water	◆Flood storage gained/lost	◆Potential for sloughing	◆Anticipated impacts to target and non-target plant species in lake	◆Use of lake water as a supply	◆Structural alteration to facilitate control
	◆Access to sediments for removal (dredging)	◆Pond bathymetry	◆Effects on peak flows	◆Potential for shoreline erosion	◆Presence of protected species	◆Depth of any wells within zone of influence	◆Pumping or alternative water moving technology
	◆Flood control	◆Area to be exposed/flooded	◆Relative area and volume of lake remaining	◆Potential for dewatering and compaction	◆Association of fauna with areas to be exposed	◆Alternative water supplies	◆Operational cost of controlling outflow
	◆Prevention of ice damage to shoreline and structures	◆Maximum/minimum volume	◆Effects on nutrient levels	◆Potential for odors	◆Potential impacts to connected wetlands	◆Downstream flow restrictions	◆Alternative water supply provision
	◆Sediment compaction	◆Timing and frequency of drawdown/flooding	◆Effects on oxygen levels	◆Access and safety considerations		◆Emergency response system	◆Monitoring program
	◆Rooted plant control	◆Outlet control features	◆Effects on pH levels			◆Possible elimination of usable open water	
	◆Fish reclamation	◆Climatological data				◆Access and safety	
	◆Flushing	◆Normal range of outflow				◆Potentially impaired appearance during drawdown	
		◆Outflow during drawdown and refill					
		◆Time to draw down or refill					

For any given technique, it is essential to know what it can and cannot do for the lake; in other words, it must be **evaluated ecologically**. Once it's determined to be applicable, its side effects relating to water quantity and quality, sediment, and system biology must be considered. Perhaps most important are its potential effects on non-target organisms, as these will affect permitting and public perception of the project.

Side effects of techniques, expected or not, positive or negative, can greatly influence practical aspects of implementation. Successful permitting and funding of the program may depend on controlling impacts on non-target organisms or public perception of risk.

The **effects on water use** must also be carefully considered. The chosen technique should further the goals set for the use of the lake, but will it do so immediately, or will there be a substantial lag time, possibly with use restrictions or other negative impacts? It cannot be assumed that long-term benefits will justify short-term impairment of uses in the public arena, although informed groups will usually make some sacrifices.

Costs are almost always a critical consideration in lake management. It has been said that "the most important elements in lake management are not phosphorus and nitrogen, but silver and gold" (Canfield, 1992). Unfortunately, many choices of techniques have been based on what was affordable at the moment, not on the long-term cost-benefit ratio.

While a technique that addresses the source of the problem usually costs more at the outset, results should last. A management program that provides at least 10 years of benefits at a high initial capital cost may not really be expensive compared to a maintenance bargain that has to be repurchased many times over a period of 10 years without ever solving the real problem. Costs should be compared on at least a 10-year basis, preferably 20 years or even longer.

Capital, operational, contractor, and monitoring costs should all be considered in comparing management options — and they should be figured for the long term.

Capital, operational, contractor, and monitoring costs should all be considered in comparing management options — and they should be figured for the long term. A single herbicide treatment that may control vegetation for a year should not be compared with a dredging program that provides several decades of control, increases water depth, and probably has other benefits. The herbicide treatment may be more appropriate in some cases, but the cost of techniques should be put on an equal scale of magnitude and longevity of benefits.

The relationship of problem and potential solution must also be evaluated in **ethical terms**. Although ethical criteria are often less clear than ecological or cost concerns, many of the evaluation points already discussed incorporate ethical judgments. Ethics deals with values and fairness, and though some will see this as highly subjective, establishing management goals must be done in an ethical manner. If society is to achieve balance and sustainability in its interactions with the environment, ethical questions must be posed and answered.

Management Options

Many of the techniques for managing and improving lakes were developed years ago, but only in the last two decades has the effectiveness of many of them been documented through applied research supported by the U.S. Environmental Protection Agency's Clean Lakes Program, corresponding state programs, the National Science Foundation, other governmental agencies, and private corporations. However, careful documentation of results, longevity, and costs continues to be a major need.

Most techniques and products described in this manual are considered to be effective in at least some circumstances; a few others are noted because of significant media coverage or theoretical appeal.

The reader is encouraged to seek additional information and opinions (see, for example, Cooke et al. 1993a; McComas, 1993; NY State Dep. Environ. Conserv. and NY Fed. Lake Assns. 1990), and to critically evaluate each option within the context of the target lake.

Lake managers should ask for scientific documentation regarding a procedure, product, or technique, especially one not described here. Discuss techniques with a lake restoration expert not financially involved in its sale or installation. Too many lake associations have spent thousands of dollars on products and procedures that don't work or are inappropriate to the problem or lake.

Funds are well spent to properly evaluate techniques with specific reference to your lake. Few lake management programs are truly inexpensive, and a proper diagnostic/feasibility assessment can save far more money than it costs.

Nuisance Algae

Excessive algal growth can become a serious nuisance in lakes. Two growth forms are most troublesome:

Techniques that reduce light and nutrients can control algae.

- Free-floating microscopic cells, colonies, or filaments — called phytoplankton — that discolor the water and sometimes form green scum on the surface. Although these algae come from a variety of algal groups, including blue-greens, greens, diatoms, goldens, euglenoids, and dinoflagellates, the blue-greens tend to cause the most problems because of their high density (blooms), taste, odor, and possible toxicity.
- Mats of filamentous algae associated with sediments and weed beds, but that often float to the surface once they reach a certain density. These are most often green algae of the orders Cladophorales or Zygnematales, or blue-green algae (more properly cyanobacteria) of the order Oscillatoriales.

Algae reproduce mainly through cell division, although resting cysts help them survive unfavorable periods. When growth conditions are ideal (warm, lighted, nutrient-rich), algae multiply rapidly and create massive blooms within a few weeks. Many algal blooms produce taste and odor problems, and their decay may decrease oxygen.

Many water treatment problems stem from algae in the water supply reservoir:

- Poor taste and odor are often associated with algal blooms and mats.
- Some common bloom-forming blue-green algae produce toxins that may kill domestic animals and be linked to certain illnesses in humans.
- The combination of algal organic matter and chlorine disinfectant can form potential carcinogenic byproducts such as trihalomethanes.

How to Control Algae

Algae need **light and nutrients** to grow. Turbidity, shading by plants, or massive algal growth itself can reduce the light; or, an essential nutrient, usually phosphorus, can be limited. Techniques (Table 7-4) such as dyes, artificial circulation, and selective plantings seek to limit light, whereas aeration, dilution and flushing, drawdown, dredging, phosphorus inactivation, and selective withdrawal are used to reduce nutrients.

Adding nutrients selectively may provide an ecologically complex solution in some cases. Altering the ratio of nutrients may encourage the growth of algae that are more amenable to other control techniques, like grazing and settling. Although theoretically sound (Tilman, 1982), this approach has rarely been used in practical lake management.

Settling, consumption by grazers, and cellular death control algae naturally. Natural processes can be accelerated by using such techniques as settling agents, biomanipulation (either encouraging grazing or adding bacteria or viruses to kill algal cells), algaecides, and mechanical removal. Unfortunately, algae are remarkably adaptable, and none of these techniques works on all algal communities.

- Many blue-greens are buoyant and resist settling.
- Nuisance green algae (Chlorococcales and Cladophorales) and certain blue-green algae (especially Aphanizomenon) are often resistant to copper, the most common algaecide, and also resist grazers.
- Only very dense algal mats can be harvested, and then with difficulty.

Filamentous algal mats are difficult to control. They usually form at the sediment-water interface or in rooted plant beds, nourished by nutrients released by decay processes in the presence of adequate light. As mat density increases, photosynthetic gases are often trapped, and the mat may float upward and expand. Neither grazing, settling, nor algaecides has much effect, and harvesting is usually not practical. The best control is to prevent mats from forming by removing sediment or treating the algae at an early stage (phosphorus inactivation or early algaecide application).

Aeration or Oxygenation

Aeration puts air into the lake, increasing the concentration of oxygen by transferring it from gas to liquid and generating a controlled mixing force. Using pure oxygen maximizes the transfer. In this chapter, we deal with aeration as a technique intended to add oxygen to deep lakes without disrupting stratification.

Aeration is most appropriately used to prevent hypolimnetic anoxia (low oxygen in the bottom layer) so that when the lake stratifies (separates into layers), minimal phosphorus, iron, manganese, and sulfides will be released from sediments.

Aeration also retards the buildup of undecomposed organic matter and compounds (e.g., ammonium) near the bottom of the lake, and can increase the amount of water available to zooplankton and fish living in the lower, colder waters.

Permits are generally required for aeration projects, but hypolimnetic aeration is among the easier lake management processes to get approved, having few adverse side effects.

Table 7-4.—Management options for control of algae.

OPTION	MODE OF ACTION	ADVANTAGES	DISADVANTAGES
PHYSICAL CONTROLS			
1. Hypolimnetic aeration or oxygenation	◆ Addition of air or oxygen at varying depth provides oxic conditions	◆ Oxic conditions promote binding/sedimentation of phosphorus	◆ May disrupt thermal layers important to fish community
	◆ May maintain or break stratification	◆ Counteraction of anoxia improves habitat for fish/invertebrates	◆ May promote supersaturation with gases harmful to fish
	◆ Can also withdraw water, oxygenate, then replace	◆ Buildup of dissolved iron, manganese, ammonia, and phosphorus reduced	◆ Permits usually required
2. Circulation and destratification	◆ Use of water or air to keep water in motion	◆ Reduces surface buildup of algal scums; promotes uniform appearance	◆ May spread local impacts
	◆ Intended to prevent or break stratification	◆ May disrupt growth of some algae	◆ May increase oxygen demand at greater depths
	◆ Generally driven by mechanical or pneumatic force	◆ Counteraction of anoxia improves habitat for fish/invertebrates	◆ May promote downstream impacts
		◆ Can eliminate local problems without obvious impact on whole lake	
3. Dilution and flushing	◆ Addition of water of better quality can dilute nutrients	◆ Dilution reduces nutrient concentrations without altering load	◆ Diverts water from other uses
	◆ Addition of water of similar or poorer quality flushes system to minimize algal buildup	◆ Flushing minimizes detention; response to pollutants may be reduced	◆ Flushing may wash desirable zooplankton from lake
	◆ May have continuous or periodic additions		◆ Use of poorer quality water increases pollutant loads
			◆ Possible downstream impacts
4. Drawdown	◆ Lowering of water allows oxidation, desiccation, and compaction of sediments	◆ May reduce available nutrients or nutrient ratios, affecting algal biomass and composition	◆ Possible impacts on contiguous emergent wetlands
	◆ Duration of exposure and degree of dewatering of exposed areas are important	◆ Opportunity for shoreline cleanup/structure repair	◆ Possible effects on overwintering reptiles or amphibians
	◆ Algae are affected mainly by reduction in available nutrients	◆ Flood control utility	◆ Possible impairment of well production
		◆ May provide rooted plant control as well	◆ Reduction in potential water supply and fire fighting capacity
			◆ Alteration of downstream flows
			◆ Possible overwinter water level variation
			◆ May result in greater nutrient availability if flushing inadequate

Table 7-4.—Management options for control of algae (continued).

OPTION	MODE OF ACTION	ADVANTAGES	DISADVANTAGES
PHYSICAL CONTROLS			
5. Dredging	◆Sediment is physically removed by wet or dry excavation, with deposition in a containment area for dewatering	◆Can control algae if internal recycling is main nutrient source	◆Temporarily removes benthic invertebrates
	◆Dredging can be applied on a limited basis, but is most often a major restructuring of a severely impacted system	◆Increases water depth	◆May create turbidity
	◆Nutrient reserves are removed and algal growth can be limited by nutrient availability	◆Can reduce pollutant reserves	◆May eliminate fish community (complete dry dredging only)
		◆Can reduce sediment oxygen demand	◆May be impacts from containment area discharge
		◆Can improve spawning habitat for many fish species	◆May be impacts from dredged material disposal
		◆Allows complete renovation of aquatic ecosystem	◆May interfere with recreation or other uses during dredging
5.a."Dry" excavation	◆Lake drained or lowered to maximum extent practical	◆Tends to facilitate a very thorough effort	◆Eliminates most aquatic biota unless a portion left undrained
	◆Target material dried to maximum extent possible	◆May allow drying of sediments prior to removal	◆Eliminates lake use during dredging
	◆Conventional excavation equipment used to remove sediments	◆Allows use of less specialized equipment	
5.b."Wet" excavation	◆Lake level may be lowered, but sediments not substantially exposed	◆Requires least preparation time or effort, tends to be least costly dredging approach	◆Usually creates extreme turbidity
	◆Draglines, bucket dredges, or long-reach backhoes used to remove sediment	◆May allow use of easily acquired equipment	◆Tends to result in sediment deposition in surrounding area
		◆May preserve aquatic biota	◆Normally requires intermediate containment area to dry sediments prior to hauling
			◆May cause severe disruption of ecological function
			◆Usually eliminates most lake uses during dredging
5.c.Hydraulic removal	◆Lake level not reduced	◆Creates minimal turbidity and impact on most biota	◆Often leaves some sediment behind
	◆Suction or cutterhead dredges create slurry which is hydraulically pumped to containment area	◆Can allow some lake uses during dredging	◆Cannot handle coarse or debris-laden materials
	◆Slurry is dewatered; sediment retained, water discharged. May involve polymer-aided settling or vacuum-aided dewatering.	◆Allows removal with limited access or shoreline disturbance	◆Requires sophisticated and more expensive containment area
			◆Requires overflow discharge from containment area

Table 7-4.—Management options for control of algae (continued).

OPTION	MODE OF ACTION	ADVANTAGES	DISADVANTAGES
PHYSICAL CONTROLS			
6. Light-limiting dyes and surface covers	♦ Creates light limitation	♦ Creates light limit on algal growth without high turbidity or great depth	♦ May cause thermal stratification in shallow ponds
		♦ May achieve some control of rooted plants as well	♦ May facilitate anoxia at sediment interface with water
6.a. Dyes	♦ Water-soluble dye is mixed with lake water, thereby limiting light penetration and inhibiting algal growth	♦ Produces appealing color	♦ May not control surface bloom-forming species
	♦ Dyes remain in solution until washed out of system	♦ Creates illusion of greater depth	♦ May not control growth of shallow water algal mats
6.b. Surface covers	♦ Opaque sheet material applied to water surface	♦ Minimizes atmospheric and wildlife pollutant inputs	♦ Minimizes atmospheric gas exchange
			♦ Limits recreational use
7. Mechanical removal	♦ Filters pumped water for water supply purposes	♦ Algae and associated nutrients can be removed from system	♦ Filtration requires high backwash and sludge handling capability for use with high algal densities
	♦ Collection of floating scums or mats with harvesters, booms, nets, or other devices	♦ Surface collection can apply on an "as needed" basis	♦ Labor intensive unless a mechanized system applied, in which case it is capital intensive
	♦ Continuous or multiple applications per year usually needed	♦ May remove floating debris	♦ Many algal forms not amenable to collection by net or boom
		♦ Collected algae dry to minimal volume	♦ May impact non-targeted aquatic life
8. Selective withdrawal	♦ Discharge of bottom water which may contain (or be susceptible to) low oxygen and higher nutrient levels	♦ Removes targeted water from lake efficiently	♦ May result in poor water quality downstream if not treated
	♦ Intake of water from low algae layer to maximize supply quality	♦ Complements other techniques such as drawdown or aeration	♦ May eliminate colder thermal layer important to certain fish
	♦ May be pumped or utilize passive head differential	♦ May prevent anoxia and phosphorus buildup in bottom water	♦ May promote mixing of some remaining poor quality bottom water with surface waters
		♦ May remove initial phase of algal blooms which start in deep water	♦ May cause unintended drawdown if inflows do not match withdrawal
		♦ May create coldwater conditions downstream	

Table 7-4.—Management options for control of algae (continued).

OPTION	MODE OF ACTION	ADVANTAGES	DISADVANTAGES
CHEMICAL CONTROLS			
9. Algaecides	◆Liquid or pelletized algaecides applied to target area	◆Rapidly eliminates algae from water column, normally with increased water clarity	◆May be toxic to non-target areas or species of plants/animals
	◆Algae killed by direct toxicity or metabolic interference	◆May result in net movement of nutrients to bottom of lake	◆May restrict water use for varying time after treatment
	◆Typically requires application at least once/yr, often more frequently		◆Increased oxygen demand and possible toxicity may result from decaying algae
			◆Nutrients may recycle, allowing other growths
9.a. Forms of copper	◆Contact algaecide	◆Effectively and rapidly controls many algal species	◆Toxic to aquatic fauna as a function of concentration, formulation, temperature, pH, and ambient water chemistry
	◆Cellular toxicant, suggested disruption of photosynthesis, nitrogen metabolism, and membrane transport	◆Approved for use in most water supplies	◆Less effective at colder temperatures or at high inorganic solids levels
	◆Applied as wide variety of liquid or granular formulations, often in conjunction with chelators, polymers, surfactants, or herbicides		◆Copper ion persistent; accumulates in sediments or moves downstream
			◆Certain green and blue-green nuisance species are resistant to copper
			◆Lysing of cells releases cellular contents (including nutrients and toxins) into water column
9.b. Forms of endothall (7-oxabicyclo [2.2.1] heptane-2,3-dicarboxylic acid)	◆Contact algaecide	◆Moderate control of thick algal mats; used where copper is ineffective	◆Non-selective in treated area
	◆Membrane-active chemical which inhibits protein synthesis	◆Limited toxicity to fish at recommended dosages	◆Toxic to aquatic fauna (varying degrees by formulation)
	◆Causes structural deterioration	◆Acts rapidly	◆Time delays on use for water supply, agriculture, and recreation
	◆Applied as liquid or granules, usually as hydrothol formulation for algae control		
9.c. Forms of diquat (6,7-dihydropyrido [1,2–2′,1′-c] pyrazinediium dibromide)	◆Contact algaecide	◆Moderate control of thick algal mats; used where copper alone is ineffective	◆Non-selective in treated area
	◆Absorbed directly by cells	◆Limited toxicity to fish at recommended dosages	◆Toxic to some zooplankton at recommended dosage
	◆Strong oxidant; disrupts most cellular functions	◆Acts rapidly	◆Inactivated by suspended particles; ineffective in muddy waters
	◆Applied as a liquid, sometimes in conjunction with copper		◆Time delays on use for water supply, agriculture, and recreation

Table 7-4.—Management options for control of algae (continued).

OPTION	MODE OF ACTION	ADVANTAGES	DISADVANTAGES
CHEMICAL CONTROLS			
10. Phosphorus inactivation	◆ Typically salts of aluminum, iron, or calcium are added to the lake, as liquid or powder	◆ Can provide rapid, major decrease in phosphorus concentration in water column	◆ May be toxic to fish and invertebrates, especially by aluminum at low pH
	◆ Phosphorus in the treated water column is complexed and settled to the bottom of the lake	◆ Can minimize release of phosphorus from sediment	◆ Phosphorus may be released under anoxia or extreme pH
	◆ Phosphorus in upper sediment layer is complexed, reducing release from sediment	◆ May remove other nutrients and contaminants as well as phosphorus	◆ May cause fluctuations in water chemistry, especially pH, during treatment
	◆ Permanence of binding varies by binder in relation to redox potential and pH	◆ Flexible with regard to depth of application and speed of improvement	◆ Floc may be resuspended in shallow areas with extreme turbulence
	◆ Potential for use on inlet streams as well		◆ Adds to bottom sediment, but typically an insignificant amount
11. Sediment oxidation	◆ Addition of oxidants, binders, and pH adjustors oxidizes sediment	◆ Can reduce phosphorus supply to algae	◆ May affect benthic biota
	◆ Binding of phosphorus is enhanced	◆ Can alter N:P ratios in water column	◆ Longevity of effects not well known
	◆ Denitrification may be stimulated	◆ May decrease sediment oxygen demand	
12. Settling agents	◆ Closely aligned with phosphorus inactivation, but can be used to reduce algae directly, too	◆ Removes algae and increases water clarity without lysing most cells	◆ May affect aquatic fauna
	◆ Lime, alum, or polymers applied, usually as a liquid or slurry	◆ Reduces nutrient recycling if floc sufficient	◆ Water chemistry may fluctuate during treatment
	◆ Creates a floc with algae and other suspended particles	◆ Removes non-algal particles as well as algae	◆ Floc may resuspend in shallow, well-mixed waters
	◆ Floc settles to bottom of lake	◆ May reduce dissolved phosphorus levels at the same time	◆ Promotes increased sediment accumulation
	◆ Re-application necessary if algal growth not controlled		
13. Selective nutrient addition	◆ Ratio of nutrients changed by additions of selected nutrients	◆ Can reduce algal levels where control of limiting nutrient not feasible	◆ May result in greater algal abundance through uncertain biological response
	◆ Addition of non-limiting nutrients can change composition of algal community	◆ Can promote non-nuisance forms of algae	◆ May require frequent application to maintain desired ratios
	◆ Processes such as settling and grazing can then reduce algal biomass (productivity can actually increase, but standing crop can decline)	◆ Can improve productivity of system without increasing standing crop of algae	◆ May have downstream effects

Table 7-4—Management options for control of algae (continued).

OPTION	MODE OF ACTION	ADVANTAGES	DISADVANTAGES
CHEMICAL CONTROLS			
14. Management for nutrient input reduction	• Generally not an in-lake process (see Chapter 6), but essential to note in any algal control program	• Acts against the original source of algal nutrition	• May involve considerable lag time before improvement observed
	• Includes wide range of watershed and lake edge activities intended to eliminate nutrient sources or reduce delivery to lake	• Decreases effective loading of nutrients to lake	• May not be sufficient to achieve goals without some form of in-lake management
	• Can involve use of wetland treatment cells or detention areas created from part of lake	• Creates sustainable limitation on algal growth	• Reduction of overall system fertility may impact fisheries
	• Essential component of algal control strategy where internal recycling is not the dominant nutrient source, and desired even where internal recycling is important	• May control delivery of other unwanted pollutants to lake	• May cause shift in nutrient ratios that favor less desirable species
		• Generally most cost effective over long term	• May cost more in the short term, as source management is generally more involved than one or a few treatments of symptoms of eutrophication
		• Facilitates ecosystem management approach which considers more than just algal control	
BIOLOGICAL CONTROLS			
15. Enhanced grazing	• Manipulation of biological components of system to achieve grazing control over algae	• May increase water clarity by changes in algal biomass or cell size distribution without reduction of nutrient levels	• May involve introduction of species
	• Typically involves alteration of fish community to promote growth of large herbivorous zooplankton, or stocking with phytophagous fish	• Can convert unwanted biomass into desirable form (fish)	• Effects may not be controllable or lasting
		• Harnesses natural processes to produce desired conditions	• May foster shifts in algal composition to even less desirable forms
15.a. Herbivorous fish	• Stocking of fish that eat algae	• Converts algae directly into potentially harvestable fish	• Typically requires introduction of non-native species
		• Grazing pressure can be adjusted through stocking rate	• Difficult to control over long term
			• Smaller algal forms may benefit and bloom
15.b. Herbivorous zooplankton	• Reduces planktivorous fish to promote grazing pressure by zooplankton	• Converts algae indirectly into harvestable fish	• Highly variable response expected; temporal and spatial variability may be problematic
	• May involve stocking piscivores or removing planktivores	• Zooplankton community response to increasing algae can be rapid	• Requires careful monitoring and management action on 1-5 yr basis
	• May also involve stocking zooplankton or establishing refugia	• May be accomplished without introduction of non-native species	• May involve non-native species introduction(s)
		• Generally compatible with most fishery management goals	• Larger or toxic algal forms may benefit and bloom

Table 7-4.—Management options for control of algae (continued).

OPTION	MODE OF ACTION	ADVANTAGES	DISADVANTAGES
BIOLOGICAL CONTROLS			
16. Bottom-feeding fish removal	◆ Removes fish that browse among bottom deposits, releasing nutrients to the water column by physical agitation and excretion	◆ Reduces turbidity and nutrient additions from this source	◆ Targeted fish species are difficult to eradicate or control
		◆ May restructure fish community in more desirable manner	◆ Reduction in fish populations valued by some lake users (human and non-human)
17. Fungal/bacterial/ viral pathogens	◆ Addition of inoculum to initiate attack on algal cells	◆ May create lakewide "epidemic" and reduction of algal biomass	◆ Largely experimental approach at this time
		◆ May provide sustained control for several years	◆ Results are uncertain
		◆ Can be highly specific to algal group or genera	◆ May promote resistant forms with high nuisance potential
			◆ May cause high oxygen demand or release of toxins by lysed algal cells
			◆ Effects on non-target organisms uncertain
18. Competition and allelopathy	◆ Plants may tie up sufficient nutrients to limit algal growth	◆ Harnesses power of natural biological interactions	◆ Some algal forms appear resistant
	◆ Plants may create a light limitation on algal growth	◆ May provide responsive and prolonged control	◆ Use of plants may lead to problems with vascular plants
	◆ Chemical inhibition of algae may occur through substances released by other organisms	◆ Shift to rooted plant dominance can improve habitat	◆ Use of plant material may depress oxygen levels
18.a. Plantings for nutrient control	◆ Plant growths of sufficient density may limit algal access to nutrients	◆ Productivity and associated habitat value can remain high without algal blooms	◆ Vascular plants may achieve nuisance densities
	◆ Plants can exude allelopathic substances that inhibit algal growth	◆ Portable plant "pods," floating islands, or other structures can be managed to limit interference with recreation and provide habitat	◆ There will be a water depth limitation on rooted plants but not algae
		◆ Wetland cells in or adjacent to the lake can minimize nutrient inputs	◆ Vascular plant senescence may release nutrients and cause algal blooms
			◆ The switch from algae to vascular plant domination of a lake may cause unexpected or undesirable changes in lake ecology, especially energy flow
18.b. Plantings for light control	◆ Plant species with floating leaves can shade out many algal growths at elevated densities	◆ Vascular plants can be more easily harvested than most algae	◆ At the necessary density, the floating plants will be a recreational nuisance
		◆ Many floating species provide valuable waterfowl food	◆ Low surface mixing and atmospheric contact promote anoxia near the sediment

Table 7-4.—Management options for control of algae (continued).

OPTION	MODE OF ACTION	ADVANTAGES	DISADVANTAGES
BIOLOGICAL CONTROLS			
18.c. Addition of barley straw	◆ Input of barely straw can set off a series of chemical reactions that limit algal growth	◆ Materials and application are relatively inexpensive	◆ Success appears linked to uncertain and potentially uncontrollable water chemistry factors
	◆ Release of allelopathic chemicals can kill algae	◆ Decline in algal abundance is more gradual than with algaecides, limiting oxygen demand and the release of cell contents	◆ May depress oxygen levels
	◆ Release of humic substances can bind phosphorus		◆ Water chemistry may be altered in other ways unsuitable for non-target organisms
			◆ Some forms of algae may be resistant and could benefit from the treatment

▾ ***How Aeration Works.*** Many methods can aerate a lake (see Fig. 7-1), but only a few maintain stratification:

- A **full lift approach,** usually driven pneumatically by compressed air, moves hypolimnetic water to the surface, aerates it, and replaces it in the hypolimnion. Return flow to the hypolimnion is generally directed through a pipe to maintain separation of the newly aerated waters from the surrounding epilimnion. To provide adequate aeration, the hypolimnetic volume should be pumped and oxygenated at least once every 60 days, preferably more frequently.

Aeration puts air into the lake, increasing the concentration of oxygen by transferring it from gas to liquid and generating a controlled mixing force.

- The **partial lift system** pumps air into a submerged chamber in which oxygen is exchanged with the deeper waters (Fig. 7-1). A housed compressor must be located on the shore, but the aeration unit itself is submerged and does not interfere with lake use or aesthetics.

- **Layer aeration** (Kortmann et al. 1994; Fig. 7-1) combines water from different, carefully chosen temperature (and thus density) regimes to form stable oxygenated layers anywhere from the upper metalimnetic boundary down to the bottom of the lake. Each layer retards the passage of phosphorus, reduced metals, and related contaminants from the layer below. Either part of or the whole hypolimnion may be aerated to the desired oxygen level (water supply may demand less oxygen than a trout refuge, for example).

Any of these three aeration systems can markedly improve lake conditions, but experience has demonstrated that effects are neither uniform nor consistent in aquatic systems:

- Zones of minimal interaction will often occur, possibly resulting in localized anoxia and phosphorus release.

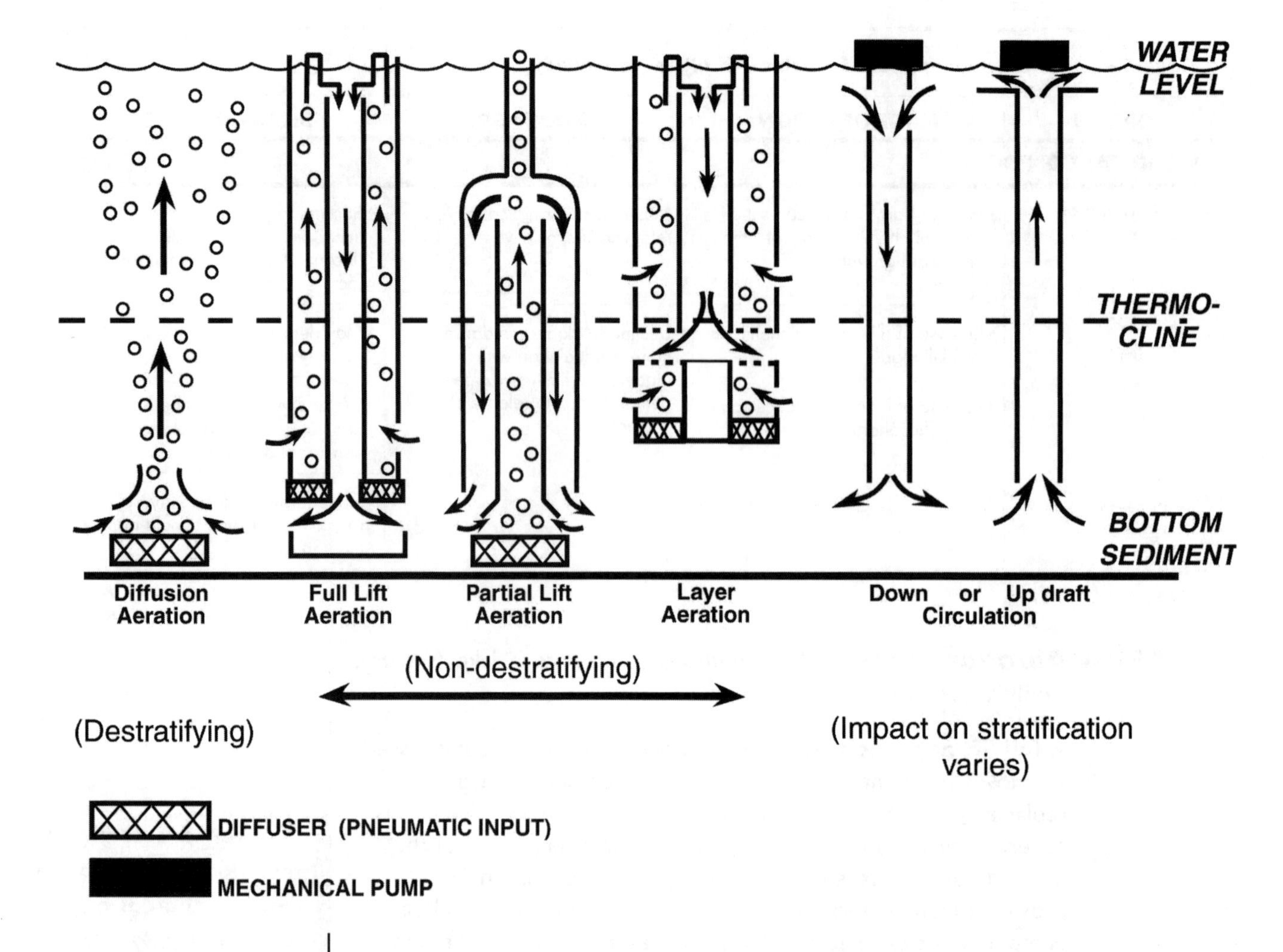

Figure 7-1.—Examples of aeration and circulation approaches.

- Partial lift systems may allow a band of anoxic water to persist near the top of the metalimnion, allowing nutrient cycling and supply to the epilimnion and discouraging vertical migration by fish and zooplankton.
- Although extremely unlikely, supersaturation of nitrogen resulting from aeration (not just oxygen is transferred if air is used) may expose fish to "gas bubble disease."

The most critical information for designing an aeration system is the amount of oxygen the lake must have. Oxygen demand is normally calculated from actual lake data. For stratified lakes, the hypolimnetic oxygen demand can be calculated as the difference in oxygen levels between the time layers formed in the lake and at some time during stratification before levels decline below 1 mg/L.

Several factors complicate the assessment of oxygen demand, among them the fact that oxygen consumption declines as oxygen supply declines. Oligotrophic lakes may need less than 250 mg/m^2/day, whereas eutrophic lakes use at least twice that (Hutchinson, 1957), and 2,000 to 4,000 mg/m^2/day have been measured in hypereutrophic lakes (Wagner, pers. obs.) An experienced professional can help you with this calculation.

▼ ***Effects on Algae.*** Aeration controls algae by reducing phosphorus. Two factors are important here: (1) Enough oxygen must be added to meet the hypolimnetic demand, and (2) there must be an adequate supply of substances that attract (bind) phosphorus (such as aluminum and iron compounds), because aeration encourages binding to remove phosphorus from the water.

Bound phosphorus does not necessarily become available after aeration ceases, but the risk of release increases, especially where iron is the dominant binder. Studies cited in Cooke et al. (1993a) show that without enough permanent binders, available phosphorus tends to decline by one- to two-thirds during aeration, but quickly rises to pre-aeration levels when treatment ceases.

Circulation and Destratification

Aeration is commonly used to mix shallow lakes (circulation), and sometimes to destratify deep lakes.

▼ ***Circulation*** mixes water to minimize stagnation in places such as coves and can eliminate or prevent thermal stratification; both conditions stimulate algal blooms, decrease oxygen, and cause sediment to accumulate. The circulatory process moves water — using surface aerators, bottom diffusers, or water pumps — to create the desired circulation pattern in shallow (usually <20 ft) lakes. Often, the effect is largely cosmetic; algae are simply mixed more evenly in the water. However, the movement may disrupt the life cycle of some algae, and limit their growth.

▼ ***Stratification*** is broken or prevented in deeper lakes by injecting compressed air into the water from a diffuser on the lake bottom or, in some instances, moving the water with a wind-driven pump (Fig. 7-1). If it's strong enough, the rising column of bubbles will mix the lake at a rate that eliminates temperature differences between top and bottom waters. Using air as the mixing force also oxygenates the water somewhat.

> **Stratification:** Distinct layers of water in a lake, separated according to temperature:
>
> - the warmest layer is nearest the surface (**epliminion**);
> - the layer that resists mixing, lies between the top and bottom layers (**metalimnion**);
> - the coldest layer is at the bottom (**hypolimnion**).

▼ ***Destratification*** may also control algae, usually through one or more of these processes:

- Mixing to the lake bottom will increase an algal cell's time in darkness, reducing the net photosynthesis and consequent algal biomass.
- Introducing dissolved oxygen to the lake bottom may prevent phosphorus release from sediments, curtailing this internal nutrient source for algae.
- Rapid circulation, air-water contact, and the introduction of carbon dioxide-rich bottom water during the initial mixing period may cause a shift to less noxious green algae from blue-greens.
- More zooplankters (which feed on algal cells) may survive because they will be mixed throughout the water column, making them less vulnerable to visually feeding fish.

Results have varied greatly:

- Problems with low dissolved oxygen have usually been solved.
- When destratification is properly used in a water supply reservoir, problems with iron and manganese can be eliminated.
- Where very small temperature differences from top to bottom have been maintained all summer, algal blooms seem to decline.
- Phosphorus and turbidity have increased; in some cases transparency has decreased.
- Surface scums have been prevented, although total algal biomass may not be reduced.
- Systems that bring deep water to the surface can be inexpensive, but unless enough water is moved to prevent anoxia near the sediment-water interface, the quality of water brought to the surface may cause deterioration of surface conditions.
- Systems that pump surface water to the bottom may improve the oxygen level near the bottom, but may also cause unfavorable circulation patterns and surface conditions.

Why do these mixing techniques sometimes fail? Either lake chemistry or equipment may be responsible:

- A lake that receives high nutrient loads from its watershed is unlikely to respond acceptably.
- If strongly stratified, a lake may be very difficult to destratify.
- Undersizing the mixing system is the major equipment-related cause of failure for this technique. Lorenzen and Fast (1977) suggested that an air flow of about 1.3 ft^3/min per acre of lake surface is required to maintain mixing within the lake.

Many engineering details must be considered in designing a circulation system, and the designer must understand the site conditions; once again, consult with a professional to design your system.

Dilution and Flushing

Algae usually do not bloom in lake waters that have minimal nutrients, particularly phosphorus. Concentrations in lake water reflect concentrations in incoming water, the flushing rate or residence time of the lake, and the net amount that settles onto sediments. While you should first try to reduce nutrients entering the lake, you can **dilute the concentration** of nutrients within the lake by adding nutrient-poor water.

When water low in phosphorus is added to the inflow, the actual phosphorus load will increase, but the mean phosphorus concentration should decrease. Lakes with low initial flushing rates are poor candidates because in-lake concentration could actually increase unless the dilution water is free of phosphorus (Uttormark and Hutchins, 1980). Internal loading is another influential factor. You must understand your lake's phosphorus budget to evaluate dilution as a potential algae control method.

Flushing can wash algae out of the lake faster than they can reproduce by adding large amounts of water, whether or not it's low in nutrients. However, complete flushing is virtually impossible in many lake systems; small, linear impoundments are the prime candidates for such treatment.

Although flushing washes out algal cells, their reproductive rate is so high (blooms form within a few weeks), that only extremely high flushing rates will be effective. A flushing rate of 10 to 15 percent of the lake volume per day is appropriate. Reliable water and nutrient budgets must be developed to evaluate flushing as an algae control technique.

Very few documented case histories of dilution or flushing exist, in part because:

- Additional water is not often available, especially water that is low in nutrients.
- Outlet structures and downstream channels must be able to handle the additional discharge.
- Qualitative downstream impacts must be considered.
- Water used for dilution or flushing should be carefully monitored prior to use in the lake.

Moses Lake in Washington state is one of the few recorded cases of successful dilution (Welch and Patmont, 1980; Cooke et al. 1993a). Low-nutrient Columbia River water was diverted through the lake, achieving water exchange rates of 10 to 20 percent per day. Algal blooms dramatically decreased, and transparency significantly improved.

Drawdown

Lowering the water level and exposing sediments to oxidize and compact them will decrease their oxygen demand and long-term phosphorus release rate. Recent research (Mitchell and Baldwin, 1998) indicates that shifts in the bacterial community during exposure may reduce phosphorus release after some drawdowns.

While the theory is attractive, in practice this approach suffers from several important limitations:

- The most problematic sediments are in the deepest part of most lakes, so the entire lake must be drained to achieve the maximum result.
- It is difficult to dewater the sediments sufficiently to get more than minor results.
- Nutrient release when the lake is refilled may actually increase until the nutrients can be flushed from the system.
- To control drawdown, the lake must have a manageable outlet structure and system hydrology normally associated only with reservoirs.

For a number of reasons, using this technique to control algae appears uncommon and unreliable:

- Drawdown has little effect on algal resting cysts.
- Chemical and physical features of the sediments influence results of drawdown.
- The effects of a drawdown are easily overwhelmed by high external phosphorus inputs.

Dredging

Removing sediment from a lake can lower in-lake nutrient concentrations and algal production by preventing the release of nutrients from the sediment. Dredging also removes the accumulated resting cysts deposited by a variety of algae. Although recolonization would probably be rapid, algal composition could change. Even where incoming nutrient loads are high, dredging can reduce the formation of benthic mats and related problems with filamentous green and blue-green algae that depend on these substrates for nutrition.

A lake can be dredged by four methods:

- **Dry excavation:** the lake is drained as much as possible; sediments are dewatered by gravity and/or pumping and removed with conventional excavation equipment such as backhoes, bulldozers, or draglines.
- **Wet excavation:** the lake is only partially drawn down (to minimize downstream flows); wet sediments are excavated by amphibious excavators or bucket dredges mounted on cranes.
- **Hydraulic dredging:** a substantial amount of water remains in the lake to float the dredge and transport the sediment. Hydraulic dredges are typically equipped with a cutterhead to loosen sediments that are then mixed with water and pumped as a slurry of 80 to 90 percent water (10 to 20 percent solids) through a pipeline extending from the dredging site to a disposal area. Polymer addition and mechanical dewatering can be used to improve handling of slurries.
- **Pneumatic dredging:** air pressure pumps sediments out of the lake at a higher solids content (50 to 70 percent), much like hydraulic dredging. This is a promising but fairly new technique that appears to require less space and facilitates drying of the solids; however, few dredges are operating in North America, and not enough experience exists for a knowledgeable analysis.

When the upper 3.3 feet of extremely nutrient-rich sediments were removed from Lake Trummen in Sweden, the total phosphorus concentration in the lake dropped sharply and remained fairly stable for at least 18 years. Phytoplankton production also declined.

Dry, wet, and hydraulic methods are illustrated in Figure 7-2, further outlined in Table 7-4, and discussed thoroughly in Cooke et al. (1993a).

Removing sediment can retard nutrient release, as illustrated by the following examples:

- When the upper 3.3 feet of extremely nutrient-rich sediments were removed from Lake Trummen in Sweden (Andersson, 1988), the total phosphorus concentration in the lake dropped sharply and remained fairly stable for at least 18 years. Phytoplankton production also declined.
- Algae decreased and water clarity increased in Hills Pond in Massachusetts after all soft sediment was removed and a stormwater treatment wetland was installed in 1994 (Wagner, 1996).
- Dredging 6-acre Bulloughs Pond in Massachusetts in 1993 has prevented thick green algal mats from forming for seven years now, despite continued high nutrient levels in urban runoff (Wagner, pers. obs.). These mats had previously begun as spring bottom growths, then floated to the surface in midsummer.

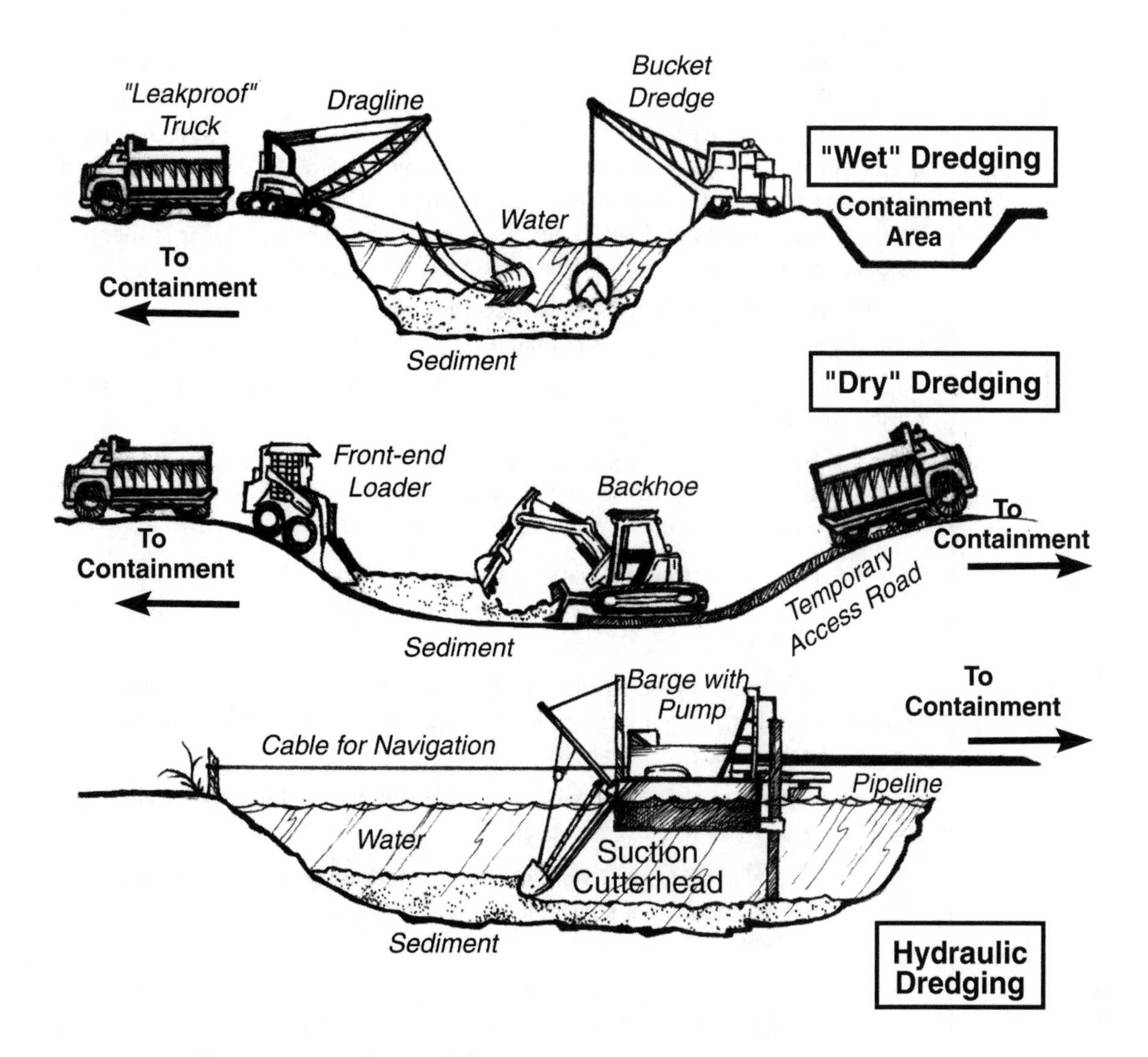

Figure 7-2.—Examples of dredging approaches.

While removing the entire nutrient-rich layer of sediment can control algae, that goal is usually secondary; dredging is most frequently done to deepen a lake, remove accumulations of toxic substances, or control macrophytes. The expense of completely removing nutrient-rich sediment and the more pressing need for watershed management are usually the primary reasons that dredging is not used more often to control algae.

Dredging may also have serious negative effects on the lake and surrounding area. Many of these problems are short-lived, and can be minimized with proper planning. It should be kept in mind, however, that dredging represents a major re-engineering of a lake, and should not be undertaken without clear recognition of its full impact, positive and negative.

Among the most serious dredging problems is not having a large enough disposal area to handle the volume of sediment and the turbid, nutrient-rich water that accompanies it. Unless the sediment-water slurry from a hydraulic dredging project can be retained long enough for it to settle, or it is treated prior to release, the nutrient-laden water will be discharged to a stream or lake. There, it

Among the most serious dredging problems is not having a large enough disposal area to handle the volume of sediment and the turbid, nutrient-rich water that accompanies it.

may deplete oxygen and cause turbidity and algal blooms. These same problems can also develop in the lake during the dredging operation, but they are usually temporary.

Containment areas for dry dredging projects are less a concern than for hydraulic projects, with wet excavation providing intermediate risks. Recent advances in polymers and mechanical dewatering now allow a hydraulically dredged slurry to be treated and loaded directly onto trucks. The cost of treatment may be offset by lower containment area costs and less material handling.

Some dredging methods create considerable turbidity, and steps must be taken to prevent the effects of contaminated sediments downstream.

Controlling inflow to the lake is another critical consideration, especially during wet or dry excavation:

- For wet excavation projects, inflows must be routed around the lake, as each increment of inflow must eventually be balanced by an equal amount of outflow, and the in-lake waters may be very turbid.
- For dry excavation, water can often be routed through the lake in a pipe or sequestered channel to prevent interaction with disturbed sediments.

Before attempting to dredge, analyze the sediments for grain size, organic content, nutrients, heavy metals, a wide variety of hydrocarbons, persistent pesticides, and other toxic materials:

- The physical and chemical nature of the dredged material will determine its potential uses; and
- Special precautions and disposal restrictions, some of them expensive, will be required if sufficient quantities of regulated contaminants are present.

Implementation and permit procedures (Chapter 8) are critical to the success of a dredging project. Failed dredging projects are common, and failure can almost always be traced to insufficient consideration of the many factors that govern dredging success:

- No technique requires more up-front information about the lake and its watershed.
- Many engineering principles are involved in planning a successful dredging project, including sediment dewatering, bulking, and transport mechanisms.
- No technique is more suitable for true lake restoration, but many potential impacts must be considered and mitigated in the dredging process.

A properly conducted dredging program removes accumulated sediment from a lake and sets it back to the time before sediment had collected.

A properly conducted dredging program removes accumulated sediment from a lake and sets it back to the time before sediment had collected. A lake can be partially dredged, but to control algae, it is far better to remove all nutrient-rich sediment, as interaction between sediments and the water column in one area can affect the entire lake.

Many benefits beyond algal control accrue from properly dredging a lake, including a deeper lake, fewer rooted plants, and less sediment-water interaction. Dredging can be very expensive, however; you should protect an investment in dredging with an active watershed management program.

Light-limiting Dyes and Surface Covers

Dyes are sometimes lumped with algaecides as management options and are often subject to the same permit process as algaecides. But they act differently; dyes reduce algae by inhibiting light penetration and resulting photosynthesis, so they are more properly classified as a physical technique.

Typically blue, inert pigments, dyes produce a pronounced, generally aesthetic color in the water column. No direct toxic effects have been reported, and organisms in the water do not absorb the color. Visibility declines in proportion to the concentration of dye.

Dyes are more effective in deep water; growths in shallow waters are unlikely to decline significantly. Dyes will not always eliminate floating scums or mats, and may actually promote surface growths. Combined with a circulation system, dyes can mask otherwise unpleasant algal blooms and improve the aesthetic appeal of ponds, reflecting pools, or similar water bodies.

Although the treatment must be repeated as the dye is flushed from the system, treatments are relatively inexpensive. The greatest negative impact is that limiting light penetration may cause thermal stratification in only 6 to 8 feet of water; the sediment-water interface may then become anoxic, creating the potential for a variety of impacts on water quality. In some cases, both aquatic vegetation and fish may be affected.

Surface covers are opaque sheets placed on the surface of a lake or reservoir to inhibit light penetration. This technique limits access on recreational lakes, but has been used in reservoirs, especially for storage of "finished" water (treated water ready for distribution). Aside from minimizing algal growth, such covers support the Federal Safe Drinking Water Act by limiting interaction with waterfowl and atmospheric deposition. Most covered storage reservoirs have roofs, but plastic covers seem to perform acceptably on others.

Mechanical Removal

Mechanical sedimentation and filtration systems routinely remove algae in drinking water treatment facilities, but such systems have not been developed for recreational lakes. Treating enough lake water to reduce algal levels in a well-mixed lake would be very difficult, as well as cost prohibitive, even for small lakes (McComas, 1993).

Algal mats can be harvested with nets, booms, or commercial macrophyte harvesters, but this can also be very expensive for an entire lake over the course of a summer. Nets or a boom system may temporarily (though inefficiently) improve small ponds. Mechanical harvesters must be properly designed to collect algal mats; although seldom used, harvesters can gather surface mats on a small scale. Since algae are mostly water, relatively little solid material is removed for each unit of effort.

Selective Withdrawal

Selective withdrawal for water supply means locating the intake at the depth where the water quality is most suited to the intended use. This requires stable vertical water density layers, and is most often used with strongly stratified lakes.

To withdraw potable water from lakes, the choice is often between high algae concentrations in the epilimnion and high iron and/or manganese in the hypolimnion. Intakes located near the thermocline (their interface) sometimes get both high algae and high metals.

A choice of intake depths is preferred, allowing the intake depth to be adjusted according to the best available water quality. For cooling water, cold hypolimnetic withdrawal is preferred, as long as it does not contain high levels of corrosive sulfides.

In managing recreational lakes, selective withdrawal is usually intended to remove the poorest quality water from the lake (normally at the bottom of the lake), discharging it at a rate that prevents anoxia near the sediment-water interface; this improves both lake conditions and discharge quality. This works in impoundments with small hypolimnia and/or large inflows, but for most lakes, the water withdrawn must be treated before discharge.

Where phosphorus has accumulated in the hypolimnion, selective discharge of hypolimnetic waters before fall turnover can reduce phosphorus. But, unless summer inflows are substantial, this may lower the lake level considerably. Selective discharge can increase the benefit of a drawdown, however, even if an outlet structure has to be retrofit; the one-time capital cost confers permanent control with minimal operation and maintenance costs.

Hypolimnetic withdrawal can reduce epilimnetic phosphorus concentrations; Nürnberg (1987) found this to be true in 17 lakes following 1 to 10 years of hypolimnetic withdrawal. To succeed, however, a withdrawal program must address the possible effects of summer drawdown, disruption of stratification, and effects on downstream water quality.

In some large western reservoirs, hypolimnetic discharges are a major outflow and actually maintain downstream coldwater fisheries. Discharged water may have to be aerated or otherwise treated, but it does remove phosphorus and other contaminants from the lake.

Algaecides

Algaecides kill algae in the lake. The oldest and still most used algaecide is copper, a cellular toxicant that comes in a wide variety of forms (Westerdahl and Getsinger, 1988a).

Copper sulfate ($CuSO_4$) is the most common and basic form; it is registered for use in potable waters, although restrictions apply in most states. Copper sulfate can be applied by towing burlap or nylon bags filled with granules (which dissolve) behind a boat, broadcasting granules, or injecting liquids. A copper slurry can be delivered to an intended depth through a weighted hose.

The method of delivery is only one factor that influences effectiveness, however. In alkaline water (150 mg calcium carbonate per liter, or more), hard water, or water high in organic matter, copper can be quickly lost from solution and thus rendered ineffective. In these cases, a liquid chelated form is often used. This formulation allows the copper to remain dissolved in the water long enough to kill algae.

Dilution is another important factor, as copper is often applied to only the upper 10 feet of water to provide a refuge for zooplankton and sensitive fish species. Vertical or horizontal mixing can rapidly decrease the treatment's effectiveness.

Most algae will be killed by doses of 1 to 2 mg $CuSO_4$/L (0.4 to 0.8 mg Cu/L) in hard water, while in soft water, doses of 0.3-0.5 mg $CuSO_4$ are usually sufficient. In most cases, cells disintegrate and release their contents into the water column. Copper sulfate usually destroys many green and blue-green algae, and nearly all diatoms, golden algae, dinoflagellates, cryptomonads, and euglenoids, although sustained control may require additional applications. Such treatments have been an important line of defense in drinking water supplies and have allowed safe swimming by increasing water clarity in many recreational lakes.

However, some planktonic forms may not be affected by copper, including certain species or strains of the filamentous blue-green algae *Aphanizomenon, Oscillatoria, Phormidium*, and *Anabaena*, and many species of the green algal order Chlorococcales. Dense algal mats, especially those formed by members of the green algal order Cladophorales, are resistant because copper cannot penetrate past the outer layer of filaments. As these are some of the most severe nuisance forms, copper treatments may eventually cause greater algal problems by not affecting these forms while reducing other algae.

Algaecides can release taste and odor agents, other organic compounds, nutrients, and toxins into the water column, where they may remain a problem.

Algaecides can release taste and odor agents, other organic compounds, nutrients, and toxins into the water column, where they may remain a problem. In killing certain species of blue-greens, algaecides may release toxins contained in their cells (Kenefick et al. 1993) that can cause human illness. Although activated carbon removes them in water treatment, simple filtration does not.

Some copper doses may also be acutely toxic to fish, although sublethal effects appear more likely (Cooke et al. 1993a). Zooplankton are especially sensitive to copper; they may die or not be able to reproduce at concentrations lower than some applied dosages. Loss of zooplankton eliminates food for many fish as well as grazing control of algae. Some doses of copper may also affect certain benthic invertebrates.

Fifty-eight years of copper sulfate treatment of several Minnesota lakes, while sometimes temporarily controlling algae, appear to have:

- Depleted dissolved oxygen;
- Increased internal nutrient cycling;
- Occasionally killed fish;
- Accumulated copper in sediments;
- Increased tolerance to copper by some nuisance blue-green algae; and
- Negatively affected fish and zooplankton.

Hanson and Stefan (1984) concluded that short-term control of algae may have been traded for long-term degradation of the lakes.

Not many alternatives to copper-based algaecides exist. **Simazine**, an organic formulation that proved highly effective against copper-resistant green algae, was voluntarily taken off the market in 1996 because of potential human health effects. **Endothall** (as the hydrothol formulation) and **diquat** are still used on hard-to-kill greens and blue-greens, but water use is restricted for multiple days after application, and diquat may be toxic to lake invertebrates. New formulations of copper are more common than new non-copper-based algaecides. Selective algaecides that take advantage of differences in cell wall composition, pigments, or food storage are just beginning to appear.

Preventive techniques are preferable to counteractions; thus, algaecide treatments should be timed to coincide with early phases of algal growth when algal

If your lake needs frequent algaecide treatment, then you should develop a more comprehensive management plan.

sensitivity is greatest and deteriorating cells minimally affect the aquatic environment. Proper timing of application requires daily to weekly tracking of algal populations; this may cost more than the actual total annual cost of the chemicals.

Given the many negative aspects of algaecides, especially those involving copper, such treatments should be used only as the last line of defense. If your lake needs frequent algaecide treatment, then you should develop a more comprehensive management plan.

Phosphorus Inactivation

Phosphorus inactivation controls algae by limiting phosphorus availability through two processes:

- Using chemicals to remove (precipitate) phosphorus from the water column; and
- Adding phosphorus binder to the lake to prevent release of phosphorus from sediments.

Phosphorus inactivation is most effective after nutrient loading from the watershed declines, because it acts only on existing phosphorus reserves in the lake.

Aluminum has been widely used for phosphorus inactivation, mostly as **aluminum sulfate** and sometimes as **sodium aluminate or polyaluminum chloride**; it binds phosphorus well under a wide range of conditions, including low oxygen. However, pH influences this process. In some cases, sodium aluminate, which raises the pH, has been successfully used in combination with aluminum sulfate, which lowers the pH (Cooke et al. 1993b). Buffering agents such as lime and sodium hydroxide may also be added to the lake before treating with aluminum sulfate.

Calcium hydroxide and **ferric chloride** successfully bind phosphorus; the former tends to raise the pH and the latter lowers the pH slightly. **Ferric sulfate** has also been applied, and lowers the pH substantially.

In practice, **aluminum sulfate** (often called **alum**) is added to the water to form colloidal aggregates of aluminum hydroxide. These aggregates rapidly grow into a visible, brownish-white floc, a precipitate that settles to the bottom in a few hours to a few days, carrying phosphorus and bits of organic and inorganic particulate matter.

After the floc settles, the water will be very clear. If enough alum is added, a thin layer of aluminum hydroxide will cover the sediments and significantly retard the release of phosphorus into the water column. In lakes where minimal nutrients enter from the watershed, this can limit algal growth for a long time.

Treating lakes with low doses of alum may remove phosphorus from the water column, but may not prevent phosphorus release from lake sediments over the long term. Determining the dosage for inactivation of sediment phosphorus depends on the form in which the phosphorus is bound (Welch and Rydin, 1999).

Good candidate lakes for phosphorus inactivation are those with low external nutrient loads and high internal phosphorus release from sediment. High alkalinity is also desirable to balance the pH when alum is used. Highly flushed impoundments are usually not good candidates because phosphorus inputs are difficult to control in such cases.

Phosphorus inactivation has succeeded in some shallow lakes, but failed where the external loads have not been previously controlled (Welch and Schrieve, 1994; Welch and Cooke, 1999). Using jar tests to evaluate appropriate dosage, successful doses have ranged from 3 to 30 g Al/m^3 (15 to 50 g Al/m^2) with pH levels remaining above 6.0. Recent studies by Rydin and Welch (1988, 1999) have produced a method for measuring available sediment phosphorus and determining the proper aluminum dose. Higher doses may be desirable if properly buffered. A 2:1 ratio of aluminum sulfate to sodium aluminate should maintain ambient pH, unless the pH is especially high because of excessive algal photosynthesis, which will be reduced by treatment.

Aluminum sulfate is often applied near the thermocline depth (even before stratification) in deep lakes, providing a refuge for fish and zooplankton that could be affected by dissolved reactive aluminum. Application methods include modified harvesting equipment, outfitted pontoon boats, and specially designed barges.

Nutrient inactivation has received increasing attention over the last decade as long-lasting results have been demonstrated in many projects, especially those employing aluminum compounds.

Nutrient inactivation has received increasing attention over the last decade as long-lasting results have been demonstrated in many projects, especially those employing aluminum compounds (Welch and Cooke, 1999). **Annabessacook Lake** in Maine suffered algal blooms for 40 years prior to the 1978 treatment with aluminum sulfate and sodium aluminate (Cooke et al. 1993a). Internal phosphorus load decreased by 65 percent, blue-green algae blooms were eliminated, and conditions have remained much improved for 20 years. Similarly impressive results have been obtained in two other Maine lakes using the two aluminum compounds together (Connor and Martin, 1989a).

Kezar Lake (New Hampshire) was treated with aluminum sulfate and sodium aluminate in 1984 after a wastewater treatment facility discharge was diverted from the lake. Both algal blooms and oxygen demand were depressed for several years, but then reappeared more quickly than expected (Connor and Martin, 1989a,b). Additional controls on external loads (wetland treatment of inflow) reversed this trend and conditions have remained markedly improved over pre-treatment conditions for 15 years. No adverse impacts on fish or benthic fauna have been observed.

Aluminum sulfate and sodium aluminate were again used with great success at **Lake Morey,** Vermont (Smeltzer et al. 1999). Treatment in late spring 1987 reduced the average spring total phosphorus concentration from 37 µg/L to 9 µg/L. Although epilimnetic phosphorus levels have varied since then, the pretreatment levels have not yet been approached. Hypolimnetic phosphorus concentrations have not exceeded 50 µg/L. Oxygen levels increased below the epilimnion, with as much as 10 vertical feet of suitable trout habitat reclaimed. Some adverse effects of the treatment on benthic invertebrates and yellow perch appear to have been temporary.

Success has also been achieved with calcium (Babin et al. 1989; Murphy et al. 1990) and iron (Walker et al. 1989) salts, but it has become clear that aluminum provides the greatest long-term binding potential for phosphorus inactivation (Harper et al. 1999). The use of **calcium** would seem to be appropriate in high pH lakes, and provides natural phosphorus inactivation in certain hardwater lakes, but has been applied on only a small scale. **Iron** seems to be most useful in conjunction with aeration systems. Aluminum salts can be used successfully in any of these cases unless toxicity becomes a problem as a consequence of pH < 6.0 or > 8.5 (7.5 if sodium aluminate is used as a buffer).

Longevity of alum treatments has generally been excellent where external inputs of phosphorus to the system have been controlled.

Longevity of alum treatments has generally been excellent where external inputs of phosphorus to the system have been controlled. As a general rule, inactivation with aluminum can be expected to last for at least three flushing cycles, with much longer effectiveness where external loading has been controlled. A review of 21 well-studied phosphorus inactivation treatments using aluminum (Welch and Cooke, 1999) indicates that effects typically last 15 years or more for dimictic (summer stratified) lakes and about 10 years for shallow, polymictic (unstratified) lakes where the technique was appropriately applied.

Despite major successes, adding aluminum salts to lakes may cause serious negative effects directly related to the alkalinity and pH of the lake water. Dosage is therefore critical. In soft (low alkalinity) water, only very small doses of alum can be added before the pH falls below 6.0. At pH 6.0 and below, $Al(OH)_2$ and dissolved elemental aluminum (Al^{+3}) become dominant: both can be toxic to aquatic life. Well-buffered, hardwater lakes can handle much higher alum doses without fear of creating toxic forms of aluminum. Softwater lakes must be buffered, either with sodium aluminate or other compounds, to prevent lowering pH while forming enough $Al(OH)_3$ to control phosphorus release.

Although pH depression is the major threat, elevated pH from over-buffering can also cause problems. **Hamblin Pond** in Massachusetts was treated with alum and sodium aluminate in 1995, after three years of pre-treatment study that demonstrated both the importance of internal phosphorus loading and limited buffering capacity (Wagner, in review). A number of problems arose during the treatment, resulting in an overdose of sodium aluminate throughout the lake. The pH rose from about 6.3 to over 9.0, resulting in a fishkill. The treatment increased the summer water transparency fivefold —and gained 10 vertical feet of coldwater fish habitat — but state agencies remain skeptical about this technique because of the fishkill.

Concerns about aluminum treatments include:

- Although the sharp increase in water transparency is usually desirable, it may allow an existing rooted plant infestation to spread into new areas or deeper water.
- The sulfates in an aluminum sulfate treatment may foster chemical reactions that disrupt the iron cycle and associated natural phosphorus binding capacity.
- Aluminum sulfate treatments that reduce the pH may cause decalcification in sensitive organisms and limit calcium control of phosphorus cycling.
- Aluminum toxicity to humans has created substantial public controversy about this treatment, but scientific investigations do not support these concerns (Krishnan, 1988).

Before using phosphorus inactivation to control algae, consult a professional who thoroughly understands your lake's chemistry.

Sediment Oxidation

Like aeration, drawdown, and phosphorus inactivation, sediment oxidation is designed to decrease phosphorus release from sediments. Called Riplox after its originator, Wilhelm Ripl, the procedure injects calcium nitrate into the top 10 inches of sediments to break down (oxidize) organic matter and promote denitrification.

If sediments are low in iron, ferric chloride or similar compounds can first be added to enhance phosphorus binding. Lime can also help raise sediment pH to the optimum 7.0–7.5.

Ripl (1976) first treated Lake Lillesjon, a 10.5-acre Swedish lake with a 6.6-foot mean depth. The treatment dramatically lowered sediment phosphorus release and lasted at least two years. A portion of a Minnesota lake was also treated, but high external loading overwhelmed the effects.

No negative impacts have been reported, but the impact on benthic communities from the chemical reactions could be severe. However, where this technique is appropriate, a significant benthic community will probably not exist prior to treatment.

Although developed in the 1970s, this technique is not widely used and needs further experimentation. Oxidation and other reactions that change sediment chemistry may be able to control internal loads of a variety of contaminants.

Settling Agents

Although the water treatment industry has long used coagulants to enhance settling and filtration, lakes have used such agents —principally alum and calcium compounds — not so much to directly remove algae as to indirectly control it by inactivating phosphorus (Babin et al. 1989; Murphy et al. 1990). Such treatments, however, do cause most algae to settle to the bottom. Various polymers could also be used, but their usage has not been documented.

The primary value of this technique is that it removes algal cells from the water column, rather than allowing them to release their contents throughout the lake. Settling may eventually result in release of cellular contents, but not rapidly and not throughout the water column.

Not all algal species settle, however; many buoyant blue-green species resist settling unless a strong floc layer develops and sweeps them out of the water column. In such cases, underdosing may not significantly reduce algal densities, or may form unsightly macroscopic clumps.

Selective Nutrient Addition

This is a theoretical approach that has not been subjected to widespread practical study. In theory, a change in nutrient ratios should drive a shift in algal composition to species better suited to the new ratio (Tilman, 1982). If algae more preferable to zooplankton become dominant, algal biomass may decline. Laboratory and some whole lake experiments support this theory, but practical applications are lacking.

In reality, competitive forces seem weak compared to predation, and changes in environmental conditions might limit competitive effects. While this approach may work under certain circumstances, it is unlikely to become a common lake management technique.

An alternative nutrient strategy involves adding nitrate to an anoxic hypolimnion (Kortmann and Rich, 1994) to limit generation of sulfides from sulfates, reduce iron-sulfide reactions, and enhance iron-phosphorus binding. Nitrogen would be released as a gas, minimizing uptake by algae. Nitrogen could also be added as aluminum nitrate or ferric nitrate to promote phosphorus binding. Cold groundwater high in nitrates might be used for this purpose. Although theoretically sound, this approach needs well-documented practical applications.

Management to Reduce Nutrient Input

Techniques in this category are largely watershed management methods (see Chapter 6). The boundary blurs, however, when part of the lake is used to treat incoming water — or when runoff is treated with phosphorus inactivators prior to discharge to a lake, where the floc then settles (Harper et al. 1999). The principles and experience set forth in Chapters 2 and 6 apply, but it is important to recognize that, with proper planning and implementation, part of the lake itself can treat watershed-based pollutants.

Enhanced Grazing

Grazing can be a powerful force in structuring the algal community. It is one of a group of procedures called "biomanipulation" that Shapiro et al. (1975) suggested could greatly improve lake quality without using expensive machines or chemicals. Biomanipulation depends on general ecological principles to manipulate the lake's biological components to produce desired conditions, and has performed satisfactorily in many systems.

Biomanipulation: The alteration of one or more biological components of an aquatic system to cause other biological components or water quality to change in a desired manner.

At times, grazing zooplankton and not the quantity of nutrients control the amount of algae in open water (McQueen et al. 1986). Even if productivity is high, grazing prevents biomass from accumulating. Zooplankters are microscopic, often crustacean animals found in every lake, but at different densities and varying sizes. A sufficient population of large-bodied herbivorous zooplankters (preferably *Daphnia*) can filter the entire epilimnion each day during the summer as they graze on algae, bacteria, and organic matter.

Although some algae are immune to grazing, continual strong grazing can reduce algae overall and increase transparency. Excessive nutrients may stimulate growth by resistant algae to overcome this effect, but, usually, large-bodied grazers will maintain the lowest possible algal biomass and highest possible clarity (Lathrop et al. 1999) for the level of fertility in the lake. If the lake is otherwise very turbid, grazing may have no observable effect.

Large-bodied grazers are not common everywhere, however.

- They won't be found in some lakes, such as Florida's subtropical lakes.
- In many others they're eaten by certain fish, including the fry of nearly every fish species and the adults of bluegill, pumpkinseed, perch, shad, alewife, shiners, and others. In lakes dominated by adult species such as largemouth bass, walleye, and northern pike, large-bodied zooplankton are more likely to be abundant because those fish have eaten the predators of the zooplankton.

Other conditions that might reduce the population of large-bodied zooplankton include:

- An anoxic metalimnion or hypolimnion, common in eutrophic lakes, that eliminates these zones as daytime refuges for zooplankton from visually feeding fish (aeration can eliminate this problem);
- The toxic effect of pesticides that enter the lake with runoff; and
- Copper sulfate used for temporary algal control can also kill zooplankton at doses below those needed for algae control.

Severe mortality of zooplankton appears responsible for the commonly observed rebound of algae following a copper treatment (Cooke and Kennedy, 1989).

Algal control by animals in the food chain is called "top-down," unlike the more common "bottom-up" algal control through nutrient limitation. Figure 7-3 depicts these food web interactions, which vary considerably over space and time.

To increase the large-bodied zooplankton population, you must reduce the number of zooplankton-eating fish (Hosper and Meijer, 1993; Dettmers and Stein, 1996). Either stock more piscivorous fish or get rid of the planktivores by using such techniques as water level drawdown, winterkill, netting, or electroshocking — even poison them as a last resort.

Removing planktivores by hook and line is usually a hopeless task. But just the act of fishing points to a potential conflict. In lakes where sport fishing is the first priority, planktivorous fish form an essential food web link in the fishery. So how do you manage both for a trophy gamefish population (which needs zooplankton-eaters) and for the clearest possible water (which needs zooplankton)? Controlling the density of stunted panfish appears to serve both goals and has improved both water clarity and fishing (Wagner, 1986).

Algae-eating fish might control algal biomass if stocked in sufficient quantities. However, no native species of fish in the United States consumes enough algae to be effective, so non-native species (e.g., *Tilapia*), probably of tropical origin, would have to be introduced. Given the track record of introduced species (Mills et al. 1994), this does not appear to be a desirable approach, and many states have banned such introductions. In addition:

- The excreted nutrients from such fish might also support the growth of as much algae as those fish could consume.
- No fish can efficiently feed on the smallest algal cells, perhaps resulting in a shift toward smaller cell size and greater turbidity per unit of algal biomass present.
- Tropical species such as *Tilapia* are unlikely to overwinter in the more northern states, limiting the duration of any effect.

Removal of Bottom-feeding Fish

Another type of biomanipulation that could improve lake transparency is eliminating fish such as the **common carp** or **bullheads** that are bottom browsers. Browsing releases significant amounts of nutrients to the water column as these fish feed and digest food. Harvesting these fish has increased clarity in some cases, but removing them can be very difficult since they tolerate very low levels of dissolved oxygen and high doses of fish poisons. Labor-intensive programs appear necessary to substantiallly reduce bottom-feeding fish populations (McComas, 1993), unless the entire fish population can be eliminated through complete drawdown, complete freezing, or extremely high doses of rotenone or other fish poisons.

Of particular interest is the **grass carp,** which has been placed in many lakes to control nuisance plants. Shallow, fertile lakes appear to have alternative stable states dominated by either rooted plants or algae (Scheffer et al. 1993). Removing the rooted plants may induce algal blooms and worsen conditions for many lake uses. The grass carp may also add enough nutrients to the water column to support algal blooms. Grass carp baits and traps are now sold, but an entire population is difficult to completely remove. If the fish are sterile, they will eventually die off, but the algae blooms may persist in the absence of vascular plants.

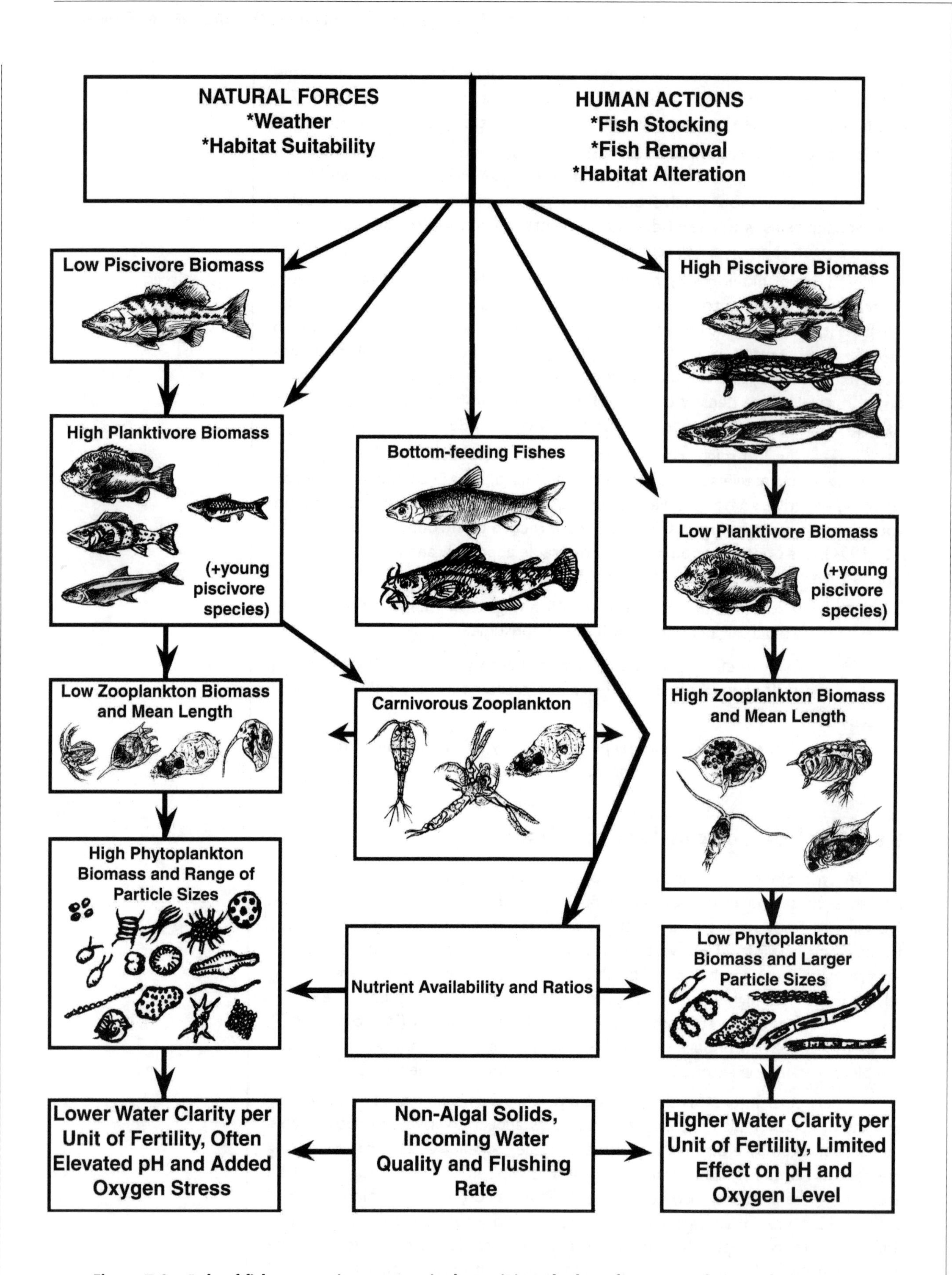

Figure 7-3.—Role of fish community structure in determining plankton features and water clarity.

Pathogens

Viral, bacterial, and fungal pathogens have each been explored as possible control methods for algae. Ideally, a lake would be inoculated with a pathogen that targets either a broad spectrum of algae, or a few especially obnoxious species. Such pathogens have been tried (Lindmark, 1979), but none has proven effective and controllable. The complexity of biological interactions appears beyond our sustained control, and although we can set processes in motion to produce desired conditions in a lake, those conditions tend to be temporary.

Competition and Allelopathy

Negative interactions between rooted plants and algae might be harnessed to control algal biomass, but they also might create a macrophyte nuisance. Commercially marketed "**nutri-pods**" incorporate rooted or floating plants into a floating structure from which excess biomass can be removed as it develops. The success of this approach has not been scientifically documented, but it has no apparent adverse ecological impacts as long as native, non-nuisance species are used.

Competition is also used at the bacterial level, using **microbial additives** developed largely in the wastewater treatment industry. According to product literature, these microbes limit the availability of nutrients essential for algal growth, thus reducing algal blooms. The products surveyed are primarily denitrifiers, removing nitrogen from the system and creating a nitrogen limitation on growth. However, little scientific documentation of this technique's effects in lakes exists, and reduction of nitrogen:phosphorus ratios favors certain nuisance-type blue-green algae.

Plantings to reduce light penetration might control algae, but also produce many negative side effects. Surface-covering growths of duckweed, water hyacinth, or water chestnut could provide such a light barrier, but at great expense to habitat and water quality.

Allelopathy: Control of one plant by another through chemical releases.

Although aging rooted plants often release nutrients that cause algal blooms, **release of allelopathic substances** during the plants' more active growth phase may inhibit algal growth. Mat-forming algae in rooted plant beds appear unaffected, but many more planktonic algal species decline when rooted plant growths are dense. This may represent a trade-off between an algal nuisance and a rooted plant nuisance, and many lakes have both.

Barley straw appears to be able to control algal densities (Barrett et al. 1996; Wynn and Langeland, 1996), and combines features of algaecides, allelopathy, and competition. Preferably added to shallow, moving water or from pond-side digesters, decaying barley straw gives off substances that inhibit algal growth, especially that of blue-green algae. Although not a thoroughly understood technique, research conducted mainly in England has demonstrated that the decomposition of the barley straw produces allelopathic compounds that act as algaecides. Microbial activity may also compete with algae for nutrients.

Doses of barley straw under well-oxygenated conditions are typically around 2.5 g/m^2 of pond surface, with doses of 50 g/m^2 or more necessary where initial algal densities are high or flow is limited. Doses of 100 g/m^2 may cause oxygen stress in the pond as decomposition proceeds, but this might be avoided by using a land-based digester into which straw is deposited and through which water is pumped as the straw decays.

Nuisance Vascular Plants

Overabundant rooted and floating vascular plants create a major nuisance for most lake and reservoir users. In extreme cases, particularly in ponds and in shallow, warm, well-lighted lakes and waterways of the southern United States, weeds can cover the entire lake surface and fill most of the water column. Obviously, plants can be both desirable and attractive, but unwanted plant (weed) infestations interfere with recreation, detract from aesthetic value, and can impair habitat as well. They can also introduce significant quantities of nutrients and organic matter to the water column, stimulating algal blooms and causing damaging fluctuations in dissolved oxygen.

Macrophytes (vascular plants and visible algal mats) are generally grouped into classes called **emergents** (such as alligatorweed and cattails), **floating-leaved** (water hyacinth and water lilies), and **submergents** (hydrilla, milfoil, and naiads), plus the mats of filamentous algae discussed in the nuisance algae section of this chapter. Understanding the factors that control plant growth is the first step in controlling weeds.

- Macrophytes reproduce by producing flowers and seeds and/or asexually from stem fragments or shoots extending from roots. The primary means of reproduction is an extremely important feature of a plant, and will greatly affect the applicability of control methods.

- Growth rates of macrophytes, especially non-native species like water hyacinth, hydrilla, and milfoil, can be very high, largely related to substrate and light conditions. Submergent plants grow profusely only when they have enough light underwater. Highly turbid lakes and reservoirs probably won't have dense beds of submerged plants. Significant reductions in algal blooms can also enhance light penetration and allow weeds to grow more extensively and densely.

- High silt loads can create a favorable plant substrate, but the silt may also cause severe turbidity, which limits growth.

- Steep-sided lakes support a much smaller plant community as a consequence of both peripheral substrate and light limitations.

- A few plants, including water hyacinth, water lettuce, duckweed, and watermeal, can float on the surface with no roots in the sediment, nearly eliminating substrate and light as key control factors.

- Most macrophytes (but certainly not all) obtain most of their nutrition from the sediment through roots. This is an important ecological feature, because they're not affected by the reduction of nutrient concentrations in the water column. When the sediments are either highly organic (very loose muck) or inorganic (rock to coarse sand), macrophytes may grow poorly because their roots can't take hold and obtain nutrients in either sediment type. In these two extremes, emergent plants may replace submergents in shallow water because their more extensive root systems are better adapted to these conditions.

Plant groupings and controls vary with geography. In southern states bordering the coast, plants grow most of the year, often rapidly, into long-lived, dense, and expansive growths of vegetation. Most non-native species are found there;

they can out-compete many more desirable native species for space. Aquatic plant management in these ecosystems often requires methods that might seem extreme in northern ecosystems.

Northern lakes and reservoirs have their share of weed infestations, but seasonal changes in light and temperature tend to limit nuisance conditions to the summer. As this corresponds with the period of greatest human use of lakes, however, plants usually need to be managed. However, opinions on the level and forms of control vary more than in southern regions.

Setting goals for rooted plant control is a critical planning step and the choice of management technique(s) depends on those goals. Ecologically, most lakes need a certain amount of plants; often a balance between rooted plants and algae must be recognized. Where fishing is the primary objective, substantial bottom coverage is desirable, with some vertical and horizontal structure created by different species of plants to enhance the habitat for different fish species or life stages. For swimming purposes, having no plants seems safer, but a low, dense cover in shallow lakes with silty bottoms can minimize turbidity, another safety concern.

A program to structure the plant community to meet clear goals in an ecologically and ethically sound manner is more appropriate, although often quite expensive.

Perhaps the simplest axiom for plant management is that if light penetrates to the bottom and the substrate is not rock or cobble, plants will grow. A program intended to eliminate all plants is both unnatural and maintenance intensive, if possible. A program to structure the plant community to meet clear goals in an ecologically and ethically sound manner is more appropriate, although often quite expensive.

Table 7-5 overviews the techniques used to control rooted plants, with notes on the action, advantages, and disadvantages of each technique. Additional details are provided in narrative form here and in Cooke et al. 1993a, and Hoyer and Canfield, 1997.

Benthic Barriers

These bottom covers work on the principle that rooted plants require light and cannot grow through physical barriers.

Benthic: Refers to life or things found on the bottom of a lake.

- ***Natural benthic barriers:*** Clay, silt, sand, and gravel have been used for many years, although plants often root in these covers eventually, and current environmental regulations usually won't approve depositing such fill. In the reverse layering technique (KVA, 1991) sand is pumped from beneath a muck or silt layer and deposited as a new layer on top. Technically, this is reorganizing the sediments, not new fill. Although expensive on a large scale and dependent on the composition of the sediment, this technique may restore the natural lake bottom without removing sediment.

- ***Artificial benthic barriers:*** Over the last three decades, various materials, including polyethylene, polypropylene, fiberglass, and nylon sheets have been developed to cover sediments. Available in both solid and porous forms, manufactured benthic barriers cover plants to limit light, physically disrupt growth, and allow chemical reactions to restrict plant development (Perkins et al. 1980).

Table 7-5.—Management options for control of rooted aquatic plants.

OPTION	MODE OF ACTION	ADVANTAGES	DISADVANTAGES
PHYSICAL CONTROLS			
1. Benthic barriers	◆Mat of variable composition laid on bottom of target area	◆Prevents plant growth	◆May cause anoxia at sediment-water interface
	◆Can cover area for as little as several weeks or permanently	◆Reduces turbidity from soft sediment	◆May limit benthic invertebrates
	◆Maintenance improves effectiveness	◆Can cover undesirable substrate	◆May interfere non-selectively with plants in target area
	◆Most often used in swimming areas and around docks	◆Can improve fish habitat	◆May inhibit spawning or feeding by some fish
1.a. Porous or loose-weave synthetic materials	◆Laid on bottom and usually anchored by sparse weights or stakes	◆Allows some escape of gases which may build up underneath	◆Allows some growth through pores
	◆Removed and cleaned or flipped and repositioned at least once per year for maximum effectiveness	◆Panels may be flipped in place or removed for relatively easy cleaning or repositioning	◆Gas may still build up underneath in some cases
1.b. Non-porous or sheet synthetic materials	◆Laid on bottom and anchored by many stakes, anchors, or weights, or by layer of sand	◆Prevents all plant growth until buried by sediment	◆Gas buildup may cause barrier to float upwards
	◆Not typically removed, but may be swept or blown clean periodically	◆Minimizes interaction of sediment and water column	◆Strong anchoring makes removal difficult and can hinder maintenance
1.c. Sediments of a desirable composition	◆Sediments may be added on top of existing sediments or plants	◆Plant biomass can be buried	◆Lake depth may decline
	◆Use of sand or clay can limit plant growths and alter sediment-water interactions	◆Seed banks can be buried deeper	◆Sediments may sink into or mix with underlying muck
	◆Sediments can be applied from the surface or suction-dredged from below muck layer (reverse layering technique)	◆Sediment can be made less hospitable to plant growths	◆Permitting for added sediment may be difficult
		◆Nutrient release from sediments may be reduced	◆Addition of sediment may increase turbidity initially
		◆Surface sediment can be made more appealing to human users	◆New sediment may contain nutrients or other contaminants
		◆Reverse layering requires no addition or removal of sediment	◆Generally too expensive for large-scale application

Table 7-5.—Management options for control of rooted aquatic plants (continued).

OPTION	MODE OF ACTION	ADVANTAGES	DISADVANTAGES
PHYSICAL CONTROLS			
2. Dredging	• Sediment is physically removed by wet or dry excavation, with deposition in a containment area for dewatering/disposal	• Achieves plant removal with some flexibility	• Temporarily removes benthic invertebrates
	• Dredging can be applied on a limited basis, but is most often a major restructuring of a severely impacted system	• Increases water depth	• May create turbidity
	• Plants and seed beds are removed and re-growth can be limited by light and/or substrate limitation	• Can reduce pollutant reserves	• May eliminate fish community (complete dry dredging only)
		• Can reduce sediment oxygen demand	• May cause impacts from containment area discharge
		• Can improve spawning habitat for many fish species	• May cause impacts from dredged material disposal
		• Allows complete renovation of aquatic ecosystem	• May interfere with recreation or other uses during dredging
			• Usually very expensive
2.a. "Dry"excavation	• Lake drained or lowered to maximum extent practical	• Tends to facilitate a very thorough effort	• Eliminates most aquatic biota unless a portion left undrained
	• Target material dried to maximum extent possible	• May allow drying of sediments prior to removal	• Eliminates lake use during dredging
	• Conventional excavation equipment used to remove sediments	• Allows use of less specialized equipment	• Risks downstream turbidity during storms
2.b "Wet" excavation	• Lake level may be lowered, but sediments not substantially dewatered	• Requires least preparation time or effort; tends to be least costly dredging approach	• Usually creates extreme turbidity
	• Draglines, bucket dredges, or long-reach backhoes used to remove sediment	• May allow use of easily acquired equipment	• Tends to result in sediment deposition in surrounding area
		• May preserve most aquatic biota	• Normally requires intermediate containment area to dry sediments prior to hauling
			• May cause severe disruption of ecological function
			• Usually eliminates most lake uses during dredging
2.c.Hydraulic removal	• Lake level not reduced	• Creates minimal turbidity and limits impact on biota	• Often leaves some sediment behind
	• Suction or cutterhead dredges create slurry which is hydraulically pumped to containment area	• Can allow some lake uses during dredging	• Requires sophisticated and more expensive containment area or advanced dewatering system
	• Slurry is dewatered; sediment retained, water discharged	• Allows removal with limited access or shoreline disturbance	• Cannot handle extremely coarse or debris-laden materials
			• Requires overflow discharge from containment area

Table 7-5.—Management options for control of rooted aquatic plants (continued).

OPTION	MODE OF ACTION	ADVANTAGES	DISADVANTAGES
PHYSICAL CONTROLS			
3. Dyes and surface covers	◆Water-soluble dye is mixed with lake water, thereby limiting light penetration and inhibiting plant growth	◆Limits light on plant growth without high turbidity or great depth	◆May not control peripheral or shallow-water rooted plants
	◆Dyes remain in solution until washed out of system	◆May achieve some control of algae as well	◆May cause thermal stratification in shallow ponds
	◆Opaque sheet material applied to water surface	◆May achieve some selectivity for species tolerant of low light	◆May facilitate anoxia at sediment interface with water
			◆Covers inhibit gas exchange with atmosphere
4. Mechanical removal	◆Plants reduced by mechanical means, possibly with disturbance of soils	◆Highly flexible control	◆May impact aquatic fauna
	◆Collected plants may be placed on shore for composting or other disposal	◆May remove other debris	◆Non-selectively removes plants in treated area
	◆Wide range of techniques employed, from manual to highly mechanized	◆Can balance habitat and recreational needs	◆May spread undesirable species by fragmentation
	◆Application once or twice per year usually needed		◆May generate turbidity
4.a. Hand pulling	◆Plants uprooted by hand ("weeding") and preferably removed	◆Highly selective technique	◆Labor intensive
4.b. Cutting (without collection)	◆Plants cut in place above roots without being harvested	◆Generally efficient and less expensive than complete harvesting	◆Leaves root systems and part of plant for re-growth
	◆May employ grinder to more completely destroy vegetation		◆Leaves cut vegetation to decay or to re-root
			◆Not selective within applied area
4.c. Harvesting (with collection)	◆Plants cut at depth of 2-10 ft and collected for removal from lake	◆Allows plant removal on greater scale	◆Limits depth of operation
			◆Usually leaves fragments which may re-root and spread infestation
			◆May impact lake fauna
			◆Not selective within applied area
			◆More expensive than cutting
4.d. Rototilling	◆Plants, root systems, and surrounding sediment disturbed with mechanical blades	◆Can thoroughly disrupt entire plant	◆Usually leaves fragments which may re-root and spread infestation
			◆May impact lake fauna
			◆Not selective within applied area
			◆Creates substantial turbidity
			◆More expensive than harvesting

Table 7-5.—Management options for control of rooted aquatic plants (continued).

OPTION	MODE OF ACTION	ADVANTAGES	DISADVANTAGES
PHYSICAL CONTROLS			
4.e.Hydroraking	◆Plants, root systems, and surrounding sediment and debris disturbed with mechanical rake; part of material usually collected and removed from lake	◆Can thoroughly disrupt entire plant	◆Usually leaves fragments which may re-root and spread infestation
		◆Also allows removal of stumps or other obstructions	◆May impact lake fauna
			◆Not selective within applied area
			◆Creates substantial turbidity
			◆More expensive than harvesting
5. Water level control	◆Lowering or raising the water level to create an inhospitable environment for some or all aquatic plants	◆Requires only outlet control to affect large area	◆May create potential issues with water supply
	◆Disrupts plant life cycle by dessication, freezing, or light limitation	◆Provides widespread control in increments of water depth	◆May have potential issues with flooding
		◆Complements certain other techniques (dredging, flushing)	◆May impact non-target flora and fauna
5.a. Drawdown	◆Lowering of water over winter allows freezing, dessication, and physical disruption of plants	◆Provides control with some flexibility	◆May impact contiguous emergent wetlands
	◆Timing and duration of exposure and degree of dewatering are critical aspects	◆Provides opportunity for shoreline clean-up/structure repair	◆May affect overwintering reptiles and amphibians
	◆Variable species tolerance to drawdown; emergent species and seed-bearers are less affected	◆May help with flood control	◆May impair well production
	◆Most effective on annual to once/3 yr. basis	◆Impacts vegetative propagation species with limited impact to seed-producing populations	◆May reduce potential water supply and fire fighting capacity
			◆May alter downstream flows
			◆May cause overwinter water level variation
			◆May cause shoreline erosion and slumping
			◆May result in greater nutrient availability for algae
5.b. Flooding	◆Higher water level in the spring can inhibit seed germination and plant growth	◆Where water is available, this can be an inexpensive technique	◆Water for raising the level may not be available
	◆Higher flows that are normally associated with elevated water levels can flush seed and plant fragments from system	◆Plant growth need not be eliminated, merely retarded or delayed	◆May cause peripheral flooding, and/or property damage
		◆Timing of water level control can selectively favor certain desirable species	◆May have downstream impacts
			◆Many species may not be affected, and some may benefit
			◆Algal nuisances may increase where nutrients are available

Table 7-5.—Management options for control of rooted aquatic plants (continued).

OPTION	MODE OF ACTION	ADVANTAGES	DISADVANTAGES
CHEMICAL CONTROLS			
6. Herbicides	◆ Liquid or pelletized herbicides applied to target area or to plants directly	◆ Wide range of control is possible	◆ May be toxic to non-target species of plants/animals
	◆ Contact or systemic chemicals kill plants or limit growth	◆ May be able to selectively eliminate species	◆ Possible downstream impacts; may affect non-target areas within pond
	◆ Typically requires application every 1-5 yrs	◆ May achieve some algae control as well	◆ May restrict water use for varying time after treatment
			◆ May increase oxygen demand from decaying vegetation
			◆ May cause recycling of nutrients to allow other growths
6.a. Forms of copper	◆ Contact herbicide	◆ Moderately effective control of some submersed plant species	◆ Toxic to aquatic fauna as a function of concentration, formulation, and ambient water chemistry
	◆ Cellular toxicant, suspected membrane transport disruption	◆ More often an algal control agent	◆ Ineffective at colder temperatures
	◆ Applied as wide variety of liquid or granular formulations, often in conjunction with surfactants or other herbicides		◆ Copper ion persistent; accumulates in sediments or moves downstream
6.b. Forms of endothall (7-oxabicyclo [2.2.1] heptane-2, 3-dicarboxylic acid)	◆ Contact herbicide with limited translocation potential	◆ Exerts moderate control of some emersed plant species, moderately to highly effective control of floating and submersed species	◆ Non-selective in treated area
	◆ Membrane-active chemical that inhibits protein synthesis	◆ Has limited toxicity to fish at recommended dosages	◆ May be toxic to aquatic fauna (varying degrees by formulation)
	◆ Causes structural deterioration	◆ Acts rapidly	◆ Time delays necessary on use for water supply, agriculture, and contact recreation
	◆ Applied as liquid or granules		
6.c. Forms of diquat (6,7-dihydropyrido [1,2-2′,1′-c] pyrazinediium dibromide)	◆ Contact herbicide	◆ Exerts moderate control of some emersed plant species, moderately to highly effective control of floating or submersed species	◆ Non-selective in treated area
	◆ Absorbed by foliage but not roots	◆ Has limited toxicity to fish at recommended dosages	◆ Sometimes toxic to zooplankton at recommended dosage
	◆ Strong oxidant; disrupts most cellular functions	◆ Acts rapidly	◆ Inactivated by suspended particles; ineffective in muddy waters
	◆ Applied as a liquid, sometimes in conjunction with copper		◆ Time delays necessary on use for water supply, agriculture, and contact recreation

Table 7-5.—Management options for control of rooted aquatic plants (continued).

OPTION	MODE OF ACTION	ADVANTAGES	DISADVANTAGES
CHEMICAL CONTROLS			
6.d. Forms of glyphosate (N phosphonom ethyl glycine)	◆Contact herbicide	◆Exerts moderately to highly effective control of emersed and floating plant species	◆Non-selective in treated area
	◆Absorbed through foliage; disrupts enzyme formation and function in uncertain manner	◆Can be used selectively, based on application to individual plants	◆Inactivated by suspended particles; ineffective in muddy waters
	◆Applied as liquid spray	◆Acts rapidly	◆Not for use within 0.5 miles of potable water intakes
		◆Low toxicity to aquatic fauna at recommended dosages	◆Highly corrosive; storage precautions necessary
		◆No time delays needed for use of treated water	
6.e. Forms of 2,4-D (2,4-dichlorophenoxy acetic acid)	◆Systemic herbicide	◆Moderately to highly effective control of a variety of emersed, floating and submersed plants	◆Has variable toxicity to aquatic fauna, depending upon formulation and ambient water chemistry
	◆Readily absorbed and translocated throughout plant	◆Can achieve some selectivity through application timing and concentration	◆Time delays necessary for use of treated water for agriculture and contact recreation
	◆Inhibits cell division in new tissue, stimulates growth in older tissue, resulting in gradual cell disruption	◆Fairly fast action	◆Not for use in water supplies
	◆Applied as liquid or granules, frequently as part of more complex formulations, preferably during early growth phase of plants		
6.f. Forms of fluridone (1-methyl-3-phenyl-5-[-3-{trifluoromethyl} phenyl]-4[IH]-pyridinone)	◆Systemic herbicide	◆Can be used selectively, based on concentration	◆Impacts on non-target plant species possible at higher doses
	◆Inhibits carotenoid pigment synthesis and impacts photosynthesis	◆Gradual deterioration of affected plants limits impact on oxygen level (BOD)	◆Extremely soluble and mixable; difficult to perform partial lake treatments
	◆Best applied as liquid or granules during early growth phase of plants	◆Effective against several difficult-to-control species	◆Requires extended contact time (40 days recommended)
		◆Low toxicity to aquatic fauna	
6.g. Forms of triclopyr (3,5,6-trichloro-2-pyridinyloxyacetic acid)	◆Systemic herbicide, registered for experimental aquatic use by cooperators in selected areas only at this time	◆Effectively controls many floating and submersed plant species	◆Impacts on non-target plant species possible at higher doses
	◆Readily absorbed by foliage, translocated throughout plant	◆Can be used selectively; more effective against dicot plant species, including many nuisance forms	◆Current time delay of 30 days on consumption of fish from treated areas
	◆Disrupts enzyme systems specific to plants	◆Effective against several difficult-to-control species	◆Necessary restrictions on use of treated water for supply or contact recreation not yet certain
	◆Applied as liquid spray or subsurface injected liquid	◆Low toxicity to aquatic fauna	
		◆Acts rapidly	

Table 7-5.—Management options for control of rooted aquatic plants (continued).

OPTION	MODE OF ACTION	ADVANTAGES	DISADVANTAGES
BIOLOGICAL CONTROLS			
7. Biological introductions	◆ Fish, insects, or pathogens that feed on or parasitize plants are added to system to effect control	◆ Provides potentially continuing control with one treatment	◆ Typically involves introduction of non-native species
	◆ The most commonly used organism is the grass carp, but the larvae of several insects have been used more recently, and viruses have been tested	◆ Harnesses biological interactions to produce desired conditions	◆ Effects may not be controllable
		◆ May produce potentially useful fish biomass as an end product	◆ Plant selectivity may not match desired target species
			◆ May adversely affect indigenous species
7.a. Herbivorous fish	◆ Sterile juveniles stocked at density that allows control over multiple years	◆ May greatly reduce plant biomass in single season	◆ May eliminate all plant biomass, or impact non-target species more than target forms
	◆ Growth of individuals offsets losses or may increase herbivorous pressure	◆ May provide multiple years of control from single stocking	◆ Funnels energy into largely unused fish biomass and algae
		◆ Sterility intended to prevent population perpetuation and allow later adjustments	◆ May drastically alter habitat
			◆ May escape to new habitats upstream or downstream
			◆ Population control uncertain unless absolutely sterile
7.b. Herbivorous insects	◆ Larvae or adults stocked at density intended to allow control with limited growth	◆ May involve species native to region, or even targeted lake	◆ Population ecology suggests incomplete control likely
	◆ Intended to selectively control target species	◆ Expected to have no negative effect on non-target species	◆ Oscillating cycle of control and re-growth likely
	◆ Milfoil weevil becoming popular, but still experimental	◆ May facilitate longer-term control with limited management	◆ Predation by fish may complicate control
			◆ Other lake management actions may interfere with success
7.c. Fungal/bacterial/ viral pathogens	◆ Inoculum used to seed lake or target plant patch	◆ May be highly species specific	◆ Largely experimental; effectiveness and longevity of control not well known
	◆ Growth of pathogen population expected to achieve control over target species	◆ May provide substantial control after minimal inoculation effort	◆ Infection ecology suggests complete control unlikely
			◆ Possible side effects not well understood
7.d. Selective plantings	◆ Establishment of plant assemblage resistant to undesirable species	◆ Can restore native assemblage	◆ Largely experimental at this time; few well-documented cases
	◆ Plants introduced as seeds, cuttings, or whole plants	◆ Can encourage assemblage most suitable to lake uses	◆ Nuisance species may eventually outcompete established assemblage
		◆ Supplements targeted species removal techniques	◆ Introduced species may become nuisances

Benthic barriers can control plant growth in small areas such as dock spaces and swimming beaches. They're also practical for creating access lanes and structural habitats. Large areas are not often treated, however, because the cost of materials and application is high and maintenance can be problematic (Engel, 1984).

Benthic barrier problems of prime concern:

- Long-term integrity of the barrier;
- Billowing caused by trapped gases;
- Accumulation of sediment on top of barriers;
- Growth of plants on porous barriers;
- Temporary decline of the benthic community because of lower oxygen and chemical changes (recovery is rapid once the barrier is removed [Ussery et al. 1997]);
- Their non-selectivity — they usually kill all plants over which they are applied; and finally,
- Market stability of the barrier materials — many types are not on the market more than 5 to 10 years.

Guidelines for successfully using benthic barriers:

- Porous barriers will be subject to less billowing, but will allow settling plant fragments to root and grow, making annual maintenance essential.
- Solid barriers will generally prevent rooting if there's no sediment on them, but will billow after enough gases accumulate; venting and strong anchoring are essential in most cases.
- Plants under the barrier will usually die completely after about a month, with solid barriers more effective than porous ones in killing the whole plant; if they're strong enough, barriers can then be moved to a new location. However, solid barriers prevent recolonization and thus might be best kept in place.
- Installation may be difficult to accomplish over dense plant growth; a winter drawdown might provide the opportunity. Late spring application also works, however, despite the presence of plants at that time. Barriers applied in early May have been removed in mid-June with no substantial plant growth through the summer (Wagner, 1991).
- Scuba divers normally apply the covers in deeper water, greatly increasing labor costs.
- Bottom barriers will usually accumulate sediment, which allows plant fragments to root. Barriers must then be cleaned, either in-place (which is very labor-intensive) or by removing them.

Despite application and maintenance issues, benthic barriers are very effective. In northern waters, benthic barriers can control milfoil locally (Engel, 1984; Perkins et al. 1980; Helsel et al. 1996), and at the same time create more edge habitat where plants grow densely.

As an example, benthic barriers have been used at Lake George since 1986 (Eichler et al. 1995). Plastic sheets initially installed over three acres of milfoil in two areas controlled milfoil there for about three years. But by 1990, milfoil had rooted in the sediment, covering portions of the barrier.

At that time, a fine mesh material and a solid sheet were installed at eight sites. Both succeeded, although when left in place without annual maintenance, recolonization was far greater on the mesh than on the solid sheet.

Both native species and milfoil recolonized areas of Lake George where benthic barriers had been removed (Eichler et al. 1995), but milfoil did not dominate for at least two growing seasons. Plant cover was sparse for at least the first month after barrier removal and typically did not exceed 74 percent after two growing seasons, providing ample opportunity for milfoil invasion.

Similar plant recolonization occurred in two swimming areas in Great Pond, Massachusetts (Wagner, 1991) where barriers were applied to improve swimming safety (not control invasive plants). In one swimming area, a plant community nearly identical to the original assemblage returned within one to two years after barrier removal. Regrowth in the second area was kept to a minimum by foot traffic; the area had been considered unusable before treatment.

Dredging

Sediment removal was described in some detail in the section on algal control, but it also works as a plant control technique in two principal ways:

- Limiting light (and growth) by increasing water depth; or
- Removing enough "soft" sediment (muck, clay, silt, and fine sand) to reveal a less hospitable substrate (typically rock, gravel, or coarse sand).

An exception is **suction dredging,** used mostly to remove specific whole plants and their seeds. Suction dredging might be considered a form of harvesting, however, as plants are extracted from the bottom by scuba divers operating the suction dredge; sediment is often returned to the lake.

The amount of sediment removed, and hence, the greater depth and light limitation are critical to long-term control of rooted, submerged plants. There appears to be a direct relation between water transparency, as determined with a Secchi disk, and the maximum depth of colonization (MDC) by macrophytes. Canfield et al. (1985) provided equations to estimate MDC in Florida and Wisconsin from Secchi disk measurements:

STATE	EQUATION
Florida	$\log MDC = 0.42 \times (\log SD) + 0.41$
Wisconsin	$\log MDC = 0.79 \times (\log SD) + 0.25$

where SD = Secchi depth in meters

For a Florida lake with a Secchi disk transparency of about 6 feet (1.8 meters), we would expect some submergent plants in 11 feet (3.4 meters) of water and more plants in progressively shallower water. A great deal of sediment (calculated from a bathymetric map) might have to be removed to create large areas of the lake with depths of 11 feet or more.

These equations also indicate that actions that greatly improve water clarity, such as erosion control or phosphorus inactivation, may encourage plant growth and distribution, increasing the depth dredging must go to limit light. Partial deepening may limit the amount of vegetation that reaches the surface, but may also favor species that tolerate low light — particularly non-native, nuisance species such as hydrilla and milfoil.

If the soft sediment supporting rooted plants is not very thick, it may be possible to remove it and expose rock ledge or cobble so as to prevent rooted plant growth. But such circumstances are rare; either the sediments grade slowly into coarser materials, or not all fine sediments can be removed from around the rock. Consequently, some regrowth is to be expected where light penetrates to the bottom. Successful dredging may keep this regrowth to only 25 percent of the pre-dredging coverage, without domination by recently invading species.

Remember, you can — and should — expect some rooted plant regrowth; it is indeed desirable for proper ecological function of the lake as a habitat and for processing of future pollutant inputs.

If you're dredging to control rooted plants — and your budget won't let you remove all the soft sediment — then dredge to the desired depth or substrate in part of the lake, rather than trying to remove some sediment throughout the lake.

If you're dredging to control rooted plants – and your budget won't let you remove all the soft sediment – then dredge to the desired depth or substrate in part of the lake, rather than trying to remove some sediment throughout the lake.

Dredging to control rooted plants has had mixed results. As with dredging for algal control, failures are invariably linked to incomplete pre-dredging assessment and planning.

- Control through light limitation appears more successful than limiting the substrate, largely because it is so difficult to remove all soft sediment from shallow areas.
- Dry dredging projects appear to remove soft sediments more thoroughly, mainly because equipment operators can visually observe the results of dredging as it takes place.
- Hydraulic dredging in areas with dense weed beds can frequently clog the pipeline to the slurry discharge area, suggesting that temporary plant control (most often herbicides or harvesting) may have to precede hydraulic dredging.

Limiting Light with Dyes and Surface Covers

The same **dyes** are used to control rooted plants as are used for algal control (see previous discussion). Dyes limit light penetration and thus restrict the depth at which rooted plants can grow, although they have little effect in shallow water (< 4 ft deep).

Dyes favor species tolerant of low light or with enough food reserves to support an extended growth period (during which a stem could reach the lighted zone). In lakes with high transparency but moderate depth and ample soft sediment, dyes may provide open water where little would otherwise exist.

Treatment must be repeated as the dye flushes out of the system. Dyes are typically permitted under the same process as herbicides, despite their radically different mode of action.

Although they could interfere with recreation, **surface covers** can be a useful and inexpensive alternative to traditional methods of weed control in small areas such as docks and beaches. They can be timed to produce results so as to affect neither summer recreation uses nor overall system ecology.

Polyethylene sheets floated on the lake surface were used by Mayhew and Runkel (1962) to shade weeds. They found that two to three weeks of cover eliminated all species of pondweeds (*Potamogeton* spp.) for the summer if the sheets were applied in spring before plants grew to maturity. Coontail was also controlled, but the generally desirable macroalga *Chara* was not.

Mechanical Removal

Mechanical management of aquatic plants is not much different from managing terrestrial plants, except for the complications imposed by the water. Indeed, mechanical weed control techniques (Table 7-5) can be thought of as:

- Mowing the lawn (cutting or harvesting);
- Weeding the garden (hand pulling); or
- Tilling the soil (rototilling or hydroraking).

▼ ***Hand pulling*** is exactly what it sounds like; a snorkeler or diver surveys an area and pulls out individual unwanted plants. This is a highly selective technique, very labor intensive — and obviously not designed for large-scale, dense beds.

But hand pulling will help keep out invasive species that have not yet become established. Hand pulling can also effectively address non-dominant undesirable species in mixed assemblages, or small patches of plants.

Hand harvesting records for Eurasian watermilfoil in Lake George in New York for 1989-91 (Darrin Freshw. Inst.1991) reveal the following:

- First-time harvest averaged 90 plants per person-hour.
- Second-time harvest (the same sites revisited the next year) averaged 41 plants per person-hour.
- Except for substantial regrowth at one site, regrowth the year after initial harvest was 20 to 40 percent of the initial density.
- Regrowth two years after initial harvest averaged less than 10 percent of the initial density.
- Although plant density and total harvest decline with successive harvesting, effort declines more slowly; harvest time per plant increases largely because of search time.
- Actual harvesting effort at 12 sites was 169 hours for first-time harvest and 90 hours for second-time harvest.

Various tools can augment hand pulling, including a wide assortment of rakes, cutting tools, water jetting devices, nets, and other collection devices. Read McComas (1993) for an extensive and enjoyable review of options.

Suction dredging can also augment hand pulling, increasing the pulling rate since the diver/snorkeler does not have to carry pulled plants to a disposal point.

▼ ***Cutting*** is also exactly what it appears to be. A blade of some kind severs the actively growing stem (and possibly much more) from its roots. Regrowth is expected, but some species regrow so rapidly that the benefits of cutting are negated in only a week or two. If the plant can be cut close enough to the bottom, or repeatedly, it will sometimes die, but this is more the exception than the rule.

Cutting is defined here as an operation that does not involve collecting the plants once they are cut, so large-scale cutting operations may affect dissolved oxygen.

Handheld cutters and mechanized barges (like those used with harvesting) cut the plants but don't collect them. Recently, barges have been developed to grind the plants to minimize their viability after cutting and processing. Dissolved oxygen may be affected as the plant biomass decays, much as with most herbicide treatments.

▼ ***Harvesting*** can collect weeds using a number of methods:

- Hand-held nets;
- Small boats towed by the person collecting the weeds;
- Small boat-mounted cutting tools that cut biomass and haul it into the boat for eventual disposal on land; or
- Larger commercial machines with numerous blades, a conveyor system, and a substantial storage area for cut plants.

Offloading accessories allow easy transfer of weeds from the harvester to trucks that haul the weeds to a composting area.

The choice of equipment is really one of scale: is your weed problem massive or confined to a small, nearshore area? Where weeds are a dominant, recurring problem, some lake associations choose to purchase and operate harvesters built to their specifications, while others prefer to contract for harvesting services.

Cutting rates for commercial harvesters tend to range from about 0.2 to 0.6 acres per hour, depending on machine size and operator ability. Even at the highest conceivable rate, harvesting is a slow process that may leave some lake users dissatisfied with progress in controlling aquatic plants.

Weed disposal is not usually a problem, in part because lakeshore residents and farmers often will use the weeds as mulch and fertilizer. Also, since aquatic plants are more than 90 percent water, their dry bulk is comparatively small. Odors can sometimes be an issue.

Key issues in choosing a harvester include depth of operation, volume and weight of plants to be stored, reliability, and ease of maintenance, along with a host of details regarding the hydraulic system and other mechanical design features.

▼ ***Rototilling and cultivation equipment*** are newer procedures with a limited track record (Newroth and Soar, 1986). A rototiller is a barge-like machine with a hydraulically operated tillage device that can be lowered 10 to 12 feet to tear out roots. If the water level in the lake can be drawn down, cultivation equipment pulled behind tractors on firm sediments can remove 90 percent of the roots.

These techniques may significantly affect non-target organisms and water quality, but can reduce severe weed infestations.

Hydrorake: A floating backhoe, usually outfitted with a York rake that looks like a farm implement for tilling or moving silage.

▼ ***Hydroraking*** can be thought of as using a floating backhoe, usually outfitted with a York rake that looks like a farm implement for tilling or moving silage. The tines of the rake attachment move through the sediment, ripping out thick root masses and associated sediment and debris. A hydrorake can be a very effective tool for removing submerged stumps, water lily root masses, or floating islands. This is not a delicate operation, however, and will create substantial turbidity and plant fragments. Hydroraking in combination with a harvester can remove most forms of vegetation found in lakes.

Most mechanical plant removal operations produce at least temporary relief from nuisance plants and remove organic matter and nutrients without adding potentially damaging substances. But, be aware that without proper planning you might face at least two problems:

- Plants often regrow very rapidly (days or weeks), especially in southern waters where midsummer growth rates of water hyacinth can exceed the rate at which they can be harvested; and
- Harvesting may reduce plant diversity and leave areas open for colonization by invasive species.

A bay of LaDue Reservoir (Geauga County, Ohio) was harvested in July 1982 by the traditional method: the operator simply mowed the weed bed like a residential lawn. Stumps of Eurasian watermilfoil plants about 0.5 to 3 inches in height were left; they completely regrew in 21 days.

In contrast, the slower method of lowering the cutter blade about 1 inch into the soft lake mud to tear out roots controlled milfoil for the entire season (Conyers and Cooke, 1983). This method has demonstrated a carry-over effect (less growth in the subsequent year), especially if the area had multiple harvests in one season. But this cutting technique won't work in stiff sediments or in water so deep the cutter bar cannot reach the mud.

Some weed species are more sensitive to harvesting than others. Nicholson (1981) has suggested that harvesting actually spread milfoil in Chautauqua Lake, New York, because the harvester distributed fragments of plants from which new growths could begin. On the other hand, milfoil has become dominant in many northeastern lakes without harvesting programs in less than five years after initial appearance (Wagner, pers. obs.)

Timely harvesting of species that re-seed can eventually limit the extent of those species, but seeds may remain viable in the sediment for many years. For example, extensive harvest of water chestnut in impounded sections of Boston's Charles River in 1996 did not seem to affect plant growth the following year. Harvesting was repeated in 1997, and 1998 growths declined substantially. Sequentially less harvesting has been necessary in 1999 and 2000, but growth from seeds continues (Smith, pers. comm.)

Few data validate the actual restorative effects of harvesting as it relates to controlling nutrients in the water column. If nutrient inputs are moderate and weed density is high, intense harvesting could remove as much as 40 to 60 percent of net annual phosphorus loading — a significant nutrient removal in many cases. On the other hand, harvesting itself can increase phosphorus concentration in the water either by disturbing the sediments or by making it easier for sediments to release phosphorus. In almost any event, nutrients in the sediment remain adequate to support dense plant growths.

Water Level Control

Historically, water level drawdown has been used in waterfowl impoundments and wetlands for a year or more, including the growing season, to improve the quality of waterfowl breeding and feeding habitat (Kadlec, 1962). It is also a common fishery management method.

Until a few decades ago, recreational lakes were lowered primarily to control flooding and allow access for cleanups and repairs; macrophyte control was a side benefit. While this technique is not effective on all submergent species, it does reduce some of the chief nuisance species, particularly those that rely on vegetative propagules for overwintering and growing (Cooke et al. 1993a). If a drawdown capability exists at your lake, lowering the water level is an inexpensive way to control some macrophytes. It will also give you the opportunity to maintain shorelines or remove nutrient-rich sediments.

If a drawdown capability exists at your lake, lowering the water level is an inexpensive way to control some macrophytes.

The ability to control a lake's water level is affected by the area's precipitation pattern, system hydrology, lake morphometry, and the outlet structure. The base elevation of the outlet or associated subsurface pipe(s) will usually set the maximum drawdown level, while the outlet's capacity to pass water and the pattern of water inflow to the lake will determine if that base elevation can be achieved and maintained. In some cases, sedimentation of an outlet channel or other obstructions may control the maximum drawdown level.

Several factors affect how well drawdown works to control plants:

- While drying of plants during drawdowns in southern areas may provide some control, the additional impact of freezing is substantial, making drawdown a more effective strategy for northern lakes during late fall and winter.
- A mild winter or one with early and continuing snow may not provide the necessary level of drying and freezing.
- High levels of groundwater seepage into the lake may also negate the desired effects by keeping the area moist and unfrozen.
- Extensive seed beds may quickly re-establish undesirable plants.
- Recolonization from nearby areas may be rapid.
- And finally, the response of macrophytes to drawdown is quite variable (Table 7-6).

Drawdown has a long and largely successful history, even if not always intended to control plants (Dunst et al. 1974; Wlosinski and Koljord, 1996). The initial winter drawdown of Candlewood Lake in Connecticut (Siver et al. 1986) reduced nuisance species by as much as 90 percent. Drawdowns in Wisconsin lakes have decreased plant coverage and biomass by 40 to 92 percent (Dunst et al. 1974). In one Wisconsin case, Beard (1973) reported that winter drawdown of Murphy Flowage opened 64 of 75 acres to recreation and improved fishing.

The effect of drawdown is not always predictable or desirable, however. Reductions in plant biomass of 44 to 57 percent were observed in Blue Lake in Oregon (Geiger, 1983) following drawdown, but certain nuisance species actually increased and herbicides were eventually applied to regain control. Drawdown of Lake Bomoseen in Vermont (Vt. Agency Nat. Resour. 1990) caused a major reduction in many species, many of which were not targeted.

Table 7-6.—Anticipated responses of some wetland and aquatic plants to winter water-level drawdown.

	CHANGE IN RELATIVE ABUNDANCE		
	INCREASE	NO CHANGE	DECREASE
Acorus calamus (sweet flag)	E		
Alternanthera philoxeroides (alligator weed)	E		
Asclepias incarnata (swamp milkweed)			E
Brasenia schreberi (watershield)			S
Cabomba caroliniana (fanwort)			S
Cephalanthus occidentalis (buttonbush)	E		
Ceratophyllum demersum (coontail)			S
Egeria densa (Brazilian Elodea)			S
Eichhornia crassipes (water hyacinth)		E/S	
Eleocharis acicularis (needle spikerush)	S	S	S
Elodea canadensis (waterweed)	S	S	S
Glyceria borealis (mannagrass)	E		
Hydrilla verticilata (hydrilla)	S		
Leersia oryzoides (rice cutgrass)	E		
Myrica gale (sweetgale)		E	
Myriophyllum spp. (milfoil)			S
Najas flexilis (bushy pondweed)	S		
Najas guadalupensis (southern naiad)			S
Nuphar spp. (yellow water lily)			E/S
Nymphaea odorata (water lily)			S
Polygonum amphibium (water smartweed)		E/S	
Polygonum coccineum (smartweed)	E		
Potamogeton epihydrus (leafy pondweed)	S		
Potamogeton robbinsii (Robbins' pondweed)			S
Potentilla palustris (marsh cinquefoil)			E/S
Scirpus americanus (three square rush)	E		
Scirpus cyperinus (wooly grass)	E		
Scirpus validus (great bulrush)	E		
Sium suave (water parsnip)	E		
Typha latifolia (common cattail)	E	E	
Zizania aquatic (wild rice)		E	

E=emergent growth form; **S**=submergent growth form; **E/S**=emergent and submergent forms

Source: Cooke et al. 1993a.

Reviewing drawdown effectiveness in a variety of lakes, Nichols and Shaw (1983) noted its species-specific effects, with both benefits and drawbacks. So, before you conduct a drawdown, do a thorough system-specific review of likely impacts.

Desirable side effects associated with drawdowns:

- The opportunity to clean up the shoreline, repair previous erosion damage, repair docks and retaining walls, search for septic system breakout, and physically improve fish spawning areas (Nichols and Shaw, 1983; Cooke et al. 1993a; Wis. Dep. Nat. Resour. 1989).
- The attendant concentration of forage fish and game fish in the same areas (Cooke et al. 1993a), although not all fishery professionals agree.
- Benefits for furbearers, since drawdowns usually help emergent shoreline vegetation (Wis. Dep. Nat. Resour. 1989), although not all wildlife biologists agree.
- The consolidation of loose sediments and sloughing of soft sediment deposits into deeper water, perceived as a benefit by shoreline homeowners (Cooke et al. 1993a; Wis. Dep. Nat. Resour. 1989).
- Encouragement of some desirable plants.

Undesirable possible side effects of drawdown:

- Loss or reduction of desirable plant species.
- Facilitation of invasion by drawdown-resistant undesirable plants; either seeds or expanding shoreline vegetation can recolonize. Cattails and rushes are the most commonly expanding fringe species (Nichols and Shaw, 1983; Wis. Dep. Nat. Resour. 1989). Drawdowns to control nuisance submergent vegetation are usually recommended for alternate years to every third year to prevent domination by resistant plant species (Cooke et al. 1993a).
- Reduced attractiveness to waterfowl (considered an advantage by some).
- Possible fishkills if oxygen demand exceeds reaeration during a prolonged drawdown.
- Changes in fish and invertebrate habitat.
- Mortality among hibernating reptiles and amphibians.
- Impacts to connected wetlands.
- Shoreline erosion during drawdown.
- Loss of aesthetic appeal during drawdown.
- More frequent algal blooms after refill in some cases.
- Reduction in water supply: processing or cooling water intakes may be exposed, reducing or eliminating intake capacity. The water level in wells with hydraulic connections to the lake will decline, as may the yield, along with changes in water quality and pumping difficulties.
- Recreational problems during the drawdown: swimming areas will shrink and beach areas will enlarge during a drawdown. Boating may be restricted both by available lake area and by access to the lake. Again, winter drawdown will avoid most of these disadvantages,

although lack of control over winter water levels can make ice unsafe for fishing or skating.

- Outlet structures, docks, and retaining walls may be damaged by freeze/thaw processes during overwinter drawdowns if the water level is not lowered beyond all contact with structures.
- Downstream flow impacts (Nichols and Shaw, 1983; Cooke et al. 1993a): flood storage and the downstream flow will both increase during a drawdown, but once the drawdown level is reached, the downstream flow should stabilize. The key to managing downstream impacts is to minimize erosion and keep flows within an acceptable natural range.
- The time it takes to refill a lake after drawdown: enough water must enter the lake to refill it in an appropriate timeframe while maintaining an acceptable downstream flow. In northern lakes, early spring is the best time for refill, with the melting snowpack and rainfall on frozen ground yielding maximum runoff.

Carefully planned fluctuations in water level can help check nuisance macrophytes and periodically rejuvenate wetland diversity. Planned disturbance is always a threshold phenomenon; a little is beneficial, too much leads to overall ecosystem decline. The depth, duration, timing, and frequency of the drawdown are therefore critical elements in devising the most beneficial program.

Herbicides

Killing nuisance aquatic weeds with chemicals is probably the oldest — and still the most widely used — control method. But few aspects of plant control breed more controversy than chemical control by herbicides, which are types of pesticides. Part of the problem stems from pesticides that have enjoyed widespread use and then been linked to environmental or human health problems, and subsequently banned. Some left long-term environmental contamination and toxicity problems behind. Books such as *Silent Spring* (Carson, 1962) and *Our Stolen Future* (Colburn et al. 1997) have raised both public consciousness and wariness of chemicals in the environment.

Killing nuisance aquatic weeds with chemicals is probably the oldest – and still the most widely used – control method. But few aspects of plant control breed more controversy than chemical control by herbicides, which are types of pesticides.

Current pesticide registration procedures are far more rigorous than in the past. While no pesticide is considered unequivocally "safe," federal pesticide regulation is based on the premise that when the chemical is used according to label instructions, it will not cause unreasonable human health or environmental effects. Federal regulations do not restrict swimming in water treated with any of the currently registered aquatic herbicies, although many applicators recommend that you should avoid contact with treated water for at least a day. Some states impose more restrictions.

Only six active ingredients are widely used in aquatic herbicides in the U.S today, with one additional ingredient awaiting approval. Westerdahl and Getsinger (1988a,b) provide a detailed discussion of herbicide properties and effectiveness on specific plants. Aquatic and terrestrial versions exist under various trade names, causing some confusion.

▼ ***Copper*** (see the algal control section) is not a primary herbicide for rooted aquatic plants, but is sometimes part of a broad spectrum formulation intended to reduce the biomass of an entire plant assemblage.

Copper formulations are also used on certain plants where use restrictions preclude the use of other herbicides. Copper concentrations should not exceed 1 mg/L in the treated waters.

- ***Endothall*** is a contact herbicide, attacking a wide range of plants at the immediate point of contact. It is sold in several formulations: liquid (Aquathol K), granular dipotassium salt (Aquathol), and the di (N, N-dimethyl-alkylanine) salt (Hydrothol) in liquid and granular forms. Effectiveness can range from weeks to months. Most endothall compounds break down readily and do not remain in the aquatic environment.

  Endothall acts quickly on susceptible plants, but does not kill roots with which it cannot come into contact, so many plants recover rapidly. Quickly dying plants can deplete oxygen if decomposition exceeds reaeration; successive partial treatments will minimize this effect. Toxicity to invertebrates, fish, or humans is not expected to be a problem at the recommended dose, yet it is not used in drinking water supplies. Depending on the formulation, concentrations in treated waters should be limited to 1 to 5 mg/L.

- ***Diquat,*** like endothall, is a fast-acting contact herbicide, producing results within two weeks. Although diquat has exhibited toxicity to aquatic fauna in the laboratory, toxicity has not been clearly documented in the field. Domestic water use is normally restricted for seven days, and this herbicide is not used in drinking water supplies. Some species have regrown rapidly (often within the same year) after treatment with diquat, while others have been controlled longer. Concentrations in treated water should not exceed 2 mg/L.

- ***Glyphosate*** is primarily a contact herbicide, but some uptake by plants has been observed. It is effective against most emergent or floating-leaved plant species, but not against most submergents. Rainfall shortly after treatment can negate its effectiveness, as does its ready adsorption to particulates in the water or to sediments. At recommended doses, it is relatively non-toxic to aquatic fauna, and degrades readily into non-toxic components in the aquatic environment. There is no maximum concentration for treated water, but a dose of 0.2 mg/L is usually recommended.

- ***2,4-D,*** the active ingredient in a variety of commercial herbicide products, has been in use for over 30 years despite claims of undesirable environmental and human health effects. This is a systemic herbicide; it is absorbed by roots, leaves, and shoots and disrupts cell division throughout the plant. Vegetative propagules such as winter buds, if not connected to the circulatory system of the plant at the time of treatment, are generally unaffected and can grow into new plants. It is therefore important to treat plants early in the season, after growth has become active but before such propagules form.

  2,4-D is sold in liquid or granular forms as sodium and potassium salts, as ammonia or amine salts, and as an ester. Doses of 50 to 150 pounds per acre are usual for submersed weeds, most often of the dimethylamine salt or butoxyethanolester (BEE) in granular form. This herbicide is particularly effective against Eurasian watermilfoil (granular BEE applied to roots early in the season) and as a foliage spray against water hyacinth. 2,4-D lasts only a short time in the water but can be detected in the mud for months.

Properly applied, granular 2,4-D generally controls nuisance macrophytes such as Eurasian watermilfoil with only sublethal damage to many native species (Helsel et al. 1996). The native community has also recovered from seed. 2,4-D has variable toxicity to fish, depending upon formulation and species. The 2,4-D label does not permit use of this herbicide in water used for drinking or other domestic purposes, or for irrigation or watering livestock. Concentrations in treated water should not exceed 0.1 mg/L.

Plastic curtains have been used to contain waters treated with 2,4-D, and only 2 to 6 percent of the herbicide escaped outside the target area (Helsel et al. 1996). This approach marks the beginning of a new wave of more selective and integrated rooted plant management.

▼ ***Fluridone*** is a systemic herbicide introduced in 1979 (Arnold, 1979). It has been widely used since the mid-1980s, although some states have been slow to approve it. Fluridone currently comes in two formulations, an aqueous solution and a slow release pellet.

This chemical inhibits carotene synthesis, which in turn exposes the chlorophyll to photodegradation. The entire plant will die with prolonged exposure to enough fluridone. Some plants, including Eurasian watermilfoil, are more sensitive to fluridone than others, allowing selective control at low dosages.

For susceptible plants, lethal effects take several weeks to several months, with 30 to 90 days given as the range of time for dieoff after treatment. Fluridone concentrations should be maintained in the lethal range for about six weeks. This may be difficult where there is substantial water exchange, but the slow dieoff rate minimizes the risk of oxygen depletion.

Fluridone has low toxicity to invertebrates, fish, other aquatic wildlife, and humans, and is not known to be a carcinogen, oncogen, mutagen, or teratogen. Substantial bioaccumulation has been noted in certain plant species, but not in animals. EPA has designated a tolerance level of 0.5 ppm (mg/L or mg/kg) for fluridone residues or those of its degradation products in fish or crayfish, and 0.15 ppm in potable water supplies; state restrictions may be lower.

If the recommended 40 days of contact time can be achieved, the liquid formulation of fluridone applied in a single treatment can be very effective. Where dilution may occur, the slow release pellet is generally used, or sequential liquid treatments are performed.

Gradual release of fluridone, which is 5 percent of pellet content, can yield a relatively stable concentration. But pellets have been less effective in highly organic, loose sediments than over sandy or firm substrates (Haller, pers. comm.), largely because of a phenomenon termed "plugging" that prevents the active ingredient from being released from the pellet. Multiple sequential treatments with the liquid formulation are more effective in areas with extremely soft sediments and significant flushing.

Eurasian watermilfoil has been controlled for several years or more without significant impact on non-target species at doses close to 0.01 mg/L (Smith and Pullman, 1997). Recently, applicators have been trying even lower doses (0.006 mg/L) with some success; most native species are minimally affected, but control of milfoil has been erratic and generally short-lived (< 2 years).

Also of interest is the first ever sequestered treatment with fluridone in 2000, whereby curtains were used to seal off target areas from the rest of the lake (McNabb, pers. comm.). Escape was low, dilution was limited, and milfoil appears to have been eliminated without impact outside the barrier.

▼ ***Triclopyr*** is currently experimental for aquatic habitats. It is highly selective and effective against Eurasian watermilfoil at a dose of 1 to 2.5 mg/L. It seems to have little or no effect on valued native species such as most monocotyledonous naiads and pondweeds (Netherland and Getsinger, 1993). It prevents synthesis of plant-specific enzymes, thereby disrupting growth processes. This herbicide is most effective when applied during the active growth phase of young plants.

Triclopyr is not known to be a carcinogen, oncogen, mutagen, or teratogen, and tested animals have been lethally affected only at concentrations over 100 times the recommended dosage rate. The experimental label calls for concentrations in potable water of no more than 0.5 mg/L, suggesting that care must be taken to allow sufficient dilution between the point of application and potable water intakes.

A herbicide treatment can be an effective short-term management procedure to rapidly reduce vegetation for periods of weeks to months. As many as five years of control have been gained with fluridone or 2,4-D. Herbicide treatments are still the best way to open the vast acreages of water infested with the exotic water hyacinth (*Eichhornia crassipes*) in Florida and other southeastern states (Shireman et al. 1982). This is a case in which chemical management becomes necessary until some other more long-term control, such as plant-eating insects, can be established.

Using herbicides to get a major plant nuisance under control is a valid element of long-term management when followed by supplementary methods for keeping such plants under control. Otherwise, herbicide treatments become simply cosmetic maintenance; such techniques tend to have poor cost-benefit ratios over the long term.

Users should follow the herbicide label directions, use only a herbicide registered by EPA for aquatic use, wear protective gear during application, and protect desirable plants. Most states require applicators to be licensed and insured.

Lake managers who choose herbicides must exercise all proper precautions. As shown in Table 7-7, effectiveness of a given herbicide varies by plant species and therefore the nuisance plants must be carefully identified. Users should follow the herbicide label directions, use only a herbicide registered by EPA for aquatic use, wear protective gear during application, and protect desirable plants. Most states require applicators to be licensed and insured.

Important questions to be answered before adopting a management program involving herbicides include:

- What are the acreage and volume of the area(s) to be treated? Proper dosage is based on this information.
- What plant species are to be controlled? This will determine the herbicide and dose to be used.

Table 7-7.—Susceptibility of common aquatic plant species to herbicides.

	CONTROLLED BY HERBICIDE APPLICATION				
	DIQUAT	ENDOTHAL	2,4-D	GLYPHOSAT E	FLURIDONE
EMERGENT SPECIES					
Alternanthera philoxeroides (alligator weed)			Y	Y	Y
Dianthera americana (water willow)			Y		
Glyceria borealis (mannagrass)	Y	N	N		
Phragmites spp. (reed grass)				Y	
Sagittaria spp. (arrowhead)	N	N	Y		Y
Scirpus spp. (bulrush)	N	N	Y	Y	Y
Typha spp. (cattail)	Y	N	Y	Y	Y
FLOATING SPECIES					
Brasenia schreberi (watershield)	N	Y	Y		N
Eichhornia crassipes (water hyacinth)	Y		Y		N
Lemna spp. (duckweed)	Y	N	Y		Y
Nelumbo lutea (American lotus)	N	N	Y	N	
Nuphar spp. (yellow water lily)	N	Y	Y	Y	Y
Nymphaea spp. (white water lily)	N	Y	Y	Y	Y
Wolfia spp. (watermeal)	Y	N	Y		Y
SUBMERGED SPECIES					
Ceratophyllum demersum (coontail)	Y	Y	Y		Y
Cabomba caroliniana (fanwort)	N	N	N	N	Y
Chara spp. (stonewort)	N	N	N	N	
Elodea canadensis (waterweed)	Y		N		Y
Hydrilla verticillata (hydrilla)	Y	Y			Y
Myriophyllum spicatum (Eurasian watermilfoil)	Y	Y	Y	N	Y
Najas flexilis (bushy pondweed)	Y	Y	N	N	Y
Najas guadalupensis (southern naiad)	Y	Y	N		Y
Potamogeton amplifolius (largeleaf pondweed)		Y	N		Y
Potamogeton crispus (curlyleaf pondweed)	Y	Y	N		Y
Potamogeton diversifolius (waterthread)	N	Y	N		
Potamogeton natans (floating leaf pondweed)	Y	Y	Y		Y
Potamogeton pectinatus (sago pondweed)	Y	Y	N		Y
Potamogeton illinoensis (Illinois pondweed)					Y
Ranunculus spp. (buttercup)	Y		Y		

Adapted from Nichols,1986. Y=Yes, N=No, blank=uncertain

Note: *Chara* spp. (stonewort) can be controlled with copper, which also enhances the performance of Diquat on *Eichhornia crassipes* (water hyacinth).

- What will the long-term costs of this decision be? Most herbicides must be reapplied annually, in some cases twice per growing season.
- How is this water body used? Many herbicides have restrictions of a day to two weeks on certain water uses following application.
- Is the applicator trained, experienced, licensed, and insured, and has a permit been obtained from the appropriate regulatory agency? All are desirable (and sometimes necessary) prior to treatment.

Shireman et al. (1982) caution that the following lake characteristics often produce undesirable water quality changes after treatment with a herbicide for weed control:

- High water temperature,
- High plant biomass to be controlled,
- Shallow, nutrient-rich water,
- High percentage of lake area treated,
- Closed or non-flowing system.

Competent applicators will be cautious in treating a lake with these conditions.

Biological Introductions

Biocontrol organisms may eventually help us achieve lasting control of nuisance aquatic vegetation. Biological control harnesses the power of biological interactions to manage plants; however, it suffers from the ecological drawback that in predator-prey relationships it is rare for the predator to completely eliminate the prey. Consequently, population cycles for both predator and prey are likely. Humans may find the magnitude of these cycles in plant populations unacceptable, so other techniques may have to be combined with biocontrols to achieve lasting, predictable results.

Biological control harnesses the power of biological interactions to manage plants; however, it suffers from the ecological drawback that in predator-prey relationships it is rare for the predator to completely eliminate the prey.

Biological controls include herbivorous fish such as the grass carp (*Ctenopharyngidon idella*), insects such as the aquatic milfoil weevil (*Euhrychiopsis lecontei*), and experimental fungal pathogens. Aside from grazing and parasitism, maintaining a healthy native community also works to limit invasive species through competition.

▼ ***Herbivorous fish*** such as the non-native grass carp (imported around 1962) voraciously consume many weeds. The grass carp grows very rapidly (about 6 pounds per year maximum; Smith and Shireman, 1983). This combination of broad diet and high growth rate can control or even eradicate plants within several seasons.

Like people, grass carp prefer certain food. They generally avoid alligatorweed, water hyacinth, cattails, spatterdock, and water lily — choosing instead species such as waterweed, pondweeds, and hydrilla. Fish will graze selectively on the preferred plant species while less preferred species, including milfoil, may increase. Overstocking, on the other hand, may eliminate all plants, contrary to the ecological axiom of encouraging population cycles. Feeding preferences are listed in Cooke and Kennedy (1989).

Not all states permit the introduction of grass carp, so consult your state fishery agency before you decide to use them. Critical controls include:

- Restrictions on the ability of the fish to reproduce (sterile triploid fish vs. reproductive diploid fish); and
- Inlet and/or outlet controls to prevent emigration.

Stocking rates are based primarily on qualitative and quantitative characteristics of the lake, with adjustment by region. Rates of up to 70 fish per acre have been used to remove dense nuisance plant growth, while rates of only 1 to 2 fish per acre have been used in lakes with a lower density of more desirable vegetation.

Stocked fish are normally 10 to 12 inches in length to protect them from predation. Stocking is typically performed on a 6-year cycle linked to fish mortality.

Although many lakes (most notably, small ponds) have successfully used grass carp, the definition of success varies.

- Introduction of 3 to 5 fish per acre into Lake Conway (Florida) greatly reduced hydrilla, nitella, and pondweeds after two years, leaving non-targeted water celery (*Vallisneria*) largely unaffected (Miller and King, 1984). But algal biomass increased, indicating that fish can affect productivity (see the algal control section of this chapter).
- In contrast, stocking about 13 fish per acre (30/acre if only vegetated acres are counted) in Lake Conroe (Texas) eliminated all submersed plants in under 2 years, increased algal biomass, and changed the algae to less desirable forms (Maceina et al. 1992).
- In small Lake Parkinson (New Zealand), grass carp eradicated the invasive, non-native Brazilian elodea (*Egeria densa*), were themselves then removed by netting and rotenone poisoning, and a native flora was naturally re-established from the existing seed bed (Tanner et al. 1990).
- The grass carp experience at Santee-Cooper (South Carolina) is an interesting story. Stocking 900,000 triploid carp eliminated all plants and stopped bass reproduction.

Using grass carp will probably drastically change the ecology of a lake. Stocked to reduce plants, grass carp typically cause a shift toward algal blooms and more turbidity – which then becomes a self-sustaining lake condition.

Failure of this technique has generally been a function of fish diet not matching targeted plant species, inappropriate stocking rates, and lack of patience (essential with biological techniques) before taking additional action.

Using grass carp will probably drastically change the ecology of a lake. Stocked to reduce plants, grass carp typically cause a shift toward algal blooms and more turbidity — which then becomes a self-sustaining lake condition. Gamefish production will suffer from this condition, which lake users may find more objectionable than the original plants.

▼ ***Non-native insects*** have historically been used to control rooted plants. Ten insect species have been imported to the United States under quarantine and have received U.S. Department of Agriculture approval for release to U.S. waters. These insects — which include aquatic larvae of moths, beetles, and thrips — have life histories specific to certain host plants and are therefore confined to the waters of southern states infested with those plants: alligatorweed, hydrilla, water lettuce, and water hyacinth (Cooke et al. 1993a).

They also appear climate-limited to southern states, their northern range being Georgia and North Carolina. Because they reproduce more slowly than their target plants, control takes awhile, although it can be accelerated by using harvesting or herbicides to reduce major plant growth, with insects concentrated on the remaining plants.

Insects have proven highly effective in controlling alligatorweed and water hyacinth; for example, Sanders and Theriot (1986) report that since 1974, the water hyacinth weevil (*Neochetina eichhorniae*) has reduced the distribution of water hyacinth in Louisiana by 50 percent.

Integrated pest management combines biological, chemical, and mechanical controls to maximize effectiveness. Insects are particularly effective when combined with other plant management techniques. Supplemental chemical or mechanical controls work best if done in early fall or winter to minimize interference with the insects.

Integrated pest management combines biological, chemical, and mechanical controls to maximize effectiveness. Insects are particularly effective when combined with other plant management techniques. Supplemental chemical or mechanical controls work best if done in early fall or winter to minimize interference with the insects.

When a Florida canal section was harvested at the peak of the growing season, both water hyacinths and weevils severely decreased (Center and Durden, 1986). Subsequent plant growth was much greater than the weevil population, greatly delaying control. Another section of the same canal was sprayed with 2,4-D at the end of the season, allowing plants and weevils to recover simultaneously and facilitating more rapid control by the weevil.

In a Florida pond, weevils gradually eliminated water hyacinth in conjunction with sequential 2,4-D treatments of sections of the pond (Haag, 1986). Chemical treatments were conducted in such a manner as to leave refuges for the weevil. Alligatorweed later invaded the pond, but was controlled by the alligatorweed flea beetle, *Agasicles hygrophila*.

Despite these successes, non-native species have a poor track record for solving biological problems (they seem to create as many problems as they solve); thus, government agencies tend to prefer alternative controls. Using native insects in a biomanipulative approach is, however, usually acceptable.

▼ ***Native insects*** — primarily the larvae of midgeflies, caddisflies, beetles, and moths — appear promising as aquatic plant controls, mainly in northern states (Cooke et al. 1993a). However, in recent years, the aquatic weevil *Euhrychiopsis lecontei* has received the most attention.

Native to North America, *Euhrychiopsis lecontei* is believed to have been associated with northern watermilfoil (*Myriophyllum sibericum*), a species largely replaced since the 1940s by non-native, Eurasian watermilfoil (*M. spicatum*). The weevil is able to switch plant hosts within the milfoil genus, although to varying degrees and at varying rates depending upon genetic stock and host history (Solarz and Newman, 1996).

Weevils do not use non-milfoil species, but can structurally damage Eurasian watermilfoil's growth points (apical meristems) and supports (basal stems) (Sheldon and O'Bryan, 1996a). Weevils feed on milfoil, lay eggs on it, and pupate in burrows in the stem.

Field observations link the weevil to natural milfoil declines in nine Vermont lakes and several lakes outside that state (no weevils were intentionally brought into any of these lakes)(Creed, 1998).

Lakewide crashes have not been observed where the weevil has been introduced into only part of the lake, although local damage has been substantial; widespread control may require more time. Like the non-native insects, the native weevil reproduces more slowly than does its host plant, so for faster results, more weevils must be stocked. Still unknown is why the weevil was able to naturally overpower the milfoil population in cases like the Vermont lakes. Longevity of control remains unknown, but classic predator-prey population cycles are expected.

One to three weevils per stem appear to collapse milfoil plants; thus, raising the necessary weevils is a major operation. The state of Vermont devoted considerable resources to rearing weevils over a two-year period, using them all for just a few targeted sites (Hanson et al. 1995).

Weevils are now marketed commercially as a milfoil control, with a recommended stocking rate of 3,000 adults per acre. Weevils are usually released from cages or onto individual stems; early research attached a stem fragment with a weevil from the lab onto a milfoil plant — a highly labor-intensive procedure.

Although an integrated milfoil management approach may be able to use weevils effectively, competing control techniques may affect their performance (Sheldon and O'Bryan, 1996b):

- Harvesting may directly remove weevils and reduce their density during the growing season.
- Because adults overwinter in debris along the edge of the lake, techniques such as drawdown, bottom barriers, or sediment removal can kill them.
- Extension of lawns to the edge of the water and application of insecticides also threaten these milfoil control agents.

▼ ***Plant pathogens*** remain largely experimental, despite a long history of research interest. Properties that make them attractive (Freeman, 1977) include:

- High abundance and diversity;
- High host specificity;
- No effects on non-target organisms;
- Ease of dissemination and self-maintenance; and
- Ability to limit host population without eliminating it.

Fungi are the most common plant pathogens, and using them to control water hyacinth, hydrilla, or Eurasian watermilfoil has been extensively evaluated (Theriot, 1989; Gunner et al. 1990; Joye, 1990). Although many problems plague this approach, combining fungal pathogens with herbicides has shown recent promise as an integrated technique (Nelson et al. 1998).

▼ ***Native plants*** provide some resistance to non-native invasions. Although invasive nuisance plant species are just what the name implies, the presence of a healthy, desirable plant community may be able to minimize or slow infestation rates. Disturbance encourages invasive species, so a stable plant community should provide a significant defense.

Unfortunately, natural disturbances abound, and almost all common plant control techniques disturb the environment. Therefore, if native species are to regain dominance after a disturbance, they may need help, such as additional seeding and planting. This is still a highly experimental procedure, but if native species are employed it should be less controversial.

Experiments conducted in Texas (Doyle and Smart, 1995) indicate that dried seeds added to an exposed area of sediment will rapidly germinate to cover the previously exposed area. However, if this is not done early enough in the growing season to allow plants to mature and produce seeds of their own, annual plants will not return the second growing season.

Transplanting mature growths into exposed areas is a better way to establish a seed-producing population. Planting cuttings gathered by a harvester (Helsel et al. 1996) did not establish native species in areas previously covered by a benthic barrier in Wisconsin.

In Lake George, New York, where the native plant community is diverse and dense, colonization by Eurasian watermilfoil has been much slower than in many other area lakes (Wagner and Clear, 1996). Although in some areas the sediment itself may be inhospitable to milfoil, when milfoil is cleared from an area and a native assemblage restored, milfoil growth greatly diminishes (Eichler et al. 1995). It would seem that establishment of desired vegetation is entirely consistent with the primary plant management axiom: if light and substrate are adequate, plants will grow. Plant control should extend beyond eradicating undesirable species to encouraging desirable plants.

Plant control should extend beyond eradicating undesirable species to encouraging desirable plants.

Sediment Buildup

Sand, silt, and organic matter from erosion, construction, shoreline collapse, urban drainage, and other sources decrease lake volume and increase the shallow water area. Not only can this interfere with recreational activities such as boating, but shallow, nutrient-rich sediments encourage the growth of nuisance aquatic plants and may contribute nutrients to the water column, thus stimulating algal blooms as well. As internal productivity increases, organic matter accumulates and the lake evolves toward an emergent wetland.

Good management of the watershed (see Chapter 6) will, of course, reduce sediment entering a lake; but how do you deal with it once it's there?

Shallowness is the primary symptom to be considered here; you can address this problem in several ways:

- **Reserve part of the lake as a detention and settling area** (cleaning it out as needed to maintain its function) to protect the remainder of the lake. Such basins will usually have to be quite large to capture the finest and most pollutant-laden material. Smaller basins to capture coarse sediments are more practical, but may not protect the lake from sedimentation impacts. This is more a preventive than restorative technique; and giving up a substantial portion of a lake is usually unacceptable.

- **Raise the water level:** Outlet limitations usually make this very difficult — and you certainly don't want to flood shoreline properties.
- **Resuspend the sediment to move it downstream:** This is not environmentally sound and is illegal in most areas.
- **Cause the sediment to decay or compact:** Drawdown sometimes compacts sediments, but usually does not significantly change water depth. Adding natural or engineered microbes, usually in conjunction with aeration, is advertised as accelerating organic sediment decay, but scientific evidence is lacking. Also, most organic mucks need a great deal of oxygen; supplying enough oxygen for decomposition almost guarantees resuspension of particles, increased turbidity, and downstream transport. And some materials simply don't decay.
- **Remove the sediment:** Dredging is a major operation, but can provide major restorative benefits if done properly.

Removing sediment is the only practical way to consistently increase water depth, and dredging has become one of the most frequently prescribed techniques. Key steps include:

- Perform a proper feasibility study of the lake and disposal sites. Consider the potential negative impacts and how to avoid them; otherwise, they can be severe. Carefully assess sediment attributes (quality and quantity).
- Design the dredging project — this involves much engineering and is best done by professionals.
- Be aware of all costs; dredging projects are expensive — but can improve water depth over the long term.
- Secure permits, which are usually extensive.

Continual sediment loading will return the lake to its pre-dredged condition, so you must control external loading — remember, watershed management is essential — to protect your investment in dredging. See the algal and rooted plant control sections of this chapter for more details on dredging.

Non-algal Color and Turbidity

Colored drinking water often indicates high concentrations of algae or humic substances. Algal control has been addressed previously in this chapter. Humic substances in drinking water are removed in a treatment facility, not in the reservoir itself. It would be difficult and expensive to sustain this treatment for an entire lake.

Humic substances: Derived from humus, the organic portion of soil resulting from decomposition of plants and animals.

Humic substances act as a natural dye, limiting the depth of rooted plant growths, and reducing phosphorus availability to algae. Consider these natural benefits before removing humic substances.

Color in drinking water drawn from a deep source (and sometimes the metalimnion) is often caused by a high concentration of iron or manganese in the raw water. These metals may also impart taste to the water. Aeration is generally used to eliminate this buildup in a reservoir, although artificial circulation might also be applicable. For more information, see the algal control section of this chapter.

Anoxia and Related Issues

Anoxia (the absence of oxygen) is a natural process common in U.S. lakes for a number of reasons:

- Decaying organic matter and sediments need oxygen — people's activities in the watershed can greatly accelerate the natural process of sediment accumulation.
- Oxygen can't enter the hypolimnion of a stratified lake from the atmosphere, so if there's not enough oxygen already in the hypolimnion to handle the oxygen demand, it may become anoxic.
- Shallow lakes may also become anoxic, especially where the sediment meets the water, if the sediment needs more oxygen than the atmosphere is delivering.

So what happens in an anoxic hypolimnion?

- Iron, manganese, ammonia, and dissolved phosphorus often accumulate to undesirable levels.
- Habitat for coldwater fish and daytime refuge for zooplankton are minimized.

And the solution?

- Adding oxygen is a logical solution to anoxia; this may involve artificial circulation (destratification) or hypolimnetic aeration.
- Removing oxygen-demanding sediments by dredging can also be effective, but is a complicated and expensive process.
- Selective discharge may prevent anoxia if water can be discharged faster than the rate of oxygen loss, but this is an uncommon situation.
- Anoxic impacts to water supplies can be minimized by elevating the raw water intake from the hypolimnion to the epilimnion. This can introduce taste and odor or other algal problems, but may be preferable to using an anoxic supply.

Review the algal control section of this chapter for a more thorough discussion of these techniques.

Acidification

Acidic waters can harm many aquatic organisms, principally by leaching sodium chloride from the body fluids of fish and other organisms. Important sport fish species may disappear at pH levels below 6.

Acidic lakes occur in two areas:

- Where the soils have little natural buffering (neutralizing) capacity; and
- Where acid rain and other human or natural processes cause acidification of water bodies.

Poorly buffered lakes subject to considerable atmospheric deposition of acid compounds have been the most widely publicized examples, but lakes subject to copper sulfate and unbuffered aluminum sulfate applications can also acidify. Acidic drainage from mines affects thousands of miles of streams and numerous lakes throughout Appalachia and in other coal and metal mining areas.

Some waters are moderately acidic because they pass through naturally acidic soils. A naturally acidic system can be severely disrupted by artificially raising the pH. Lakes dominated by these waters possess an adaptive ecology that should not be confused with that of lakes impaired by cultural acidification.

Clearly, controlling atmospheric emissions is the best way to counteract acidification — attacking the problem at its source. But where damage from cultural acidification must be mitigated on an in-lake basis, acidity is most often managed by adding neutralizing materials. Olem (1990) describes methods of counteracting acidification:

- **Limestone**, a natural mineral, is often the major component of surface water buffering systems that neutralize acidity. Limestone works on lakes just like common antacid tablets work on our stomachs. Calcium carbonate is the active ingredient in both.

  And, because it is used extensively for agricultural liming, limestone is easily available at a low cost. It's usually applied by boat (helicopter if the lake is not accessible by boat); a limestone-water slurry is spread over the lake surface.

  When added to surface water, limestone dissolves slowly, gradually increasing the pH. It is often desirable to add enough limestone so that some settles to the bottom of the lake. This "sediment" dose continues to slowly dissolve over time.

  Limed water bodies typically increase in pH to levels between pH 7 and 9. These pH levels are best for growth and reproduction of many aquatic organisms, and reduce the concentration of toxic forms of aluminum. The effects typically last about twice the lake detention time.

- **Injection of base materials into lake sediment** is an experimental procedure that has been applied to only a few lakes (Lindmark, 1985; Willenbring et al. 1984) but shows promise for those with short detention times. The technique injects neutralizing materials such as limestone, hydrated lime, or sodium carbonate into the sediments of acidic lakes, to gradually change lake pH and increase acid neutralizing capacity in the water column during spring and fall lake turnover. This treatment should last about five to seven times longer than adding limestone to the lake surface.

- **Base injection** may release phosphorus from the sediments to the water column, thus increasing productivity. In low fertility systems commonly associated with low pH, this can benefit fish. But it can also disrupt the benthic community and increase turbidity, and is expected to cost more than liming lake water directly. This technique is generally limited to small, shallow lakes with soft organic sediments; they must be accessible by roads adequate for transporting materials and application equipment.

- **Pumping naturally alkaline ground water** into a few lakes has also been attempted.

Toxic Substances

Substances that cause negative reactions in exposed organisms are called "toxic". An overused and frequently misused term in recent years, it is a generic label put on a wide variety of natural and anthropogenic compounds.

Lakes often become sinks for toxic materials generated in the watershed — such as DDT and PCBs — but only a few toxic substances originate in the lake itself. These include ammonia, hydrogen sulfide, methyl mercury, and algal toxins.

Disinfection byproducts (DBPs) result from reaction of organic molecules with disinfection chemicals (mainly chlorine) in water treatment to form potentially toxic compounds. The organic molecules may come from the watershed, primarily as decaying plants, or from vascular plants and algae in the reservoir. The concentration of these organic molecules is expected to be higher in more eutrophic water bodies. Consequently, watershed management and most in-lake techniques for controlling algae and vascular aquatic plants can limit DBP formation (Cooke and Carlson, 1989).

The initial source of **mercury** in many lakes is not known, but its widespread distribution suggests airborne dispersion; waste incinerators and fossil fuels are often blamed. Once in the aquatic ecosystem, mercury can be transformed into a highly toxic substance. While dilution, flushing, or dredging can improve the situation, methyl mercury contamination has no simple in-lake solution.

Artificial circulation or aeration can usually control **ammonia** and **hydrogen sulfide**. Most algal control techniques can minimize **algal toxins**, which are most often produced by blue-greens but can come from other algae as well.

Natural uptake, adsorption, and settling usually remove or inactivate toxic substances introduced to a lake. The sediments in urban and agricultural lakes often contain substantial quantities of a wide variety of potentially toxic compounds, but these compounds usually interact very little with the overlying water column. Removing the sediment can be desirable, but disposal of contaminated dredged material can be complicated and very expensive.

Where a drinking water supply is involved, raw water is commonly treated with coagulants, adsorption, settling, and/or filtration to minimize the amount of potentially toxic substances in water sent on to consumers. In general, treated water supplies are very safe for consumption.

Pathogens

Controlling the biological pathogens (viruses, bacteria, fungi, and protozoa) of concern to humans is usually a matter for watershed management. Such pathogens come mainly from waste materials, especially animal wastes, so management of wastewater treatment plants, on-site (septic) disposal systems, and concentrated animal feedlots is important. Stormwater runoff is often a source of pathogens as well, and requires attention in watershed management. Exceptions would include direct inputs from humans, waterfowl, beavers, and other wildlife or livestock using the lake.

Pathogen: A specific causative agent of a disease.

Pathogens of special concern in water supplies today include the protozoans *Giardia* and *Cryptosporidium*. The former is associated with a variety of wildlife, but most notably beaver, while the latter is linked to livestock, especially young cows. Both produce debilitating symptoms in humans, and *Cryptosporidium* has been fatal in some cases (Fayer, 1997). Risk of infection in recreational lakes is very low; current regulations concentrate on contamination of drinking water supplies.

Pathogens are not easy to measure for several reasons:

- Fecal coliforms, bacteria that inhabit the intestinal tract of warmblooded animals, are often used to indicate pathogen presence, but these bacteria are not usually pathogenic themselves.
- Other forms of bacteria can be assessed by culture techniques, but this process is not rapid or completely reliable.
- Detection of *Giardia* and *Cryptosporidium* depends mainly on visual examination of samples with a microscope, and is not extremely reliable (LeChevallier et al. 1997).

Most pathogens live only briefly in well-oxygenated waters, although some do form cysts that can become active pathogens after ingestion. Aeration and circulation techniques, discussed previously in this chapter, sometimes remedy this problem in the lake. Drinking water subject to possible pathogenic sources should be treated by filtration and disinfection. Although risk cannot be eliminated in water supplies, proper treatment (as stipulated under the Safe Drinking Water Act) will minimize it.

Undesirable Fisheries

Fish production is directly related to lake or reservoir fertility. This relationship is also the source of many fishery problems. While nutrient-rich waters may produce a larger fish population, they may also promote intense algal blooms, anoxia in deeper waters, or serious fish imbalances (predators vs. prey, bottom vs. water column feeders, or age and size classes).

At the other end of the spectrum, lake or reservoir fertility may be so low that fish grow and reproduce poorly, and stocking efforts will usually fail if the lake is not fertilized. Lake Mead, in Nevada and Arizona, is a case in point (Axler et al. 1988).

Thus, both low and high fertility situations have become targets for management where fishing is a high priority. Some may say it's unethical to drastically change a natural system to meet fishery goals, but management may be the only way to maintain the system where the aquatic habitat has been created by humans.

Fishery issues include:

- **Small population size** — not enough fish, usually as a result of lack of fertility, but also from overfishing, disease, anoxia, or toxic substances;
- **Undesirable size distribution** — stunted fish populations, usually as a result of low mortality rates and overcrowding, leading to sub-lethal food shortages and slow growth;
- **Predator-prey imbalances** — not enough predators to keep the prey species in check (leading to an undesirable size distribution) or not enough prey to support the desired level of gamefish;
- **Non-native or disruptive species** — presence of enough of an undesirable species to impact the desirable species (lamprey damage to trout) or quality of the fishing experience (turbidity created by carp); and

- **Poor fish condition** — low weight to length ratios, tainted flesh (taste), contaminated tissue, poor appearance (parasites).

Just as with lake restoration, the condition of the fish community must first be diagnosed before beginning a fish management program. This usually involves fish sampling to assess the condition of the present fish community. Before attempting to diagnose a fishery condition, consult with state fisheries professionals. Fish management should also involve an understanding of the physical, chemical, and biological features of the lake aside from fish; a diagnostic/feasibility study is a useful tool in virtually every aspect of lake management.

Fishery management in lakes and reservoirs falls into three broad categories (Baker et al. 1993):

- **Habitat management** — alteration of the physical and/or chemical features of the lake to suit a fish species or group. Aeration to reduce anoxia, liming to reduce acidity, dredging to expose gravel substrate, placement of artificial reefs, and structuring the plant community by removal or plantings are all examples.

- **Fish manipulation** — stocking or removing fish of selected species or sizes to enhance the food base or remove competitive or predatory limits on a species or group. Removal of planktivores (sunfish) or rough fish (carp), stocking of top predators (pike or bass), or addition of prey species (minnows, alewife, shad) come under this heading.

- **Managing fishing pressure** — setting limits on the types, sizes, or numbers of fish that can be removed to minimize human disruption of the fish assemblage. Fishing regulations should not only enhance equity of opportunity for anglers, but should act as a tool to maintain or improve fishery characteristics over time.

All the habitat management techniques except artificial reefs have been discussed elsewhere in this chapter. **Artificial reefs** are somewhat controversial, in that it is not clear they help fish production. Certain species will benefit, but the concentration of those fish around reefs may favor anglers more (Bortone, 1998).

▼ ***Fish manipulation*** typically revolves around stocking or removal. Removal has been practiced on large and small scales by a variety of methods (McComas, 1993). Piscicides, such as rotenone, were widely used in the mid-1900s, but are less popular today as a consequence of unintended side effects. Still, chemical control is generally more effective than physical techniques such as netting or angling. Drawdown can eliminate most species of fish, but complete drawdowns are uncommon and also have serious side effects. Biological control, which generally involves adding larger predator species, has become more popular over the last quarter century. Fish stocking, while indiscriminate in decades past, is now somewhat more tightly controlled and normally requires a permit from the state fish agency.

In either case, the goal is to alter the ratio of predators to prey, or large fish to small fish, or one species to another, so that one or more specific species will become more abundant. An exception would be put-and-take stocking, in which catchable-sized fish are placed in a lake for immediate angling enjoyment and may have little long-term impact on the fish community.

Fish manipulation requires careful planning, along with followup monitoring and usually repeat manipulations on an annual to five-year basis. Its primary negative impacts are:

- Loss of native species or genetic stocks: stocking of hatchery-reared trout has become more controversial because of its effect on native genetic diversity.
- The ever-present potential for biological interactions to affect the lake with unexpected severity: for example, alewife or shad stocked to provide food for gamefish can produce a trophy fishery, but can also greatly reduce zooplankton and the average zooplankter body length and algae grazing potential (Wagner and Carranza, 1986; Brandt, 1998). Alewife and shad can discourage recruitment of gamefish through food competition with juveniles and possibly direct predation on eggs and fry. Adult gamefish growth has therefore been enhanced at the expense of algal biomass control, water clarity, and gamefish recruitment.

Managing fishing means managing people, which requires a firm scientific basis and the ability to both communicate and enforce decisions.

In artificial or intensely managed systems, however, fish manipulation may be necessary to achieve fish production and fishing goals.

Managing fishing means managing people, which requires a firm scientific basis and the ability to both communicate and enforce decisions. Controls range from closed seasons to size or catch restrictions to encouragement for keeping certain types or sizes of fish (Baker et al. 1993). Management options include:

- When fish are more susceptible to fishing pressure at certain times, they may need to be protected by a closed season.
- Where fishing pressure is intense, fish must be protected until they reach a reproductive age if the population is to be naturally sustained, and limits on fish size and catch per angler may be necessary to ensure continued fishing.
- Where small and large fish need protection, a slot limit can be instituted, allowing only an intermediate size class to be taken.
- If there are too many fish of a given species for the system to support, thinning by angling is possible, although this has rarely reduced the stock to the desired level by itself (McComas, 1993).

Negative aspects of regulations are mainly related to angler dissatisfaction with being regulated, not with impacts on the lake itself.

Fishing regulations tend to maintain, not change fisheries. Regulations are generally intended to minimize undesirable shifts in population characteristics. Negative aspects are mainly related to angler dissatisfaction with being regulated, not with impacts on the lake itself.

Some fish management techniques are intended to enhance aquatic conditions, not angler success:

- Removing bottom-feeding fish from shallow lakes is a prime example (see algal control techniques). Fish are often removed to increase zooplankton size and biomass, so they will graze on algae and thus improve water clarity.

- Grass carp are stocked to reduce vascular plant growth, often to the detriment of other fish species and with large increases in algal biomass.

Animal Nuisances

Many animals annoy other lake users, but those that affect humans get the most attention in management programs: various insects, waterfowl, aquatic invertebrates, and rodents. In some cases, management for human protection or comfort is consistent with other lake goals and sound ecological principles, but in many cases we are fighting nature to meet human goals.

▼ ***Insects*** that carry diseases or are simply annoying (mosquitoes, midges, blackflies, other biting insects) have little effect on water quality and may actually fuel fish production, but are often targeted because of their negative interactions with humans.

- **Chemical solutions:** pesticide applications that may also affect other non-target organisms.
- **Physical solutions:** drying up possible breeding areas (which may or may not affect associated lakes), dredging to deepen the habitat, and flushing to control nuisance populations.
- **Biological solutions:** funneling production of problem insects into other biological components of the system, most often fish.

As with algal control, humans have technological superiority but an evolutionary disadvantage; complete control of insect pests is rarely achieved — lake users must accept some level of co-existence.

As with algal control, humans have technological superiority but an evolutionary disadvantage; complete control of insect pests is rarely achieved – lake users must accept some level of co-existence.

▼ ***Waterfowl*** become nuisances when they become abundant enough to add nutrients to the lake or interfere with recreation either directly or by raising bacterial counts (and thus forcing beach closures). Canada geese are the most cited offenders, but a variety of ducks and seagulls can also cause problems. Control depends largely on the priority of lake uses.

- Waterfowl breeding and migration areas are best used to enjoy the diversity of waterfowl they attract, so controls should be aimed at helping the waterfowl: preventing diseases and managing the population. Recreation is secondary in these areas, unless hunting is a priority use.
- Where the lake is used primarily for recreation or water supply, taking action against larger populations of waterfowl may be justified.

Waterfowl control may involve scare tactics, including exploders, shellcrackers, noisemakers, scarecrows, and balloons; or dogs, chemical repellents, or habitat manipulation to keep them away from certain areas (U.S. Dep. Agric. 1986; McComas, 1993). Most scare tactics will work for some time, but waterfowl will eventually get used to them and become immune to the technique. Dogs are an exception if they remain aggressive. Chemical repellents are inconsistent, and must be reapplied frequently.

Habitat manipulation can succeed, and is ecologically consistent with most other management goals, but takes time and often requires more effort. In the case of geese, plant barriers between water and land will often discourage geese from using an area. Dense stands of cattail, iris, or other stiff plants can physically impede goose movement, and tall grasses limit forage and visibility. Fencing at the edge of the water also inhibits use by geese. Wires, fishing line, or other overhead physical barriers bar both geese and seagulls from targeted areas.

Direct population control is also possible where laws permit, and includes shooting, trapping with relocation, and shaking of eggs (addling). Eliminating supplemental feedings is appropriate, and food poisoning has been employed in extreme cases.

▼ ***Aquatic invertebrate nuisances*** include certain larval invertebrate stages that cause swimmer's itch, leeches, other invertebrates that bite or infect humans, and non-native zebra mussels that drastically alter the flow of energy in a lake. Although copper sulfate or other pesticides kill them, these treatments may harm the aquatic environment. Methods that use the ecology of the problem species against it are much preferred. Several fairly simple techniques can help alleviate this problem:

- Swimming areas can be altered to favor humans and exclude most nuisance invertebrate species by limiting plants and soft substrates, placing barriers around the edge of the swimming area, and changing water chemistry.
- A few salt blocks can increase the salt concentration to a level leeches find objectionable.
- Limiting snails and waterfowl in swimming areas by physically removing them or changing their habitat can minimize the number of larval invertebrates responsible for swimmer's itch.

▼ ***The zebra mussel problem*** deserves special mention as this non-native species expands at tremendous rates in suitable waters, with significant effects:

- Because they filter the water, zebra mussels greatly enhance water clarity, reducing food resources for zooplankton and fish.
- More energy is funneled through benthic pathways, favoring different species of fish than usually found in the affected lakes.
- Intake or discharge pipelines can become clogged.
- Bottom coverage creates a hazard for swimmers and waders, who can be cut on the sharp edges of an open shell.
- Growth on boats, mooring lines, and buoys creates hazards and maintenance problems.

Zebra mussels may be controlled biologically through predation by certain species, including the freshwater drum, but this has not been widely successful. Several other approaches have been tried:

- Standard treatment chemicals such as copper or chlorine can be effective, but in many cases the mussels can't be exposed to the chemicals long enough to kill them.

- Additives to paints, such as hot sauces, can minimize the attractiveness of surfaces for colonization, but are not completely effective for an extended period of time.
- Physical removal is possible but labor-intensive.
- Some treatment facility operators limit control efforts, preferring co-existence strategies such as oversized pipes (Sarrouh, 1998).

▼ ***Beaver, muskrat, and nutria*** become nuisances through their own efforts at manipulating their habitat. Burrows and dams create physical impediments to human use, and consumption of shoreline vegetation promotes erosion. Other than direct trapping, control methods consist largely of counteracting the tendency of these animals to adjust their environment (McComas, 1993):

- Remove dams — but you can also foil animals through tubes or sluices designed to prevent intentional clogging.
- Discourage burrowing with flagstone, riprap, or wire mesh.
- Protect trees and shrubs with wire barriers.
- Use unpleasantly scented (usually non-toxic) chemical repellents.

User Conflicts

User conflicts result when one use of the lake negatively affects one or more other uses. These may be between human uses such as water supply or recreation and non-human uses such as habitat for fish and wildlife. Or perhaps human user groups have different priorities for lake use. Conflicts between non-human user groups also exist, but tend to get minimal attention from lake managers.

The most controversial user conflicts in recent years relate to:

- Provision of water supply from reservoirs versus downstream flow for habitat maintenance;
- Use of outflow for power generation versus desired water level for other uses; and
- Motorized watercraft versus virtually all other lake uses.

Lake and reservoir management includes planning to meet the needs of water users (Chapter 3), and conflicts must be anticipated and addressed. Politics and economics will figure into conflict resolution as much as science, and the ability to effectively deal with people is as important as expertise in any of these fields.

Lake and reservoir management includes planning to meet the needs of water users, and conflicts must be anticipated and addressed. Politics and economics will figure into conflict resolution as much as science, and the ability to effectively deal with people is as important as expertise in any of these fields.

Resolving conflicts over flows and water levels tends to revolve around water quantity allocations that bring about specific downstream flows or in-lake water levels at specified times of the year (Blaha and LoVullo, 1997; Moyle et al. 1998). You must thoroughly understand the water budget of the lake or reservoir, its morphometry and storage capacity, and the operating limits of the control structures that hold or discharge water. You must also understand the configuration and resource sensitivity of the downstream channel. You may then be able to craft a management plan that satisfies all uses, or at least minimizes conflicts. Sometimes, however, you will simply have to establish clear priorities for lake use and manage accordingly.

Instead of going to the extreme and banning a use, look for the win-win; you may be able to satisfy everyone's "needs" if you all understand the nature of the conflict.

Most conflicts involving recreational lake use can be managed by allocating either space or time (Jones, 1988; Wagner, 1990, 1994). Instead of going to the extreme and banning a use, look for the win-win; you may be able to satisfy everyone's needs if you all understand the nature of the conflict. Compromise by restricting activities to certain spaces and/or times.

Motorized watercraft interfere with many other lake uses, mostly because they're noisy and disturb bottom sediments in shallow areas. Personal watercraft have received much attention for their noise impact. Try restricting the speed of watercraft operating within some defined distance from shore and zoning certain portions of the lake for non-motorized use. You might also restrict the hours of engine operation or allow use of motors on only odd-numbered days or at certain times. Restricting engine size has been tried, but the connection between the engine size and impacts is not so clear. Some approaches, such as limiting parking space at the boat launch, combine space and time restrictions on boaters into density regulation.

It should be possible to resolve most recreational conflicts and achieve balanced resource use if all parties are committed to a fair allocation for each use.

You must understand the motivation and use pattern of each user group if you are to develop a mutually satisfying solution. It is also important to understand the lake resource, as some features will predispose the lake to certain impacts. Space zoning tends to work better on large lakes, while time zoning is often necessary on small lakes. As with flow and water level conflicts, sometimes decisions must be made on the basis of use priorities, but it should be possible to resolve most recreational conflicts and achieve balanced resource use if all parties are committed to a fair allocation for each use.

Cost of Lake Management

The cost of managing a lake or reservoir depends on several factors (Table 7-8), and the process of estimating costs is highly lake-specific.

- The cost of chemicals depends on dose and volume to be treated, each of which requires information about the system to be treated. Costs are typically reported on a per acre basis, however, obscuring some details.

- Harvesting costs depend on the machine to be used, the type and density of the plants to be harvested, and the location of the offloading point relative to the area targeted for harvesting. Contractor estimates will incorporate costs for mobilization and equipment maintenance that will not be obvious in an areal (per acre) cost. Estimates derived by lake associations or towns that have purchased a harvester should also include maintenance expenses, as well as salaries and benefits for operators and insurance costs. Again, a simple cost per unit of area harvested obscures the details and limits the comparative value of the estimate.

Cost should reflect all aspects of the project, from engineering and permitting through actual dredging to shoreline and containment area restoration, and will be highly lake-specific.

- Estimating dredging cost is quite complicated, and involves substantial knowledge of lake and sediment features, access and equipment considerations, dewatering capability, containment area characteristics, and ultimate disposal of dredged materials. The cost of each program element is calculated and summed to derive the project cost, but for comparison among dredging projects this value is often divided by the quantity of sediment to be removed to yield a cost per unit volume (typically per cubic yard). This cost value should reflect all aspects of the project, from engineering and permitting through actual dredging to shoreline and containment area restoration, and will be highly lake-specific.

For any given lake and problem, comparison of costs for alternative management methods should be based on a careful accounting of all expenses associated with each method over the intended duration of benefits, recognizing how the benefits vary for different techniques. Don't, for example, compare the cost of a single herbicide treatment with a dredging project that should provide multiple benefits for decades. Similarly, nutrient inactivation may control algae to the same extent as aeration for less long-term cost, but will not increase hypolimnetic oxygen levels as much. Comparison of contract harvesting versus purchase and local operation of a harvester should be based on the cost over the expected lifetime of the harvester (typically 10 to 15 years).

Separate costs into capital and operating expenses, as high capital methods may require different approaches to funding and implementation, and may ultimately affect the choice of technique. Estimators should also note any factors that cause uncertainty of cost values, and express estimates as cost ranges for each method of managing a given problem in a given lake. Cost ranges for alternative methods can then be compared directly and in light of other non-cost issues that affect selection of management methods, such as variable benefits, permitting requirements, and social acceptability.

While the lake-specific approach is recommended for serious comparisons and detailed budgeting, it does require much information you may not have in the early planning stages of lake or reservoir management. Thus, you may want to make general cost comparisons among techniques or get an initial rough estimate for discussion purposes. Most lake-related texts cite specific cases and associated costs, but extrapolating them to your case may be difficult. Where appreciable cost data are available (e.g., dredging, nutrient inactivation, herbicide treatment), the range is generally large as a consequence of lake- or region-specific features.

Comparing Costs

With all of these caveats in mind, Table 7-8 compares costs (in 2001 U.S. dollars) among selected techniques based on clearly outlined assumptions, a uniform unit of application (per acre), a common target area (100 acres), and an extended period of benefit (20 years). The ranges in Table 7-8 are not all-inclusive, and some assumptions may not hold true for even most possible cases. Cost estimates are based on interviews with lake management practitioners across the country, but are adjusted to reflect the added costs of design, permitting, and monitoring for each technique.

Preventing lake and reservoir problems would seem preferable to in-lake management wherever possible. . . . Very few techniques provide lasting relief at a consistently low cost.

Consider, for example, the costs Table 7-8 gives for addressing two primary problems: excessive algal growths and aquatic macrophyte infestation. It is obvious that addressing these problems on an in-lake basis is expensive when considered over an extended period of time, no matter which technique is employed. Preventing lake and reservoir problems would seem preferable to in-lake management wherever possible.

Very few techniques provide lasting relief at a consistently low cost. If algae can be controlled with selective withdrawal through an existing structure (either

Table 7-8.—Cost of selected management options within lakes and reservoirs.

TECHNIQUE	ASSUMPTIONS	COST RANGE ($) PER ACRE TREATED, WITHOUT CONSIDERATION OF LONGEVITY OF EFFECTS	COST RANGE ($) FOR A HYPOTHETICAL 100-ACRE TARGET AREA OVER A HYPOTHETICAL 20-YEAR PERIOD
ALGAL CONTROL			
Herbicide Treatment with Copper	Copper sulfate powder/crystal	50–150	100,000–1,200,000
	Chelated formulations	150–300	300,000–2,400,000
Artificial Circulation	Shallow water circulation	300–5000	70,000–350,000
	Destratifying diffusion	500–7000	90,000–400,000
Aeration	Full or partial lift, prevention of anoxia	800–2000	180,000–300,000
	Full or partial lift, DO>5 mg/L	1000–3000	280,000–400,000
	Layer aeration, prevention of anoxia within layer	500–1000	120,000–180,000
	Layer aeration, DO>5 mg/L within layer	700–1200	180,000–240,000
Selective Withdrawal	Necessary structures in place	<100	50,000–100,000
	Structural alteration and/or treatment of discharge required	1000–3000	200,000–1,000,000
Bottom Sealing	Artificial covers	5,000–60,000	1,000,000–6,000,000
	Reverse layering	20,000–40,000	2,000,000–4,000,000
Sediment Treatment with Riplox	No major obstructions to bottom treatment	8,000–12,000	1,600,000–2,400,000
Dredging	Average sediment depth = 2 ft	15,000–50,000	1,500,000–5,000,000
	Average sediment depth = 5 ft	25,000–80,000	2,500,000–8,000,000
Nutrient Inactivation	Alum with no buffering, external load controlled	500–700	50,000–140,000
	Buffered alum treatment, external load controlled	600–1000	60,000–200,000
Dilution	Water readily available	500–2500	1,000,000–5,000,000
	Piping, pumping, or structural alteration necessary, and/or water treatment necessary	5000–25,000	5,000,000–25,000,000
Flushing	Water readily available	500–2500	1,000,000–5,000,000
	Piping, pumping, or structural alteration necessary	5000–10,000	5,000,000–10,000,000
Dye Addition	Detention time >1 month	100–500	200,000–1,000,000
Partitioning for Pollutant Capture	5-acre detention pool	10,000–40,000	75,000–200,000
	5-acre constructed wetland	15,000–75,000	75,000–275,000
Biomanipulation of Fish/Zooplankton	Predator stocking	500–1500	200,000–600,000
	Planktivore removal	1000–5000	400,000–2,000,000

Table 7-8.—Cost of selected management options within lakes and reservoirs (continued).

TECHNIQUE	ASSUMPTIONS	COST RANGE ($) PER ACRE TREATED, WITHOUT CONSIDERATION OF LONGEVITY OF EFFECTS	COST RANGE ($) FOR A HYPOTHETICAL 100-ACRE TARGET AREA OVER A HYPOTHETICAL 20-YEAR PERIOD
MACROPHYTE CONTROL			
Cutting	No collection of vegetation	200–400	400,000–800,000
Hand Pulling	With removal	100–500	200,000–1,000,000
Harvesting	Moderately dense, submersed vegetation	200–600	400,000–2,400,000
	Very dense or difficult to cut/handle	1000–1500	2,000,000–4,000,000
Hydroraking or Rotovation	Softer submersed vegetation	2000–4000	2,000,000–4,000,000
	Emergents and root masses	6000–10,000	3,000,000–6,000,000
Suction Dredging/ Harvesting	Primarily plants removed	5000–10,000	2,500,000–6,000,000
Benthic Barriers	Includes installation and removal/annual maintenance	20,000–50,000	4,000,000–7,000,000
Water Level Control	Necessary structures in place	<100	50,000–100,000
	Structural alteration required	1000–2000	200,000–600,000
Dredging	Average sediment depth = 2 ft	20,000–50,000	1,500,000–5,000,000
	Average sediment depth = 5 ft	40,000–80,000	2,500,000–8,000,000
Dye Addition	Detention time >1 month	100–500	200,000–1,000,000
Herbicide Treatment with Diquat	Liquid application	200–500	400,000–2,000,000
Herbicide Treatment with Endothall	Liquid application	400–700	800,000–2,800,000
Herbicide Treatment with Glyphosate	Surface spray application	500–1000	1,000,000–2,000,000
Herbicide Treatment with 2,4-D	Granular formulation, 100 lbs/acre	300–800	300,000–1,600,000
Herbicide Treatment with Fluridone	Liquid formulation, single treatment	500–1000	200,000–1,000,000
	Liquid formulation, 3 sequential treatments	1000–2000	500,000–1,000,000
	Pellet formulation	800–1200	400,000–1,200,000
Herbivorous Fish	Grass carp stocked	50-300	25,000–150,000
Herbivorous Insects	Selected insects stocked	300–3000	150,000–1,500,000

Cost estimates are based on prices obtained from cooperating lake management practitioners across the USA, adjusted to the units of measurement for this table. While cost ranges do not include all possible situations, costs associated with each technique are intended to reflect all directly associated expenses, including design, permitting, capital cost, operating cost, and monitoring. Assumptions made to construct this direct comparison table may not hold true in all possible cases. Costs are given in 2001 U.S. dollars.

for water supply intake or by discharge of poor quality water), costs are minimized. Likewise, if external loading has been reduced to an acceptable level, inactivation of phosphorus with aluminum can provide long-term control at a cost lower than for most competing techniques. However, both of these situations could be considered "special cases" that do not include commonly encountered costs for structural alterations and watershed management.

Certain types of rooted aquatic plants can be controlled by drawdown using an existing outlet structure at a low long-term cost, but not all macrophytes will be controlled and not all lakes have outlet structures that facilitate such drawdowns. Herbivorous fish also offer a potentially low cost, long-term control option, but they don't consume all plant species with equal preference, control over the fish is limited once stocked, and use of non-native species is illegal in some states. Both drawdown and herbivorous fish can produce undesirable effects on a lakewide scale, and counteracting these effects can raise costs considerably.

In most cases, it makes more sense to narrow the range of techniques based on non-cost factors, then develop a lake-specific cost comparison of the most appropriate methods.

With few exceptions, the cost of addressing excessive algal and macrophyte growths is as variable within techniques as among techniques. Again, the difficulty of making generalized cost comparisons among techniques is underscored. In most cases, it makes more sense to narrow the range of techniques based on non-cost factors, then develop a lake-specific cost comparison of the most appropriate methods.

Long-term cost might best be minimized by first using one or more applicable and cost-effective lakewide techniques to control a major problem, then using different techniques on a more local or infrequent basis to prevent its recurrence.

Permitting Lake Management

Nearly all in-lake management techniques require some form of permitting or agency approval (see Chapter 8). Such processes add time and cost to the project, but they underscore the importance of considering all possible impacts and minimizing undesirable effects. Regulatory processes may include:

- General federal or state environmental review (NEPA or state equivalent)
- Environmental impact reporting (if required as a result of NEPA)
- Clean Water Act Section 401 (water quality certification)
- Clean Water Act Section 404 (wetlands permit)
- State and local wetlands protection statutes
- Federal Endangered Species Act consultation
- State protected species statutes
- Fish and wildlife permit/notification
- Discharge permits (NPDES [see Chapter 6] and/or state equivalent)
- Aquatic structures permit
- Chemical application permit
- Dam safety/alteration permit
- Drawdown permit

NEPA: The National Environmental Policy Act of 1969 that established a national policy for the environment, created the Council on Environmental Quality, and directed that every recommendation or report on proposals for legislation and other major federal actions significantly affecting the quality of the human environment include a detailed statement on the environmental impact of the proposed action. For the complete act, see Council on Environmental Quality, 1991.

- Dredging permit
- Waste disposal permits
- Water diversion/use permit

Always consult local and state regulatory authorities before implementing a lake or reservoir management plan, and build an appropriate lead time into your program to acquire permits. Small projects or those involving limited impacts in minimally sensitive systems may require only a month or two to gain approval, while large-scale restoration and management efforts may require more than a year to get permitted.

References

Andersson, G. 1988. Restoration of Lake Trummen, Sweden: effects of sediment removal and fish manipulation. *In* G. Balvay, ed. Eutrophication and Lake Restoration. Water Quality and Biological Impacts. Thonon-les-Bains, Switzerland.

Arnold, W.R. 1979. Fluridone, a new aquatic herbicide. J. Aquat. Plant Manage. 17:30-3.

Axler, R. 1988. Fish aid — the Lake Mead fertilization project. Lake Reserv. Manage. 4:125-35.

Babin, J., E. Prepas, T. Murphy, and H. Hamilton. 1989. A test of the effects of lime on algal biomass and total phosphorus concentrations in Edmonton stormwater retention lakes. Lake Reserv. Manage. 5:129-35.

Baker, J.P., H. Olem, C.S. Creager, M.D. Marcus, and B.R. Parkhurst. 1993. Fish and Fisheries Management in Lakes and Reservoirs. EPA 841-R-93-002. Terrene Inst. U.S. Environ. Prot. Agency, Washington, DC.

Barrett, P., J. Curnow, and J. Littlejohn. 1996. The control of diatoms and cyanobacterial blooms in reservoirs using barley straw. *In* J. Caffrey and P. Barrett, eds. Management and Ecology of Freshwater Plants. Proc. 9th Int. Symp. Aquatic Weeds. Kluwer Academic Publishers, London.

Beard, T.D. 1973. Overwinter Drawdown: Impact on the Aquatic Vegetation in Murphy Flowage. Wis. Tech. Bull. 61. Wis. Dep. Nat. Resour., Madison.

Benndorf, J. 1989. Food web manipulation as a tool in water quality management. JWSRT Aqua. 38:296-304.

Blaha, D.W. and T.J. LoVullo. 1998. Sebago Lake: source protection through water level management. *In* Watershed Management: Moving from Theory to Implementation. A Conf. Denver, CO. May 1998. Water Environ. Fed., Alexandria, VA.

Bortone, S. 1998. Artificial reef management perspective. Fisheries 23:6-10.

Brandt, S. 1998. Impacts of alewife introduction on plankton and water quality. Presented at New Engl. Conf. N. Am. Lake Manage. Soc. Univ. New Hampshire, Durham.

Canfield, D.E. 1992. Address to the North American Lake Management Society, Cincinnati, OH.

Canfield, D.E., K. Langeland, S. Linda, and W. Haller. 1985. Relations between water transparency and maximum depth of macrophyte colonization in lakes. J. Aquat. Plant Manage. 23:25-8.

Carson, R. 1962. Silent Spring. Houghton Mifflin Co., New York.

Center, T. and W. Durden. 1981. Release and establishment of *Sameodes albiguttalis* for the biological control of water hyacinth. Environ. Entomol. 10:75-80.

Colburn, T., D. Dumanoski, and J.P. Myers. 1997. Our Stolen Future. NAL/Dutton, New York.

Connor, J. and M. Martin 1989a. An assessment of sediment phosphorus inactivation, Kezar Lake, NH. Water Resour. Bull. 25:845-53.

————. 1989b. An Assessment of Wetlands Management and Sediment Phosphorus Inactivation, Kezar Lake, NH. New Hampshire Dep. Environ. Serv., Concord.

Conyers, D.L. and G.D. Cooke. 1983. A comparison of the costs of harvesting and herbicides and their effectiveness in nutrient removal and control of macrophyte biomass. *In* Lake Restoration Protection and Management. EPA 440/5-83-001. U.S. Environ. Prot. Agency, Washington, DC.

Cooke, G.D. 1999. Ecosystem rehabilitation. Lake Reserv. Manage. 15:1-4.

Cooke, G.D. and R.E. Carlson. 1989. Reservoir Management for Water Quality and THM Precursor Control. Am. Water Works Assn. Res. Found., Denver, CO.

Cooke, G.D. and R.H. Kennedy. 1989. Water quality management for reservoirs and tailwaters. Report 1. In-lake reservoir water quality management techniques. Tech. Rep. E-89-1. U.S. Army Corps Eng., Vicksburg, MS.

Cooke, G.D., et al. 1993a. Restoration and Management of Lakes and Reservoirs. Lewis Publishers, Boca Raton, FL.

———. 1993b. Effectiveness of Al, Ca, and Fe salts for control of internal phosphorus loading in shallow and deep lakes. Hydrobiologia 253:323-35.

Council on Environmental Quality. 1991. Environmental Quality: 21st Annual Report. Washington, DC.

Creed, R. 1998. A biogeographic perspective on Eurasian watermilfoil declines: Additional evidence for the role of herbivorous weevils in promoting declines? J. Aquat. Plant Manage. 36:16-22

Darrin Freshwater Institute. 1991. Hand Harvesting Eurasian Watermilfoil in Lake George. DFWI Rep. 91-7. Rensselaer Polytechnic Institute, Troy, NY.

Dettmers, J. and R. Stein. 1996. Quantifying linkages among gizzard shad, zooplankton, and phytoplankton in reservoirs. Trans. Am. Fish. Soc. 125:27-41.

Doyle, R.D. and R.M. Smart. 1995. Restoration and rehabilitation of Texas reservoirs: The role of native aquatic macrophytes. Presented at 15th Annu. Symp. N. Am. Lake Manage. Society, Toronto, Ontario.

Dunst, R. et al. 1974. Survey of Lake Rehabilitation Techniques and Experiences. Tech. Bull. 75. Wis. Dep. Natural Resour., Madison.

Eichler, L.W., R.T. Bombard, J.W. Sutherland, and C.W. Boylen. 1995. Recolonization of the littoral zone by macrophytes following the removal of benthic barrier material. J. Aquat. Plant Manage. 33:51-4.

Engel, S. 1984. Evaluating stationary blankets and removable screens for macrophyte control in lakes. J. Aquat. Plant Manage. 22:43-8.

Fayer, R. ed. 1997. *Cryptosporidium* and Cryptosporidiosis. CRC Press, Boca Raton, FL.

Freeman, T.E. 1977. Biological control of aquatic weeds with plant pathogens. Aquat. Botany 3:175-84.

Garrison, V. and E. Smeltzer. 1987. Lake observer survey for Vermont lay monitoring program. Vt. Dep. Water Resour. Environ. Eng,. Montpelier.

Geiger, N.S. 1983. Winter drawdown for control of Eurasian water milfoil in an Oregon oxbow lake. *In* Lake Restoration, Protection and Management. EPA 440/5-83-001. Washington, DC.

Gunner, H., Y. Limpa-amara, B. Bouchard, P. Weiterstein and M. Taylor. 1990. Microbiological Control of Eurasian Watermilfoil. Tech. Rep. A-90-2. U.S. Army Corps Eng., Vicksburg, MS.

Haag, K. H. 1986. Effective control of water hyacinth using Neochetina and limited herbicide application. J. Aquat. Plant Manage. 24:70-5.

Hanson, M.J. and H.G. Stefan. 1984. Side effects of 58 years of copper sulfate treatment in the Fairmont Lakes, Minnesota. Water Res. Bull. 20:889-900.

Hanson, T.C. Eliopoulos, and A. Walker. 1995. Field Collection, Laboratory Rearing and In-Lake Introductions of the Herbivorous Aquatic Weevil, *Euhrychiopsis lecontei,* in Vermont. Year 2. Vt. Dep. Environ. Conserv., Montpelier.

Harper, H.H., J.E. Herr, and E. Livingston. 1999. Alum treatment of stormwater: the first ten years. Pages 159-80 *in* W. James, ed. Modeling Urban Water Systems. Computational Hydraulics Int., Guelph, Ontario.

Heiskary, S. 1989. Integrating ecoregion concepts into state lake management programs. Pages 89-100 *in* Enhancing the States' Lake Management Programs. Northeast Ill. Plann. Commiss., Chicago.

Hosper, H. and M-L. Meijer. 1993. Biomanipulation, will it work for your lake? Ecol. Eng. 2:63-72.

Hoyer, M.V. and D.E. Canfield, eds. 1997. Aquatic Plant Management in Lakes and Reservoirs. N. Am. Lake Manage. Soc., Aquat. Plant Manage. Soc., U.S. Environ. Prot. Agency, Washington, DC.

Hutchinson, G.E. 1957. A Treatise on Limnology. Vol. I, Part 2. Chemistry of Lakes. John Wiley and Sons, New York.

Jones, W.W. 1988. Balancing competing uses for water resources — a Griffey Lake example. Lake Reserv. Manage. 4:73-80.

Joye, G.F. 1990. Biocontrol of the aquatic plant Hydrilla verticillata (L.f.) Royce with an endemic fungal disease. Unpubl. rep. Environ. Lab. Waterways Experiment Station, U.S. Army Corps Eng., Vicksburg, MS.

Kadlec, J.A. 1962. Effects of a drawdown on a waterfowl impoundment. Ecology 43:267-81.

Kenefick, S.L., S.E. Hrudey, H.G. Peterson, and E.E. Prepas. 1993. Toxin release from *Microcystis aeruginosa* after chemical treatment. Water Sci. Tech. 27:433-40.

KVA. 1991. Reverse Layering, An Alternative Approach to Dredging for Lake Restoration. Tech. Rep. Res. Demon. Prog. Mass. Dep. Environ. Prot., Boston.

Kortmann, R. and P.H. Rich. 1994. Lake ecosystem energetics: The missing management link. Lake Reserv. Manage. 8:77-97.

Kortmann, R., G.W. Knoecklein, and C.H. Bonnell. 1994. Aeration of stratified lakes. Lake Reserv. Manage. 8:99-120.

Krishnan, S. 1988. The case against aluminum. Can. Res. 21(3):32-5.

Lathrop, R.C., S.R. Carpenter, and D.M. Robertson. 1999. Summer water clarity responses to phosphorus, *Daphnia* grazing, and internal mixing in Lake Mendota. Limnol. Oceanogr. 44:137-46.

LeChevallier, M., W. Norton, and T. Atherholt. 1997. Protozoa in open reservoirs. J. Am. Water Works Ass. 89:84-96.

Lindmark, G. 1979. Interaction between LPP-1 virus and Plectonema boryanum. Ph.D. Diss., Univ. Lund, Sweden.

———. 1985. Sodium carbonate injected into sediment of acidified lakes: a case study of Lake Lilla Galtsjon treated in 1980. Lake Reserv. Manage. 1:89-93.

Lorenzen, M.W. and A.W. Fast. 1977. A Guide to Aeration/Circulation Techniques for Lake Management. EPA-600/3-77-004. U.S. Environ. Prot. Agency, Washington, DC.

Maceina, M., M. Cichra, R. Betsill, and P. Bettoli. 1992. Limnological changes in a large reservoir following vegetation removal by grass carp. J. Freshw. Ecol. 7:81-95.

Mayhew, J.K. and S.T. Runkel. 1962. The control of nuisance aquatic vegetation with black polyethylene plastic. Proc. Iowa Acad. Sci. 69:302-7.

McComas, S. 1993. Lake Smarts: The First Lake Maintenance Handbook. Terrene Inst., U.S. Environ. Prot. Agency, Washington, DC.

McQueen, D.J., J.R. Post, and E.L. Mills. 1986. Trophic relationships in freshwater pelagic ecosystems. Can. J. Fish. Aquat. Sci. 43:1571-81.

Mills, E.L., J.H. Leach, J.T. Carlton, and C. Secor. 1994. Exotic species and the integrity of the Great Lakes. BioScience 44:666-76.

Miller, A.C. and H.R. King. 1984. Large-scale Operations Management Test of Use of the White Amur for Control of Problem Plants. Rep. 5: Synthesis Rep. Tech. Rep. A-78-2. U.S. Army Corps Eng., Vicksburg, MS.

Mitchell, A. and D. Baldwin. 1998. Effects of desiccation/oxidation on the potential for bacterially mediated P release from sediments. Limnol. Oceanogr. 43:481-87.

Moyle, P., M. Marchetti, J. Baldrige, and T. Taylor. 1998. Fish health and diversity: justifying flows for a California stream. Fisheries 23:6-15.

Murphy, T., E.E. Prepas, J.T. Lim, J.M. Crosby, and D.T. Walty. 1990. Evaluation of calcium carbonate and calcium hydroxid treatment of prairie drinking water dugouts. Lake Reserv. Manage. 6:101-8.

Nelson, L., J. Shearer, and M. Netherland. 1998. Mesocosm evaluation of integrated fluridone-fungal pathogen treatment of four submersed species. J. Aquat. Plant Manage. 36:73-7.

Netherland, M.D. and K.D. Getsinger. 1993. Control of Eurasian watermilfoil using triclopyr. Down Earth 48:1-5.

Newroth, P. and R. Soar. 1986. Eurasian watermilfoil management using newly developed technologies. Lake Reserv. Manage. 2:252-57.

Nichols, S.A. 1986. Community manipulation for macrophyte management. Lake Reserv. Manage. 2:245-51.

Nichols, S.A. and B.H. Shaw. 1983. Review of management tactics for integrated aquatic weed management. *In* Lake Restoration, Protection and Management. EPA 440/5-83-001. U.S. Environ. Prot. Agency, Washington, DC.

Nicholson, S.A. 1981. Changes in submersed macrophytes in Chautauqua Lake, 1937-75. Freshw. Biol. 11:523-30.

Nürnberg, G. 1987. Hypolimnetic withdrawal as a lake restoration technique. J. Environ. Eng. Div. Am. Soc. Civil Eng. 113:1006-17.

N.Y. State Dep. Environ. Conserv. and N.Y. Federation of Lake Associations. 1990. Diet for a Small Lake. Albany, NY.

Olem, H. 1990. Liming Acidic Surface Waters. Lewis Publishers, Chelsea, MI.

Perkins, M.A., H.L. Boston, and E.F. Curren. 1980. The use of fiberglass screens for control of Eurasian watermilfoil. J. Aquat. Plant Manage. 18:13-19.

Ripl, W. 1976. Biochemical oxidation of polluted lake sediment with nitrate — a new lake restoration method. Ambio 5:132-135.

Rohm, C.M., J.M. Omernik, and C.W. Kiilsgaard. 1995. Regional patterns of total phosphorus in lakes of the northeastern United States. Lake Reserv. Manage. 11:1-14

Rydin E. and E.B. Welch. 1998. Aluminum dose required to inactivate phosphate in lake sediments. Water Res. 32:2969-76.

———. 1999. Dosing alum to Wisconsin lake sediments based on *in vitro* formation of aluminum bound phosphate. Lake Reserv. Manage. 15:324-31.

Sanders, D. and E. Theriot. 1986. Large-scale Operations Management Test of Insects and Pathogens for Control of Water Hyacinth in Louisiana. Tech. Rep. A-85-1. U.S. Army Corps Eng., Vicksburg, MS.

Sarrouh, S. 1998. A strategy for co-existing with zebra mussels. Opflow 24:6-7.

Scheffer, M., S.H. Hosper, M-L. Meijer, B. Moss, and E. Jeppeson. 1993. Alternative equilibria in shallow lakes. Trends Ecol. Evol. 8:275-79.

Shapiro, J., V. LaMarra, and M. Lynch. 1975. Biomanipulation: an ecosystem approach to lake restoration. *In* P.L. Brezonik and J.L. Fox, eds. Symp. Water Quality Manage. Biological Control. Univ. Florida, Gainesville.

Sheldon, S. and L. O'Bryan. 1996a. The life history of the weevil, *Euhrychiopsis lecontei*, a potential biological control agent of Eurasian watermilfoil. Entomolog. News 107:16-22.

————. 1996b. The effects of harvesting Eurasian watermilfoil on the aquatic weevil *Euhrychiopsis lecontei*. J. Aquat. Plant Manage. 34:76-7.

Shireman, J., W. Haller, D. Canfield, and V. Vandiver. 1982. The Impact of Aquatic Plants and Their Management Techniques on the Aquatic Resources of the United States: An Overview. EPA 660/3-83-098. U.S. Environ. Prot. Agency, Washington, DC.

Siver, P.A., A.M. Coleman, G.A. Benson, and J.T. Simpson. 1986. The effects of winter drawdown on macrophytes in Candlewood Lake. Conn. Lake Reserv. Manage. 2:69-73.

Smeltzer, E., R.A. Kirn, and S. Fiske. 1999. Long-term water quality and biological effects of alum treatment of Lake Morey, VT. Lake Reserv. Manage. 15:173-84.

Smith, C.S. and G.D. Pullman. 1997. Experience using Sonar A.S. aquatic herbicide in Michigan. Lake Reserv. Manage. 13:338-46.

Smith, C. R. and J. Shireman. 1983. White amur bibliography. Misc. Paper A-83-7. U.S. Army Corps Eng., Jacksonville, FL.

Solarz, S. and R. Newman. 1996. Oviposition specificity and behavior of the watermilfoil specialist *Euhrychiopsis lecontei*. Oecologia 106:337-44.

Tanner, C., R. Wells, and C. Mitchell. 1990. Re-establishment of native macrophytes in Lake Parkinson following weed control by grass carp. N.Z.J. Mar. Freshw. Res. 24:181-6.

Theriot, E. 1989. Biological control of aquatic plants with plant pathogens. *In* Proc. Workshop on Management of Aquatic Weeds and Mosquitoes in Impoundments. Water Resour. Res. Inst. Univ. North Carolina, Charlotte.

Tilman, D. 1982. Resource Competition and Community Structure. Princeton Univ. Press, Princeton, NJ.

U.S. Department of Agriculture. 1986. Animal Damage Control. USDA Animal Plant Health Inspection Serv., Amherst, MA.

Ussery, T., H. Eakin, B. Payne, A. Miller, and J. Barko. 1997. Effects of benthic barriers on aquatic habitat conditions and macroinvertebrate communities. J. Aquat. Plant Manage. 35:69-73.

Vermont Agency of Natural Resources. 1990. The Lake Bomoseen Drawdown. Waterbury, VT.

Wagner, K.J. 1986. Biological management of a pond ecosystem to meet water use objectives. Lake Reserv. Manage. 2:53-61.

————. 1990. Assessing the impacts of motorized watercraft on lakes: Issues and perceptions. *In* Enhancing the States' Lake Management Programs, Northeast Ill. Plann. Commiss., Chicago.

————. 1991. Implementation Project for the Management of Great Pond, Eastham, Mass.: Aquatic Plant Control Program. BEC, Inc., East Longmeadow, MA.

————. 1994. Of hammocks and horsepower: The noise issue at lakes. Lake Line 14:24-28.

———————. 1996. Summary Report on the Restoration of Hills Pond in Menotomy Rocks Park, Arlington, Mass. Fugro East, Northborough, MA.

———————. In review. The restoration of Hamblin Pond, Barnstable, Mass., by phosphorus inactivation. Lake Reserv. Manage.

Wagner, K.J. and C. Carranza. 1986. Diagnostic/Feasibility Study of Quaboag and Quacumquasit Ponds. BEC, Inc., East Long Meadow, MA.

Wagner, K.J. and T. Clear. 1986. Lake George Milfoil Control Evaluation. Fugro East, Northboro, MA.

Wagner, K.J. and R.T. Oglesby. 1984. Incompatibility of common lake management objectives. *In* Lake and Reservoir Management. EPA 440/5-84-001, U.S. Environ. Prot. Agency, Washington, DC.

Walker, W., C.E. Westerberg, D.J. Schuler, and J.A. Bode. 1989. Design and evaluation of eutrophication control measures for the St. Paul water supply. Lake Reserv. Manage. 5:71-83.

Welch, E.B. and G.D. Cooke. 1999. Effectiveness and longevity of phosphorus inactivation with alum. Lake Reserv. Manage. 15:5-27.

Welch, E.B. and E. Rydin. 1999. Dosing alum to Wisconsin lake sediments based on *in* vitro formation of aluminum bound phosphorus. Lake Reserv. Manage. 15:324-31.

Welch, E.B. and G.D. Schrieve. 1994. Alum treatment effectiveness and longevity in shallow lakes. Hydrobiologia 276:423-31.

Welch, E.B. and C.R. Patmont. 1980. Lake restoration by dilution: Moses Lake, Washington. Water Res. 14:1317-25.

Westerdahl, H.E. and K.D. Getsinger, eds. 1988a. Aquatic Plant Identification and Herbicide Use Guide. Vol. 1. Aquatic Herbicides and Application Equipment. Waterways Experiment Station, U.S. Army Corps Eng., Vicksburg, MS.

———————. 1988b. Aquatic Plant Identification and Herbicide Use Guide. Vol. 2. Aquatic Plants and Susceptibility to Herbicides. Waterways Experiment Station, U.S. Army Corps Eng., Vicksburg, MS.

Willenbring, P., M. Miller, and W. Weidenbacher. 1984. Reducing sediment phosphorus release rates in Long Pond through the use of calcium nitrate. *In* Lake and Reservoir Management. EPA 440/5-84-001. U.S. Environ. Prot. Agency, Washington, DC.

Wisconsin Department of Natural Resources. 1989. Environmental Assessment of Aquatic Plant Management (NR 107) Program. Wis. Dep. Natural Resour., Madison.

Wlosinski, J.H., and E.R. Koljord. 1996. Effects of Water Levels on Ecosystems: An Annotated Bibliography. USGS LTRMP 96-T007. Environ. Manage. Tech. Center, Onalaska, WI.

Wynn, E. and K. Langeland. 1996. Barley straw for algae control. Aquatics 18:4-6.

Chapter 8

Developing and Implementing a Management Plan

Planning is an essential aspect of managing your lake or reservoir — indeed, planning and management cannot be separated. The investments you make in developing and implementing your management plan will pay off in terms of reaching your goals.

Why Plan?

If you follow a plan, you will reach your goals.

It can't be stated more simply than that. People concerned about lakes often say they can't afford to plan. "We are only volunteers" or "We have just so much money, so we want to spend it doing something, not planning." But many lake associations waste money on their lake because they did not have a plan.

People who don't prepare financial plans say, "I can't afford to do that right now." Yet, people who prepare and follow financial plans are better off.

Businesses that prepare and follow business plans are more likely to succeed. Indeed, most banks won't loan money to businesses without business plans.

Without a plan, you will not reach your goals.

Planning is **doing**. Planning goes hand in hand with management; it provides the framework for what you do. With planning, management is more focused, meaningful, and effective.

Crises often spark management actions that mistakenly substitute for planning. The discovery of a new exotic species or a proposed development on the lakeshore often incites quick organization and action — but almost always outside of a meaningful planning context. After the urgency disappears, the organization and the effort dissolve.

These situations argue for advanced planning — often the crisis could have been avoided and effectively managed in the context of a plan.

"When good things happen to us it is often the result of planning. The better the planning, the better the result."

From, "Planning Ahead . . . . Strategically." Olsen Thielen *Advisor,* 1998.

Imagine a complicated computer factory with white-coated technicians emerging from rooms sealed with positive air hatches as they move products along an ultra-sophisticated production line. This is the end result of many decades of research and development and millions of dollars of capital expenditures.

Now imagine your favorite lake during a summer's sunrise. The 'technicians' are the many plants and animals of an ultra-sophisticated biological production line: the end result of millions of years of evolutionary 'research.'

Just as the factory, the lake did not just appear and it needs even more research and attention than the computer factory, believe it or not. Otherwise, how will we be able to maintain its engines of life? There are now too many contaminants and unidentified influences from all of the people technicians in this factory — which might be fatal in the computer analogy.

What to do? We need to institute careful planning and attention to detail for our lakes to maintain their engines of life that we have come to appreciate. This will take resources — people, ingenuity, money, and commitment — to accomplish. *–Bruce Wilson, Minnesota Pollution Control Agency. An analogy illustrating the importance of planning for lakes. December 1998.*

What is a Management Plan?

Your management plan should answer the question:

▼ ***Who will do what by when and with what expected result?*** We can take this question apart to form the elements of the plan.

- The **who** refers to the members of the lake community who are willing to invest in the management of their lake. Including those people who are unwilling or who are coerced into action will work against sustaining a meaningful management program.
- The **what** refers to management actions directed toward a meaningful and measurable objective. Here we must exercise care regarding the source of funding. Outside monies — those coming from outside the lake community — may tend to direct management actions to goals not embraced by the community; or, these monies may make your community complacent by delaying their investment in the lake.
- The **when** is important because without a deadline, there is no commitment.
- Specifying an **expected result** forces an ongoing and critical evaluation, which is essential to a long-term effort. The evaluation must be objective and measurable, and it requires flexibility in the management program.

Because lake and reservoir management is not an exact science, ongoing evaluation is essential, and should, as a rule-of-thumb, be up to 20 percent of the management effort. Lake McCarrons, Minn., went for 10 years without any critical evaluation; then, the community learned that the restoration program had not improved the lake in any way (Metropolitan Council, 1997). Not attaining the expected result does not represent a failure of the management program, but rather an opportunity to adjust and improve.

▼ ***'Plan' is a verb, not a noun.*** Okay, it's really both, but the point is the outcome of planning is not just a report, it **must** include action — and that presents unique challenges. Unlike personal financial plans, which involve one person or a couple, lake management plans must satisfy multiple, often conflicting objectives. This means that preparing a lake management plan requires a greater investment of time and resources. And stakeholder participation and conflict management will be an important aspect of developing and implementing the plan.

"Failures don't plan to fail; they fail to plan."

"Be like a postage stamp; stick to it until you get there."

"A dream is just a dream. A goal is a dream with a plan and a deadline."

Harvey Mackay's Column-ending 'morals'; United Features Syndicate

The Planning Cycle

Planning for lake and reservoir management is a thoughtful, intentional, and systematic process that results in action. Investment in and attention to the details of preparing and implementing a management plan will result in attaining your management goals. Planning and management go hand-in-hand in an ongoing cycle.

The two main ingredients in an effective management plan are **discipline** and **clear vision**. Discipline keeps you faithful to and focused on the planning process. Clear vision helps you develop reasonable goals and objectives.

When Do You Start Planning?

Now, any time, always. Planning is an integral part of management and begins and continues with management as long as anyone cares about the lake or reservoir.

The basic steps:

- Step 1. Get started, make a commitment.
- Step 2. Analyze the situation, take stock.
- Step 3. Set directions — vision, goals, and objectives.
- Step 4. Evaluate alternative strategies and actions.
- Step 5. Take action, implement.
- Step 6. Monitor and evaluate progress.
- Step 7. Repeat steps 2 — 6.

Steps 1 through 4 comprise the plan development phase, and steps 5 through 7, the plan implementation phase (Kehler et al. [no date]). Although variations of this planning model exist (see Chapters 3 and 6), the most important consideration is to adhere to the basic elements (Fig. 8-1).

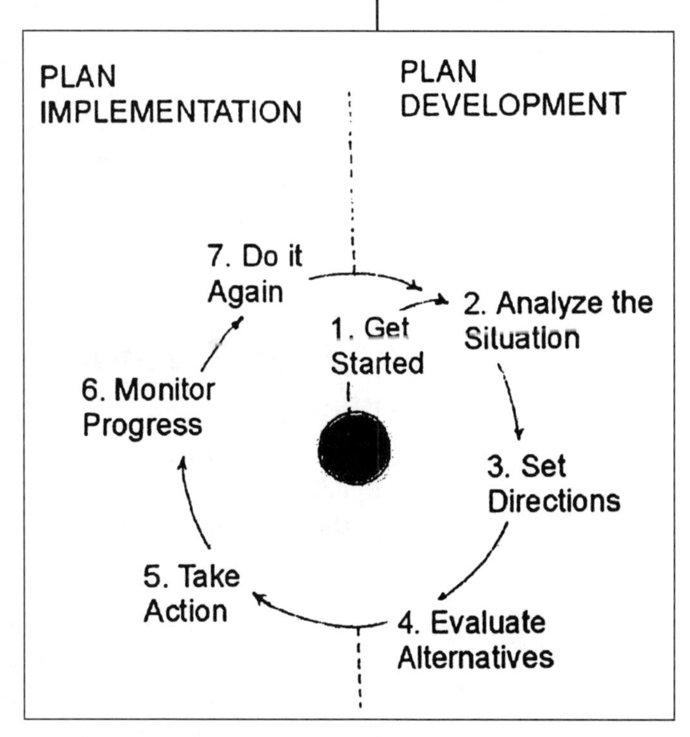

Figure 8-1.—The planning cycle.

Steps I — 6 will normally take from 6 months to a year to complete, require the involvement of dozens of people, and may include a professional planner or facilitator.

The amount of time required to complete the planning cycle depends on the complexity of your lake. A small pond in a condominium development requires much less formal preparation than a more significant resource like Lake Tahoe. Then, to be meaningful and lasting, some formal management entity must be empowered to oversee the ongoing management of the lake or reservoir. That organization should repeat the planning cycle every three to five years as a routine part of its ongoing management.

Step 1: Get Started

Getting started may be the most difficult step. Most people are not inclined to plan. That's why people often do not prepare financial plans, save for their children's college, write wills, or prepare for retirement.

Lake management planning involves many people — sometimes hundreds — with potentially conflicting interests. Getting started requires a high level of organization, which is especially difficult since many participants are volunteers.

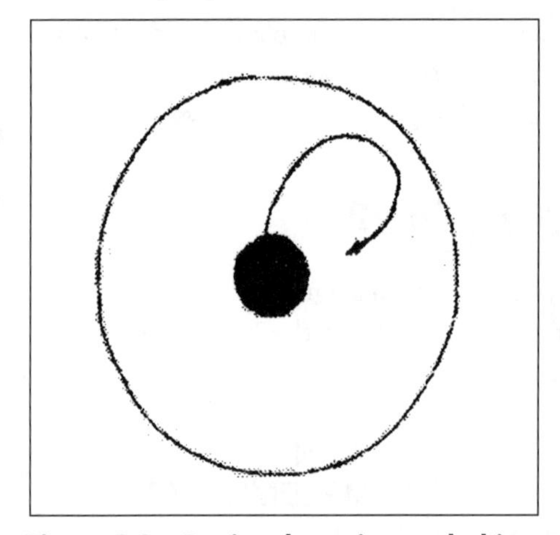

Figure 8-2.—Depicts the trajectory lacking the energy to attain orbit, and therefore, falls short.

So, the hurdle of 'getting started' is real. At a minimum, you will need enough money, technical support, and organizational effort to carry the initial planning cycle. You may need help from outside sources, but long-term management will ultimately rely upon local resources.

Start-up resources are akin to the booster rocket that is required to break the earth's orbit (Fig. 8-2): it must be large enough to get the capsule into orbit, but once in orbit, the booster is no longer needed. The capsule (management planning) can then sustain its orbit under gravity (local resources).

Step 2: Analyze the Situation

Before a group can agree on how to manage the lake, they must find consensus on where to start. A **situation analysis** —a snapshot of the physical, social, economic, and political environment surrounding the lake or reservoir — prompts this discussion and sets the initial direction for developing the management plan.

The situation analysis will include:

- **A community survey** to evaluate perceptions — and a "wish list" — of lake conditions, changes in lake conditions, uses and values of the lake, conflicts regarding the use of the lake and lakeshore, and concerns about ongoing management of the lake. See Chapter 3 for survey methods.

- **A physical inventory** to summarize the chemical condition of the lake, trophic state, fish and aquatic plant communities, shoreline plant and animal communities, lake level fluctuations, pollution sources, watershed delineation, estimates of pollution sources and amounts, and land use. A physical inventory is not a detailed diagnostic or engineering study.

- An assessment of the **human environment** that includes a list of management and regulatory authorities, population projections, planning and zoning matters, business and commercial interests, management and control activities, and lake uses.

- A stab at a **collective vision** for the lake. The person preparing the situation analysis normally gets a good sense of the community's collective desires. It is useful to prompt the group with an independent view of their situation. This will be the point of departure between the matter-of-fact situation analysis and the remainder of the planning process.

When you've completed it, summarize the situation analysis in a brief, non-technical report that sets the initial directions for developing the management plan.

Step 3: Set Directions

According to *Top 10 Watershed Lessons Learned* (U.S. EPA, 1997), the best plans have "... clear visions, goals, and action items." Getting from a broad vision to discrete, meaningful, doable actions requires a good deal of time and energy from the group developing the management plan.

The management plan of the Blue Water Commission, the committee that developed the plan to protect Lakes Nokomis and Hiawatha in Minneapolis, Minn. (Osgood, 1998), contained numerous goals, objectives, and actions that were organized in a readable and logical format. The following example shows how the vision translated into goals, objectives, and actions (one each shown) for Lake Nokomis:

- **Vision:** *Lakes Nokomis and Hiawatha are the focal points for our community and should be a showcase for Minneapolis. The lakes are valuable natural assets that must be protected and improved for the benefit of our neighborhoods and the city as a whole.*

- **Goal #1 (of 19):** *Eliminate nuisance algae blooms.*

- **Objective** (one of several objectives to address this goal): *Prevent 294 pounds of phosphorus per year from entering Lake Nokomis.*

- **Action 1c:** *The Minnehaha Creek Watershed District will build three wet detention ponds (specific designs and locations provided in the plan).*

Visions — general statements of where the lake community wants to go and what it will accomplish over a given time span (usually 5 to 10+ years). Visions should be comprehensive enough to capture the thrust of the planning process.

Goals — more specific than visions, break into logical pieces what is needed to obtain vision, refer to components of overall effort, sometimes quantifiable.

Objectives — steps to achieve the goals, describe types of management or activities and are quantifiable where possible.

Action Items — explain who is going to do what, where, and when; they generally articulate how to implement the objectives and should be quantified if possible; benchmarks of existing conditions and/or indicators should be developed for action items.
— from *Top 10 Watershed Lessons Learned, U.S. EPA, 1997*

Step 4: Evaluate Alternative Strategies and Actions

The temptation at this point is often to go right to a solution without carefully evaluating all options within the framework established in the first three steps.

Don't just pick a solution that looks good at first. Be thorough, prudent, and systematic. The final management plan will probably contain actions that, in total, will address several problem areas. The initial evaluation needs to recognize how these actions might fit together to complement rather than conflict with the plan's objectives.

Step 5: Take Action

Now the community is ready to begin the implementation phase of lake management — at some point, someone has to take the prescribed actions. The management plan per se will not take action, it will spell out *who will do what by when and with what expected result*; this is the point where planning stops and implementation begins. Recall that both are critical and ongoing aspects of the planning cycle (Fig. 8-1).

Much attention and energy needs to be invested in 'smell checking' possible actions. It is easy to make recommendations that contain the word **should**, but these are meaningless without an agreement as to whether the action **will** occur. Consider a recommendation that reads:

> ***The Pretty Lake Watershed District should implement a phosphorus fertilizer ban.***

This recommendation is basically one party directing another party to take a specific action. What if the second party — in this case, the Pretty Lake Watershed District — does not agree with the action? Will they do it? Probably not.

So, rather than assuming someone else (here, the Watershed District) will take action, that "someone else" should be included in the planning process. That way, they will cease to be "someone else" but become part of the entire effort to manage the lake wisely. Then, when everyone has agreed to the recommendation, the plan will read:

> ***The Pretty Lake Watershed District will (or has agreed to) implement a phosphorus fertilizer ban.***

Will, not should. All parties have reached agreement and now we know the action will occur.

Step 6: Monitor and Evaluate Progress

Like children in school, a lake project must be monitored and evaluated. Although you expect certain outcomes, environmental management is not a precise science. You can expect a financial investment of $100 at 5 percent interest to return $105 in a year, but lake science is much less exact, and therefore much less certain. The environmental situation is also constantly changing. So, monitoring

Case Study: Lake McCarrons

Lake McCarrons is an urban lake in Roseville, Minnesota. A restoration project was implemented in 1985 to reduce algae blooms in the lake.

In 1996, 10 years after the project began, lake residents complained of persistent algae blooms and obnoxious aquatic plants. A situation analysis revealed that:

1. Lake phosphorus concentrations had not declined. Likewise, algae levels and water clarity had not improved. Thus, the project had not met the goal of lowering phosphorus and reducing algae blooms.

2. While the lake had been continually monitored, progress had not been evaluated.

3. Nuisance aquatic plants were still a problem.

This illustrates the importance of two critical steps in the planning process. First, the community was not involved in setting the initial directions for the plan (Step 3). If they had been, the plan would have recognized their concern with nuisance aquatic plants. Second, after a decade of monitoring, it was clear the lake's condition had not improved, but since no critical evaluation was built into the planning process, the plan has never been modified to better address its goals (Step 6). —*Osgood (1996a) and Metropolitan Council (1997).*

and evaluating progress help you track what's really going on so you can make adjustments in the management program.

Finally, over a longer time, the perceptions and expectations of those who use and value the lake may also change. At some point, normally after five to 10 years, it is time to begin another planning cycle. With sound monitoring and evaluation in hand, the subsequent planning cycles should be more focused than the initial effort.

Monitoring and evaluation are designed to assure performance of the management program, to determine whether it's meeting your stated goals and objectives. Mid-term failure to meet your objectives does not imply a failure of the management program, but instead offers an opportunity to better understand, then make adjustments.

Given the inherent uncertainty in environmental management and the ongoing need for up-to-date information, monitoring and evaluation are essential in managing lakes and reservoirs. They should comprise about 20 percent of the management effort.

Planning should never stop.

Step 7: Do it Again

Planning should never stop. After the initial effort, the emphasis on planning will fade; but it should never die. After awhile the basis for your plan will change as the environment changes, technologies change, human needs and expectations change, the economy changes, even your lake changes. Just as we periodically review our business or retirement plans to keep them up to date, so must we review our lake management plan.

Sustaining the Management Effort

To be effective and sustainable, a lake planning and management program must be organized and focused. It takes **work** to keep organizations focused and it takes **energy** to sustain a desired condition in a lake or reservoir.

To realize your lake management goals, you must sustain your management efforts by:

- Building a dynamic, locally-based management organization that focuses only on managing the lake or reservoir.

Without formal organization, a management effort may be unfocused, ineffective, and short-lived. Organizations built around crisis issues or loosely put together tend not to last. On the other hand, organizations that are formally organized and focused will:

- Focus on comprehensive concerns instead of single issues
- Be anticipatory instead of reactive
- Be well connected with their community
- Have dynamic leadership instead of burning out their leaders
- Focus on programs instead of projects
- Tend to have volunteer workers instead of volunteer leaders
- Seek solutions instead of blame
- Be oriented toward problem-solving instead of problem-finding

To sustain a credible and long-term management effort, the organization must:

- Be directed by leaders with a close connection to the lake community
- Have a clear vision
- Have a sound development strategy

Build a Local Management Organization

The community — those groups, organizations, users, and individuals closest to the resource — is in the best position to manage the lake for the long run. Therefore, you must either identify or create a local organization to assume responsibility for the management plan.

Always remember, while lake ecosystems can function very nicely without people, it is precisely the human integration that prompts the need for management.

This local lake management organization must engage the lake community in meaningful management actions for the long run. But maintaining the organization requires **work:** time, resources, and money.

Work that must be done by people — the members of the organization. Always remember, while lake ecosystems can function very nicely without people, it is precisely the human integration that prompts the need for management. Build your organization on the solid base of those who use and value your lake: year 'round and seasonal residents, local businesses, people who treasure it for a variety of reasons.

You can't expect everyone in your watershed to be interested in your lake, however: the geography of the lake community may not coincide with the watershed boundary. Thus, the watershed approach to lake management, while an important element of the physical system, may be less significant in the social context (see also Osgood, 1999b). Keep this in mind as you develop your lake management plan.

The point is that no 'one-size-fits-all' organization suits all lakes and reservoirs; the organization must be tailored to the unique circumstances of the resource.

The watershed approach to lake management, while an important element of the physical system, may be less significant in the social context (see also Osgood, 1999b). Keep this in mind as you develop your lake management plan.

Key Characteristics of a Successful Watershed Organization

- Full-time employees
- Established office space and equipment
- Access to water quality information
- High-level concern for water quality
- Water quality monitoring program in place
- Interest in (and encouragement of) citizen participation
- Presence of a public outreach program

Source: from *Environmental Ground*, no date

Involve the Whole Lake Community

The lake community is the people, groups, and organizations that use and value the lake. Sometimes called stakeholders, these people should be called upon to develop and implement the management plan. They are, after all, the community for whom the lake is being managed and to whom you will turn to sustain the management effort.

The lake community is the people, groups, and organizations that use and value the lake.

Identifying the lake community is critical. Select people with a stake in the resource. Take care not to leave key players out of the planning process, but avoid too large a number — many interests will have no stake in the resource and can dilute the planning effort. In both cases, the planning suffers.

The following categories of people, groups, and organizations may be involved:

- **People**. Lakeshore owners, neighbors, residents, and others, including members of related lake associations.
- **Groups**. Lake-related businesses, chambers of commerce, and lake user groups (for example, anglers, water skiers, swimmers, divers, wildlife observers).
- **Organizations**. Governmental agencies, water utilities, tribal governments, municipalities, special units of government, religious organizations, and environmental and conservation organizations.

Not all of these people and organizations need to be involved. Remember to include all with a stake in the resource, but no more. When assembled, the stakeholders should be engaged to act in their mutual interests. Their management actions will result from a desire to act, rather than from an outside directive; and their motivation will come from their 'ownership' or value of the resource.

Consider Hiring a Professional Planner or Facilitator

The role of the planner is not so much to write the plan, as to direct its development and facilitate agreements required to implement it. The planner should be familiar with lake management, versed in planning, and skilled in conflict resolution. It is important that the planner not be associated with or attached to any of the interests of the stakeholders — the planner must be neutral. This frees the planner to facilitate all discussions and negotiations even-handedly and fairly, thereby assuring that the plan works for the community.

Tips for selecting and hiring a planner/facilitator:

- Know what qualifications you need, and
- Check references.

Focus on the Whole Ecosystem

Here 'ecosystem' is considered in the broadest sense: comprised of the lake or reservoir, the people, and the watershed — plus all their physical, social, and political interactions. These elements are the heart, soul, and backbone of a meaningful management program.

- The lake (heart) is the primary management focus.
- The people (soul) are the lake community. They provide the vision and goals. After all, the people are the ones who care most about the lake and for whom the lake is being managed.
- The watershed (backbone) is the basic management unit.

Here 'ecosystem' is considered in the broadest sense: comprised of the lake or reservoir, the people, and the watershed – plus all their physical, social, and political interactions.

The lake or reservoir, of course, is the main focus of the management plan. The lake ecosystem is much more complicated than the unpleasant consequences of too much phosphorus, the classic management focus. Fish and aquatic plant communities are key elements of the lake ecosystem that require equal attention for sound and sustainable ecosystem management. The interactions of these ecosystem elements are complex, but critical to sound management (Jeppesen et al. 1997; Moss et al. 1996; Osgood, 1996b; and other chapters in this manual).

First, identify the needs and interests of the lake community. Human needs and interests can be translated into environmental attributes and then into lake-specific objectives (see Chapter 3). Be careful when using lake standards or criteria as a guide for management objectives. Because many in the lake community may not understand how they were developed or applied, using standards or criteria may be a distraction. Better to first tie a management need to a human need (Walesh, 1999).

Those needs can differ drastically, as illustrated by two lake associations in the Twin Cities (Minnesota) metro area. The Mooney Lake Association's main management objective is to keep the Secchi disk transparency greater than 3 feet. When the lake clarity falls below this, they apply copper sulfate and the clarity increases to about 4 feet. This level of clarity is in the lowest 10 percentile for all lakes in the metro region (Osgood, 1988), and well below the region's water quality standard, but it exceeds the needs of the Mooney Lake Association (Osgood, 2000b). Indeed, when the Secchi disk reaches 4 feet, the president of the Mooney Lake Association says the lake 'sparkles.'

In contrast, the Christmas Lake Association aims to keep its lake's clarity above 17 to 20 feet (Osgood, 1995). Christmas Lake is among the clearest lakes in the region — well above the top 10 percentile. The 'normal' clarity for lakes in this region is 7 to 10 feet, but that has no relevance for either association. Achieving a 'normal' lake clarity would be unthinkable for the Mooney Lake Association and a disaster for the Christmas Lake Association. Here, each organization manages its lake to meet its unique needs and interests.

Watershed management is to lake management what paying your bills is to financial management. Both are essential elements of management, but they alone will probably not help you achieve your goals.

Learn from this example that societal or regulatory 'standards' or 'norms' may not jibe with the needs and interests of the lake community. You best discover what your community wants — and make sure that thinking is included in the goals and objectives of the lake management plan.

Watershed management is **not** lake management. This is an important distinction because we have discovered that many lakes are not sensitive to phosphorus reductions through watershed improvements (see Carpenter et al. 1998, 1999; Newton and Jarrell, 1999; and Osgood, 2000a). Watershed management **is** important in three ways:

- First, it's critical in preventing lake eutrophication problems;
- Second, it helps sustain most lake management goals; and
- Third, watershed management also addresses other kinds of pollution.

So, without a sound watershed management element in your lake management plans, other strategies and actions are less effective.

Sound watershed management can be approached in three ways (see Chapter 6):

- **Point Source Controls**. Controlling the pollution discharged (through pipes) by cities and industry almost always improves lake conditions (Edmondson, 1991).
- **Nonpoint Source Controls**. Controlling pollution carried by runoff improves a lake only when carried out as a long-term, comprehensive program. Although sustained effects have seldom been demonstrated, nonpoint source controls stabilize the watershed and minimize the effects of changing land use.
- **Education**. Education is often used to encourage behavioral changes (a significant factor in controlling nonpoint pollution). To be effective, education programs must address a planning objective with measurable outcomes. Education can also be a marketing and community-building tool; both are important for implementing the management plan (Osgood, 1999c).

Developing the Management Plan

To use your resources wisely in developing a management plan, you must know the essential elements involved and how to put them together.

Getting Organized

▼ ***Develop a Work Plan:*** Planning takes time, effort, and resources. If you have never prepared a management plan for your lake, much of the initial management budget will be spent up front on developing the plan.

Use the sample work plan (Table 8-1) to guide the development of the plan and suggest reasonable expectations for completing it. Keep in mind that there is no one model: the right model for your lake or reservoir must be tailored to the specific needs of your lake community.

▼ ***Expect Conflict.*** Conflict is a normal part of human interaction. Any time you gather people together to discuss their interest in managing a shared resource, you can expect differing views and positions regarding the importance, use, value, or investment in that resource. Even when you might think everyone concurs in their desire to manage or protect a lake, conflict will erupt over almost every part of developing a management plan.

Rather than working to **avoid** conflict, it is more effective to **manage** conflict. Look for differing needs and interests, recognize them, and address them openly and respectfully.

You would be wise to look for a facilitator skilled and conversant in strategic planning, lake management, and the governmental and regulatory system of lake management.

▼ ***Use a Facilitator.*** A neutral facilitator skilled in conflict resolution or mediation should be used to facilitate the discussion needed to develop the management plan (Ury, 1993; Fisher et al. 1991). The role of the facilitator is to (1) guide the plan development process in a positive, productive direction, (2) assure that all participants have an equal chance to voice their interests, and (3) identify and help resolve conflict when it occurs. The facilitator will work with the committees, staff, and others to help them come to consensus regarding significant issues.

You would be wise to look for a facilitator skilled and conversant in strategic planning, lake management, and the governmental and regulatory system of lake management.

▼ ***Look for Consensus.*** Robert's Rules of Order can be an impediment to open discussion and reaching agreements. Voting and other democratic-like methods (for example, nominal group processes [see Chapter 3]) may work against consensus. Regardless of how the process is structured, the outcome of a vote fosters a 'win-lose' environment. While these processes are quantitatively objective and satisfying, they are psychologically demeaning, especially for the losers. Indeed, those in the minority often believe that, if only they had better made their case, the vote would have swung the other way. In the extreme, the minority may feel that the majority decision was coerced.

You can't always obtain balanced representation, especially with a range of issues. Thus, unless there is unanimous agreement, there are always 'losers.' While voting, Robert's and other structured processes have their place (see Chapter 3). Seeking consensus is best when **all** parties need to support and implement the management plan.

Table 8-1.—Management Plan Development — Sample Work Plan

ACTIVITY	TIMELINE	AGENDA	OUTCOME
1. Initial meeting	Month 1	Assemble interested parties to discuss the need for managing your lake.	A resolution to proceed.
2. Organize	Months 2–3	Set up a more formal organization. Identify committees & committee chairs. Assign tasks and deadlines. Contact agency staff, local authorities, and business representatives. Identify funding needs & sources.	Deadline for making decisions to proceed.
3. Develop work plan & secure funding	Month 3	1. Develop a work plan and timetable. If you are unsure how to develop the work plan, ask for proposals. 2. Secure funding. The first planning cycle will require resources.	A draft work plan.
4. Hire a planner or facilitator	Month 4–5	Prepare a list of qualifications and scope of work. Seek candidates.	Hire the planner or facilitator.
5. Community survey	Months 6–8	Develop, implement, and evaluate a community survey.	An indication of community needs, interests, and values.
6. Inventory of lake condition	Months 6–8	Gather & summarize all available physical, chemical, biological, land-use, and demographic information available for your lake and watershed.	An objective picture of the environmental condition of the lake and its environs.
At this point, it may be necessary to implement a diagnostic study. If it is apparent that the problems — either real or perceived — lack sufficient physical data for satisfactory resolution, then a one- or two-year diagnostic study may be needed. If, on the other hand, environmental monitoring is required to simply better define an identified problem, the diagnostic work may be a management action that is an outcome of the planning process.			
7. Preliminary problem definition	Months 6–8	Based on input from the community survey and the inventory of lake condition, identify main problems to be addressed in the management plan.	Written summary of problems to be addressed.
8. Form advisory committee	Months 6–8	Based on input from the survey and other sources, identify community representatives to serve on an advisory committee. The size of the advisory committee should be 20 to 30. Extend invitations.	– Advisory committee roster. – A specific charge.
9. Form technical advisory committee	Months 6–8 (technical input may be needed earlier in the process)	Invite technical regulatory and governmental representatives to serve on a technical advisory committee.	– Technical advisory committee roster. – A specific charge.
10. Plan development	Months 9+ (6–12 months) (4–12 meetings)	**Advisory Committee Agenda** – Introductions – Review survey & inventory summaries – Educate committee – Develop vision & goals – Evaluate alternative strategies & actions – Decide on management actions – Develop a monitoring & evaluation plan – Commit to action	Management plan

Seeking consensus allows open expression of all points of view and agreement on action. This rule for consensus is useful:

> ***If on a particular issue, committee members disagree, the members do agree that all have been given an equal chance to be heard and the consensus agreement represents the best solution at that time.***

Consensus-building is procedurally more cumbersome to administer and usually more time-consuming, but the result is almost always more palatable and will lead to willing future action and participation. Consensus also allows for future reconsideration. If new information becomes available, the monitoring program finds an unexpected result, or the needs of the community change, the issue as well as the management response can be re-opened.

Implementing the Work Plan

The following steps (see Table 8-1) will be required to develop the management plan. The timelines presented here are meant as guidelines as well as benchmarks for realistic expectations. Because the ultimate success and effectiveness of your management plan depend to a great degree on community buy-in, it is better to err on the side of taking the suggested amount of time rather than looking for short cuts.

1. Initial Meeting (Month 1)

Most likely, the group has been brought together by a common concern for their lake or reservoir. The only important outcome of the initial meeting is some commitment to proceed with developing a lake management plan. And this is enough.

2. Organize (Months 2–3)

A more formal oversight committee, sometimes referred to as a Steering Committee (Interagency Lake C.C., 1996), should be formed to coordinate the development of the management plan. This group will be responsible for securing and administering funding and implementing the work plan to develop the management plan; they may even be responsible for adopting and implementing the management plan.

Identifying working committees and setting them about their tasks is done at this time. Working committees may include:

- Finance and Administration
- Planning
- Program Management
- Communications and Community Relations

3. Develop Work Plan and Secure Funding (Month 3)

A detailed work plan should be developed to guide the development of the management plan. The work plan has several purposes:

- Project realistic timelines and work loads for developing the plan
- Guide the planner/facilitator
- Serve as a communication tool

Also, funding needs should be identified and funding secured by this time. You shouldn't proceed beyond this point without enough money to complete the development of the management plan.

4. Hire Planner/Facilitator (Months 4–5)

If, in preparing the work plan, the group decides it needs professional assistance, you or one of your committees can develop a scope of work for the planner or facilitator. With that in hand, you can seek candidates to hire.

Although your funders or Steering Committee may dictate how you go about the search, you might want to issue a Request for Proposal, particularly if you are having difficulty developing your scope of work or your work plan. An RFP, as it's known, asks candidates to tell you how they would approach this project; that will give you several independent ways to think about the project. Be sure, however, to include your budget in the RFP so responders can stay within your financial limits.

Look for planners or facilitators who are qualified and experienced in:

- Lake or reservoir management
- Lake and watershed management and regulatory agencies
- Local government
- Mediation and facilitation

The planner or facilitator should also have demonstrated the ability to:

- Be balanced and neutral
- Solve problems and find solutions
- Write and communicate clearly and effectively
- Work with diverse interests

Be sure to ask for and check references!

Although it may be tempting to select someone from a local agency or the community, think twice! It is critical that the planner be able to facilitate the planning process as a neutral participant — a local person may have a real or perceived allegiance to one of the interested parties. Your planner/facilitator must also be experienced with other planning situations to provide broader perspective and 'out-of-the-box' thinking.

5. Community Survey (Months 6-8)

Use the Community Survey to get an overview of the issues and concerns that must be addressed in the management plan. The response will also help you identify individuals who can best articulate concerns for their community and who might be good candidates for the Advisory Committee.

The results of the survey are not meant to give you definitive or quantitative answers about management issues, but instead should serve as a point of departure for the work of the Advisory Committee.

The sample (on the following page) is open-ended to allow for full expression of community concerns. The planner or survey analyst must carefully summarize and interpret the results — maybe even list the written responses. Remember, you're not looking for definitive results; the purpose of the survey is to guide the long-term development process. (Chapter 3 gives you a more quantitative survey instrument.)

Deciding who to survey is a critical step. Target individuals and interest groups who clearly have a stake in the lake and its management; this will give you better responses (as high as 80 percent is not unusual) than random samples. Also, a survey designed with specific knowledge of the lake and its management issues will engage respondents and encourage them to participate in developing the management plan.

When the debate and deliberations are conducted in a positive, respectful atmosphere, the solutions and agreements reached are likely to be sustainable.

Distribute between 100 to 300 surveys. Because the results will not be analyzed statistically, this number is determined more by the way it represents the community. Asking for a short response deadline (1-2 weeks) helps prevent the forms from getting 'lost in the in-box.' And enclosing a return envelope increases the returns.

Sometimes follow-up phone interviews can be useful. You will find that people who might not ordinarily come forward will, when personally contacted, discuss — sometimes at length — important concerns. Listening always pays off in terms of developing a management plan that connects with the community.

Finally, develop a candidate list for the Advisory Committee based on interest expressed by the respondents. The Advisory Committee is the key to connecting with the community that will be called upon to implement the plan.

Look for strong spokespeople who represent a cross-section of the community as well as those who may espouse single issues or may otherwise be detractors. The Advisory Committee is the place to bring diverse views together and to work for their resolution. When the debate and deliberations are conducted in a positive, respectful atmosphere, the solutions and agreements reached are likely to be sustainable.

6. Inventory of Lake Condition (Months 6-8)

Gather all available information about your lake and prepare an inventory. Every objective bit of information is fair game here. This includes physical, chemical, social, historical, political, commercial, and management information. Using the responses from the community survey, you can gauge what information is more or less important to your community.

The inventory should include both historic and contemporary information and be reported in a concise and readable format. This should not be a detailed

Sample Community Survey

I. Personal Information

a. How long have you lived in or conducted business in the Pretty Lake community?

b. The next phase of the lake planning project will involve an advisory committee to help focus the management plan. Are you interested in serving on the Advisory Committee?

c. Are you aware of anyone else whose input would be important? If so, please give their name.

d. Please include your name, address, and phone.

II. Uses and Values of Pretty Lake

a. Describe your recreational use(s) of Pretty Lake.

b. What about Pretty Lake is valuable to our community?

c. What about Pretty Lake and its environs are of aesthetic or environmental value to you?

d. What uses conflict with or detract from how you use or value Pretty Lake?

e. What uses should be regulated?

III. The Condition of Pretty Lake

a. How do you perceive the overall water quality of Pretty Lake?

b. How do you perceive the quality of the fishery in Pretty Lake?

c. How do you perceive the quality of aquatic plants in Pretty Lake?

d. How does the condition of Pretty Lake add to or detract from how you use or value the lake?

IV. Issues and Concerns

a. What are important issues and concerns for the management of Pretty Lake?

b. What is the most important issue?

c. What areas of conservation are important to you?

d. What concerns do you have about the management (or lack of management) in Pretty Lake?

V. Other Comments. Please take this space for any other comments.

____

____

____

____

____

engineering report, but rather a basic summary that describes the condition of the lake in the context of past changes, regional perspectives, and predominant uses and values.

Include a summary of the following categories, including only the detail necessary to characterize the lake's condition in a format that the Advisory Committee can use:

- **Physical and Chemical.** Lake basin dimensions, basic chemistry, oxygen regime, mixing patterns, nutrient concentrations, historic comparisons.
- **Lake Condition and Trophic State.** The three trophic state variables — phosphorus, chlorophyll, and Secchi disk — and their relevance.
- **Fish.** Population assessments, gamefish, stocking, and management programs.
- **Aquatic Plants.** Predominant species, their distribution, importance to the lake ecosystem, and management and control programs.
- **Shoreline Environment.** Stabilization structures, areas of erosion, plant communities.
- **Exotic Species.** Those that are present or may infest the lake.
- **Watershed.** A map and description that includes points of inflow and point sources of pollution.
- **Land Use and Population.** Historic, present, and projected land use as well as population and settlement patterns.
- **Lake Uses.** Recreational, wildlife, aesthetic, and commercial categories.
- **Planning and Zoning.** A zoning map that describes planned uses and development patterns.
- **Conservation Holdings, Parks, and Public Lands.**
- **Commercial and Business Interests.** Businesses in the community that receive some income from the lake, waterfront commercial establishments, or the operational plan of a reservoir.
- **Management and Control.** Individuals, agencies, and organizations with management, permitting, and control authority, including past, present, or proposed control efforts.
- **Diagnostic-Feasibility Studies.** A federal Section 314 (Clean Lakes) or similar study either completed or currently ongoing.

Because the Advisory Committee will use this report to get up to speed quickly, it should contain enough information in a readable format to help them understand the lake's condition. And they should be able to read and understand it in less than an hour.

7. Preliminary Problem Definition (Months 6-8)

Before the first Advisory Committee meeting, prime their strategic thinking by offering your (or your planner's) assessment of the main problems confronting this community. This shows that you are in tune with the community and will get your planning effort off to a good start. Remember, however, to present this as your best shot — the final assessment is the Advisory Committee's to make. See example from White Bear Lake, Minn. (below).

Directions for Management Planning

Based on the survey and the assessment of lake conditions, several concerns require attention by the Advisory Committee:

- White Bear Lake is a highly valued resource, providing aesthetic, recreational, commercial, and wildlife qualities that ought to be protected and preserved. But certain lake uses threaten or conflict with these values:
  —Surface uses that lead to congestion and are noisy
  —Commercial activities, public access, over-development
  —Ice fishing houses (litter/sanitation), motor boats (gas/oil)
  —Uses that are unsafe or incompatible with other uses
- Aquatic plants are generally not a nuisance and serve the lake by providing fish and wildlife habitat and water quality benefits. Eurasian watermilfoil has not yet become a nuisance, but may someday.
- Pollution from the watershed — such as stormwater runoff, lawn chemicals, and on-site septic systems — is or could become excessive.
- The history of large lake level fluctuations is a concern. No plan is in place to control lake levels.
- There is concern regarding the leadership and direction provided by the White Bear Lake Conservation District.

Preliminary problem definition from Osgood (1999a)

8. Form an Advisory Committee (Months 6-8)

The Advisory Committee connects the planning effort with the lake community. Although they will be responsible for developing a meaningful management plan for your lake that addresses real needs and values, the ultimate solutions to managing your lake have more to do with connecting with the community than with a particular engineering or biological approach. Thus, it is essential that the Advisory Committee represent the entire community — who they are and how they are selected is critical.

A committee of 20 to 30 people is recommended. More than 30 may pose a challenge for effectively managing the plan development process. However, if you are flooded with willing participants, it is a worthwhile trade-off to be as inclusive as possible. At a minimum, invite people representing:

- Lakeshore homeowners
- Community residents
- Businesses

- Elected and appointed officials
- Users and user groups
- Community organizations
- Environmental and conservation organizations
- Historic societies

The Steering Committee is sometimes tempted to invite people with technical training to sit on the Advisory Committee because they "get it." This temptation stems from the concern that a citizens committee will not come up with the correct approach, so "techies" will help them along. Experience has shown the opposite to be true — citizens demonstrate an impressive grasp of relevant lake management issues as well as the ability to focus their attention on positive, workable solutions. Thus, resist this temptation; the Advisory Committee can turn to the Technical Advisory Committee when they need such advice.

The basic charge to the Advisory Committee is to develop the management plan and to recommend its adoption and implementation. The simple rationale for this is that the committee and the community they represent will be called upon to implement the actions they recommend in the management plan, so they should be the ones to direct the process. Prepare a written charge and ask the committee if this is agreeable at their first meeting.

Don't be concerned that the Advisory Committee will "take over" — in fact, any inclinations they have to direct the process should be encouraged. The Advisory Committee will appreciate the support of the Steering Committee and the planner/facilitator.

The basic charge to the Advisory Committee is to develop the management plan and to recommend its adoption and implementation.

While the committee members were selected on the basis of how they represent various interests, you should make it clear that they are not expected to represent any specific interest, but instead represent their own best judgment.

9. Form a Technical Advisory Committee (Months 6–8)

The Technical Advisory Committee provides technical and regulatory oversight to the development of the lake management plan. This committee interacts with the Advisory Committee as needed to assure the technical soundness and adequacy of the planning process by making it consistent with lake management science and regulatory requirements.

The Technical Advisory Committee may meet regularly or in response to the needs of the Advisory Committee. It is important to clearly distinguish the functions of the two.

10. Plan Development (Months 9+)

The main work — and the fun work — begins here. The agenda for the Advisory Committee is to develop the management plan according to the steps in the planning cycle (see Blue Water Commission example). This process may take from six to 12 months and involve four to 12 meetings. Meetings will normally be a minimum of two hours in duration, but should not be longer than four hours.

Determine the agenda in advance and keep to it. This is important because it demonstrates your appreciation and respect for the significant commitment by these volunteer members. Remember, planning is an ongoing process and it is okay to save some concerns for another time.

Now it is a matter of completing the management plan. The basic strategy is to start big and work smaller. This means begin with a simple problem statement, agree upon a vision, set goals, evaluate and decide upon management actions with measurable objectives, and provide for ongoing monitoring and evaluation.

Elements of the management plan, taken directly from the planning cycle model (see Fig. 8-1), are repeated here to emphasize how to use the work plan to complete the management plan. These steps also serve as a framework for the contents of your management plan document.

- **Step 2: Analyze the situation, take stock.** This is a report that has been prepared in advance and presented to the Advisory Committee before their first meeting. The report is the basis for the initial problem and vision statements — the Advisory Committee must agree on both early in the process.

The basic strategy is to start big and work smaller. This means begin with a simple problem statement, agree upon a vision, set goals, evaluate and decide upon management actions with measurable objectives, and provide for ongoing monitoring and evaluation.

Blue Water Commission
MEETING SCHEDULE

(Two-hour meetings, except day-long workshop)

Nov. 12, 1997	**Introductory meeting.** Welcome; committee charge; committee structure & process.
Nov. 15	**Workshop.** Introduction to lakes and watersheds; historical context; preliminary goal setting; vision.
Dec. 2	**Resource speakers.** Lake condition; What is the deal with phosphorus?
Dec. 16	**Resource speakers.** Aquatic plants; fish; fish contamination.
Jan. 6, 1998	**Resource speakers.** Urban watershed management; watershed management authorities.
Jan. 20	**Resource speakers.** Setting realistic goals; education programs.
Feb. 3	**Goals.** Report from the Technical Advisory Committee; preliminary discussion of goals.
Feb. 17	**Goals.** Continued discussion.
Mar. 10	**Diagnostic study.** Presentation of the diagnostic study results by the Technical Advisory Committee.
Mar. 24	**Management approaches.** Report from the Technical Advisory Committee.
Apr. 7	**Management approaches.** Setting targets and considering management alternatives.
Apr. 21	**Consideration of management targets and actions.** Review draft chapters of report.
May 13	**Consider Draft Report.** Review Feasibility Report and recommended management actions.
May 27	**Final meeting.** Wrap up.

(from Osgood, 1998)

- **Step 3: Set directions — vision, goals, and objectives.** Again, work from big (vision) to small (objectives). This is a systematic, deliberative process. Resist the temptation to talk about goals before there is consensus on the vision. Likewise, don't move on to management actions and objectives until there is consensus on goals. The facilitator and the Technical Advisory Committee play key roles in making sure the discussion stays on track and the outcomes are reasonable.

- **Step 4: Evaluate alternative strategies and actions.** This process assesses the feasibility and balance of all possible approaches and technologies that can help reach your management objectives. Evaluate feasibility on technical, financial, political, and other factors. Also, consider fees, permits, or other requirements. Weigh all alternatives, one against another, to assure they work to complement rather that conflict with a common goal. Normally, management actions fall in the categories of plans, policies, programs, or projects.

- **Step 5: Take action; implement.** The management plan prescribes management actions. Be sure that a recommended action will take place before putting it in the plan by answering such questions as, "Is there a reliable funding source available?" or "Has the agency expected to implement the recommendation expressed the willingness to do so — do they have the authority?"

- **Step 6: Monitor and evaluate progress.** A program to monitor the performance of management actions is essential to the long-term success of your management program. The purpose of the monitoring program is to track progress as well as to have objective information that may be required to adjust your management program. You can always expect to make adjustments when you're working in the natural environment. See Wedepohl et al. (1990) for a technical reference and Simpson (1991) for guidance for volunteers.

- **Step 7: Do it again.** Provide for ongoing planning. At a minimum, a management organization should be identified or created to oversee the ongoing management of your lake.

Implementing the Plan

Meaningful, effective, and positive management action requires a plan — which you now have — and an organization with the resources to carry it out.

The Management Organization

The management organization will make the plan work and take over the ongoing planning cycle (see Fig. 8-1). It should be locally based so it can focus community resources on implementing the management plan.

Types of Organizations

To find the right organization to manage your lake, consider several factors:

- The geographic extent of your lake and watershed
- The existence of an appropriate organization
- The management and oversight needs identified in your lake management plan
- The resources available in your community
- How to most effectively sustain your management program

Management organizations include these categories:

- **Governmental:** Various intragovernmental programs that can fold the administration of a lake management plan into an existing framework; special units of government or assessment districts; or special legal arrangements.
  - ⇒ Existing programs in a federal, state, or local agency
  - ⇒ Lake Improvement District
  - ⇒ Watershed District
  - ⇒ Soil and Water Conservation District
  - ⇒ A joint powers agreement
- **Non-governmental:** Organizations organized formally or informally with powers and authorities ranging from voluntary compliance to legal incorporation.
  - ⇒ Nonprofit Corporation
  - ⇒ Lake Association
  - ⇒ Partnerships

From the Management Plan of the Blue Water Commission (Osgood, 1998)

Action #10. Identify or create an entity to champion the goals and recommendations of the Blue Water Commission.

The principals agree to do this by:

- Formally organizing with a mission compatible with the Blue Water Commission's goals.
- Including in the mission coordinating, facilitating, and advocating the Blue Water Commission's recommendations.
- Identifying and securing funding necessary to protect the lakes.
- Providing the resources needed to sustain their effort.
- Monitoring, evaluating, and reporting progress toward accomplishing their goals.
- Amending this plan in response to evolving community values, developments in lake and watershed management technologies, and changes in the environment in and around the two lakes.

Keys for Effective Organizations

To be effective and long-lasting, your lake management organization should be locally-based and focused only on managing the lake: this requires clear vision and a sound organizational and development strategy. You must pay attention to your management organization and invest in its success.

To be effective and long-lasting, your lake management organization should be locally-based and focused only on managing the lake: this requires clear vision and a sound organizational and development strategy. You must pay attention to your management organization and invest in its success.

In their book, *Profiles of Excellence: Achieving Success in the Non Profit Sector,* the authors cite four "Hallmarks of Excellence" for building effective nonprofit organizations (Knauft, et al. 1991; see also Hummel, 1997):

1. **The Primacy of Mission.** A clear sense of mission with goals to carry out that mission are essential. Everyone in the organization should be aware of and agree with the mission, and make the mission come alive as they go about their respective jobs.

2. **Effective Leadership.** "The best leaders embody the organization's mission — they can clearly articulate it and translate it to others with a sense of excitement." There are many leaders in an organization, ranging from the Board and Chief Executive Officer to staff and volunteers. Normally the CEO is looked upon for day-to-day leadership in carrying out the organization's mission.

3. **A Dynamic Board.** A board should represent the interface between the community and the organization. A dynamic board represents the diversity of the community, has varied skills, relates well among its members, and is committed to the organization.

4. **Strong Development Program.** The ability to attract and sustain a sound and diverse financial basis to run the organization takes work. This work is most effective when shared between the board and staff. Relying exclusively on one or the other is risky.

   The board is critical to the success of your management organization. Their role should include:

   - Budget and Finance
   - Strategic and Annual Plans
   - Fundraising
   - Human Resources
   - Community Relationships
   - Program Evaluation
   - Board Development
   - Advising Staff

Funding and Assistance for Lake Management Organizations

The basic funding for operating and administering the organization must come from the local community.

To sustain a meaningful and long-lasting management effort, your lake management organization will require funding and assistance. As a practical matter, the basic funding for operating and administering the organization must come from the local community.

Numerous funding programs and sources are available for start-up funds, planning and feasibility studies, special projects, and implementation programs. Use this list as a starting point:

- ***Federal Agencies.*** Funding programs, appropriation levels, and program requirements change frequently. U.S. EPA's *Catalog of Federal Funding Sources for Watershed Protection*, now in its second edition (1999), continues to offer the most up-to-date information on federal funding — check EPA's web site: www.epa.gov\owow\watershed\wacademy\fund.html.
  - **U.S. Department of Agriculture:** Grants and loans are available from most USDA agencies, including:
    - ⇒ Farm Services Agency
    - ⇒ Cooperative Extension
    - ⇒ Farmer's Home Administration
    - ⇒ Forest Service
    - ⇒ Natural Resources Conservation Service
  - **The Economic Development Administration of the Department of Commerce** makes loans and grants.
  - **U.S. Environmental Protection Agency** has numerous programs authorized through the Clean Water Act. Historically, the Clean Lakes Program used Section 314 grants administered by states to help public lakes; today, that funding can come through Section 319 (nonpoint source). Contact your state environmental agency for information.
  - **The Department of Housing and Urban Development** supports a broad range of planning and management activities.
  - **Department of Interior** agencies make grants available through:
    - ⇒ Office of Surface Mining Reclamation and Enforcement
    - ⇒ Bureau of Reclamation
    - ⇒ Fish and Wildlife Service
  - **U.S. Geological Survey** helps states through cooperative programs.
- ***State and Local Agencies.*** Many state and local governments supplement or complement federal funding programs. Your Technical Advisory Committee members should know about these sources.
- ***Other Sources.***
  - **Foundations.** These are organizations — both private and corporate — set up to give away money. They're listed by type of giving in *The Foundation Directory* (for information, see the Foundation Center's web site at www.fdncenter.org). You could set up your own foundation to fund your management program; but this will dilute the focus of your management program, and might eventually sever the connection between the community and the lake they want to protect.

- **Corporations.** These are usually direct giving programs, usually with a strong tie to the community served by the business.

- **Membership Dues.** Member dues normally fund the costs of supporting your members, such as mailings, newsletters, etc., but not programs or projects. Multiple membership levels — such as becoming a Bald Eagle for $5,000 — is one way to increase giving levels, nurture individual gifts, and begin to support management programs.

- **Membership Organizations.** Service clubs, fraternal organizations, chambers of commerce, women's and men's groups, and many other groups often contribute financially to a resource they value.

- **Individuals.** Substantial gifts usually require a great deal of time invested in nurturing the individual. This can include planned giving through estate planning.

- **Volunteers.** Volunteers will likely be a significant source of assistance. Be sure that you have a volunteer coordinator to assure that volunteers are used effectively.

Other Considerations for Management Organizations

Other considerations include administrative and oversight functions required to manage programs, personnel, contractors, grants, and so on. If an existing organization is going to implement the management plan, then they will probably already have these systems in place.

If, however, you are creating a new organization or assigning an existing organization significant new responsibilities, then you need to make sure it has the capacity and resources to do the work. These areas should receive particular attention:

If, however, you are creating a new organization or assigning an existing organization significant new responsibilities, then you need to make sure it has the capacity and resources to do the work.

- **Personnel Management**. This includes hiring (and firing), supervising, managing, directing, and coordinating staff.

- **Managing Consultants and Contractors**. This involves writing requests for proposals or qualifications, interviewing, managing and directing, oversight, and contract management.

- **Permits and Regulations**. Many watershed and lake management projects require permits or have regulatory restrictions. As you develop the management plan, the Technical Advisory Committee should be advising you on applicable regulations and permits.

- **Community Relations**. Staying connected with the community who use and value the lake managed is key to effective long-term management.

What Next?

Go. Do not feel you must abide by every detail in this chapter as you prepare your management plan. Take what works and what your lake needs, then run with it. As my father used to say, "*Let's do something, even if it's wrong.*" With common sense and these guidelines, planning for the management and protection of your lake or reservoir cannot go wrong.

References

Carpenter, S.R., N.F. Caraco, D.L. Correll, R.W. Howarth, A.N. Sharpley, and V.H. Smith 1998. Nonpoint pollution of surface waters with phosphorus and nitrogen. Ecol. Appl. 8:559-68.

Carpenter, S.R., D. Ludwig, and W.A. Brock. 1999. Management of eutrophication for lakes subject to potentially irreversible change. Ecol. Appl. 9(3):751-71.

Edmondson, W.T. 1991. The Uses of Ecology: Lake Washington and Beyond. University of Washington Press, Seattle and London.

Environmental Ground. No date. The State of Watershed Water-Quality Management in Minnesota. Environmental Ground, Inc., and the Minnesota Association of Watershed Districts.

Fisher, R., W. Ury, and B. Patton. 1991. Getting to Yes: Negotiating Agreement Without Giving In. Penguin Books, New York.

Jeppesen, E., J.P. Jensen, M. Sondergaard, T. Laurridsen, L. J. Pedersen and L. Jensen. 1997. Top-down control in freshwater lakes: The role of nutrient state, submerged macrophytes and water depth. Hydrobiologia 342/343: 151-64.

Hummel, J.M. 1997. Starting and Running a Nonprofit Organization. 2nd ed. University of Minnesota Press, Minneapolis.

Interagency Lake Coordinating Committee. 1996. Developing a Lake Management Plan. Minnesota Board of Water and Soil Resources, Minnesota Department of Natural Resources, Minnesota Pollution Control Agency, Minnesota Department of Agriculture, St. Paul.

Kehler, R., A. Ayvazuan, and B. Senturia. No date. Thinking Strategically: A Primer on Long-range Strategic Planning for Grassroots Peace and Justice Organizations. The Exchange Project of the Peace Development Fund, Amherst, MA.

Klessig, L., B. Sorge, R. Korth, M. Dresen, and J. Bode. No date. A Model Lake Plan for a Local Community. Wisconsin Lakes Management Program, Madison. (Reprinted in this manual as Appendix 3-A.)

Knauft, E.B., R.A. Berger, and S.T. Gray. 1991. Profiles of Excellence: Achieving Success in the Nonprofit Sector. Jossey-Bass Publications, San Francisco.

MacKay. H. Column. United Features Syndicate, Chicago.

Metropolitan Council. 1997. Lake McCarrons wetland treatment system — Phase III study report. Metro. Counc. Environ. Serv. Publ. No. 32-97-026, St. Paul.

Moss, B., J. Madgwick, and G. Phillips. 1996. A Guide to the Restoration of Nutrient-enriched Shallow Lakes. Braods Authority, Norwich, Norfolk, UK.

Newton, B.J. and W.M. Jarrell. 1999. Procedure to estimate the response of aquatic systems to changes in phosphorus and nitrogen inputs. Natl. Water Climate Center. U.S. Dep. Agric., Washington, DC.

Olsen Thielen Co. Ltd. 1998. Planning ahead . . . Strategically. O&T Advisor: May/June. www.olsen-thielen.com

Osgood, D. 1988. The Limnology, Ecology and Management of Twin Cities Metropolitan Area Lakes. Metro. Counc. Publ. No. 590-88-123, St. Paul, MN.

———. 1992. Managing Minnesota's Lakes: A Report of the Minnesota Lake Management Forum. Freshwater Foundation, Wayzata, MN.

———. 1995. Christmas Lake Management Plan. Christmas Lake Ass. and Ecosystem Strategies, Shorewood, MN.

———. 1996a. Lake McCarrons: Strategic Management Plan. Prepared for Ramsey County and the Lake McCarrons Neighborhood Association by Ecosystem Strategies, Shorewood, MN.

———. 1996b. The ecological basis for lake and reservoir management. Lake Line 16(2).

———. 1998. Blue Water Commission: Report and recommendations for the management of Lake Nokomis and Lake Hiawatha. May 1998, Minneapolis.

———. 1999a. White Bear Lake Management Plan: Report and recommended actions of the Advisory Committee. White Bear Lake Conservation District and Ecosystem Strategies, Shorewood, MN.

———-. 1999b. The phosphorus paradigm. Lake Line 19 (2).

———-. 1999c. Putting education in lake plans. Lake Line 19(3-4):50-1.

———. 2000a. Lake sensitivity to phosphorus changes. Lake Line 20(3):9-11.

———. 2000b. Mooney Lake Management Plan. Mooney Lake Ass. and Ecosystem Strategies, Shorewood, MN.

Simpson, J.T. 1991. Volunteer Lake Monitoring: A Methods Manual. EPA 440-4-91-002. Office of Water, U.S. Environ. Prot. Agency, Washington, DC.

Ury, W. 1993. Getting Past No: Negotiating Your Way From Confrontation to Cooperation. Bantam Books, New York.

U.S. Environmental Protection Agency. 1990. Monitoring Lake and Reservoir Restoration: Technical supplement to the Lake and Reservoir Restoration Guidance Manual. EPA 440/4-90-007. Washington, DC.

———. 1997. Top 10 Watershed Lessons Learned. EPA 840-F-97-001. Office of Water, Washington, DC.

———. 1999. Catalog of Federal Funding Sources for Watershed Protection. EPA 841-B-99-008. Office of Water, Washington, DC.

Walesh, S.G. 1999. DAD is out, POP is in. J. Am. Wat. Resour. Ass. 35:535-44.

Wedepohl, R.E., D.R. Knauer, G.B. Wolbert, H. Olem, P.J. Garrison, and K. Kepford. 1990. Monitoring Lake and Reservoir Restoration. EPA 440/4-90-007. Prep. by N. Am. Lake Manage. Soc. for U.S. Environ. Prot. Agency, Washington, DC.

Wilson, B. 1998. An analogy illustrating the importance of planning for Lakes. Minnesota Pollution Control Agency, St. Paul.

CHAPTER 9

Lake Protection and Maintenance

When it comes to lake protection and maintenance, the old adage, "an ounce of prevention is worth a pound of cure," is good advice. If your lake association has recently completed a restoration effort, you know that preventing pollution and its negative effects on the lake is easier to achieve, and much less expensive. If your lake is still in good shape then this chapter can help you keep it that way or even improve it.

People are attracted to water like magnets to steel. Fishing, swimming, boating, hiking, watching the sun set over the water, just sitting on the shore — all make waterfront property very desirable, and increasingly valuable. But enjoying a lake also means caring for it; with the pleasure offered by the lake comes the responsibility for protecting those uses. Although this chapter describes many approaches for assuming that responsibility, the key to lake protection and maintenance is public involvement and organization.

Forming and Enhancing Lake Organizations

Lake Associations: Roles, Benefits, and Activities

Whose responsibility is it to protect and maintain the water quality of a lake — lake residents, lake users, government, or local businesses? Not surprisingly, the answer is all of the above. Each of these lake interests contributes in different ways to protecting the lake.

A lake association is one of the most important tools available to lakefront residents. Many lake associations are organized in response to a lake crisis such as nuisance weeds, declining fisheries, or foul odors. People find they can accomplish more as an organized group than they can individually, and this rationale holds true for lake protection and maintenance.

Lake associations play many different roles:

- They bring together a group, usually of lake property owners, to maintain the lake.

Participants in a watershed tour listen attentively to their guide.

The Walloon Lake Association in the northwest Lower Peninsula of Michigan has a governmental affairs committee that meets regularly to discuss local government activities and how they may affect the lake and its watershed. They also organize special breakfast meetings, inviting all local government officials to discuss topics of interest to the lake association such as stormwater management or innovative zoning approaches. This committee has fostered excellent communication between the association and government officials. In fact, some of the lake association members are now members of planning commissions or township boards.

- They monitor and maintain water quality.
- They monitor activities that may harm the lake.
- They influence and participate in local government activities.
- They are an information source and can play an important social role.
- They advocate proper management and prevent activities that harm the lake.
- They educate lake residents and the public.
- And, perhaps most important, they inform and involve people in lake management decisions.

The more informed people are about lake problems, alternative management techniques, and watershed dynamics, the more thoughtful their decisions will be when selecting and implementing appropriate protection and maintenance procedures.

As with any organization, volunteers must be involved from the beginning. Many lake associations have retired professionals, such as lawyers or accountants who can help with setting up the bylaws, filing for nonprofit tax-exempt status, and electing a board.

Most lake association funding comes from memberships of lake residents and lake users. Building and maintaining a membership can involve significant work such as developing mailing lists of lake property owners and listening and acting on members' concerns. Sponsoring special projects such as a water quality study may require funding above and beyond membership contributions; this means special fundraising. Many references are available to help nonprofits with fundraising.

Getting involved in local and statewide government is a very valuable role for lake associations. Zoning/planning commissions make land-use decisions that can affect lakes. If lake association members attend these meetings they can influence decisions in a way that benefits the lake. Road and transportation commissions, county boards, drain commissioners, and other governmental offices also make decisions that can affect lake management.

It is even more important to foster these relationships where lake residents are primarily seasonal, as it helps permanent residents value the help of lakeshore property owners year-round.

Getting to know state and federal representatives is also important. State and federal government programs fund most large lake projects. Meeting with those people who represent the lake's district can help insure that funding, programs, and staff are available to help with lake concerns.

Isn't meeting with elected officials lobbying, and therefore illegal under nonprofit tax-exempt law? The answer is no. Most lake associations educate rather than lobby. They meet with their representatives to inform them about the association's goals and activities and share the association's concerns and desires for legislation.

If, however, a nonprofit organization classified as 501(c)(3) decides to lobby, it must file a 501(h) with the IRS. That gives the organization the ability to spend no more than 20 percent of its time and budget on lobbying. Lobbying is defined as communications intended to influence specific legislation. In other words, to

Bylaws, incorporation, tax-exempt status — these legal buzzwords will become everyday words to members of a new lake association. A steering committee should develop the articles of incorporation, bylaws, and application for tax-exempt status. For assistance, contact other lake associations and attorneys who live on the lake. Tips on developing these legal documents:

- **Articles of Incorporation:** Legal papers that form an organization filed with a state. Makes the group a legal entity and may protect individual members from liability.
- **Bylaws:** Details that state how the organization functions; e.g., its purpose, board structure, membership and annual meeting requirements, what makes a quorum, how decisions are made.
- **Tax-exempt Status:** The Internal Revenue Service reviews applications, articles of incorporation, and bylaws to determine if an organization should be granted this status; the most relevant to lake associations is the 501(c)(3). This exempts the organization from having to pay some taxes and allows it to solicit tax-deductible contributions. For an organization to receive and maintain tax-exempt status, its activities must be beneficial to society and serve a function that would otherwise have to be done by government. For example, a lake association with a goal and activities directed toward protecting property values for riparians will not likely receive tax-exempt status. However, a lake association with a goal and activities directed toward protecting water quality for public and private recreational uses would most likely receive the status. A tax-exempt status may bring discounts from certain retailers and other service providers, and also makes an organization eligible for many grants or other funding sources. Being a 501(c)(3) tax-exempt organization limits the amount of lobbying a group can do.

discuss a bill or a ballot proposal with a government official is lobbying. Educating government officials on lake management issues is not considered lobbying.

The opportunities for lake associations are virtually limitless. It may be as simple as holding informal meetings of homeowners to share information about the lake or as complicated as monitoring the passage of enabling legislation to form special districts to protect and improve lakes. Some ideas your lake association may want to try:

Education

- Publish a newsletter for lake residents.
- Distribute educational materials about shoreline property management.
- Sponsor field trips to explore the lake's ecosystem by canoe or motorboat.
- Organize a watershed tour to promote understanding of polluted runoff and watershed features.
- Work with local schools to sponsor a special program for students on your lake.

Boyne City, Mich., students experience what happens on Lake Charlevoix as they use EnviroScape® (a watershed model) to learn how to prevent water pollution.

Water Quality Monitoring

- Participate in volunteer lake monitoring programs.
- Organize volunteers to implement followup monitoring from restoration activities.
- Develop a water quality monitoring program for the larger streams that enter your lake.
- Work with government, universities, and other groups to accomplish special lake studies.
- Encourage all lake residents to monitor for exotic species.

Recreation

- Work with fish specialists to conduct a creel survey.
- Conduct boat survey of residents and public users.
- Maintain bulletin boards at public access sites with information on special rules, fishing regulations, etc.

Advocacy

- Organize lake association members to attend local government meetings to monitor activities and influence decision-making.
- Meet with state and federal representatives to let them know about the lake association's goals, any needs for improved administration of existing laws, new legislation, funding, etc.
- Write letters to inform government representatives about lake issues.
- Comment on proposed activities that may harm the lake and its ecosystem, such as dredge and fill permits.

Community and Social

- Organize a picnic or barbeque that follows an association meeting.
- Coordinate a sailing club or fishing contest.
- Sponsor special events to recognize volunteers and celebrate successes.

Learn from History Tip: There once was a beautiful lake in Michigan where many of the residents thought the lake level was too high and others thought it was too low. Each group formed their own lake association: the High Level Association and the Low Level Association. They fought expensive court battles until a legal decision determined in favor of the low lake level. A few years later both lake associations dissolved and now the lake is without any organized group working to protect it. The moral of the story — don't let one issue dictate the entire purpose of your association; work together, even if you disagree about what is best for the lake.

Lake Districts

Some states give lake districts authority to raise money and implement projects to restore and improve lakes. The activities they can fund vary with each state's enabling legislation. Wisconsin lake districts, for example, have the power to tax, levy special assessments, borrow and bond to raise money, and implement projects such as weed control to protect and improve their lakes.

Generally, lake district boards must have representatives from all local units of government (county, township, and city or village), the lake association or a lake resident, the road commission, the drain commissioner, and the state environmental department that oversees lake activities.

Nine Steps to Forming a Lake Association

If your lake does not have a lake association, identify several people who share your interest and form one! An active lake association will do wonders for your lake.

1. **Develop a steering committee:** Invite interested individuals to a meeting to discuss organizing a lake association. If your lake currently has a problem and people have different ideas on how to solve it, make sure all views are represented.

2. **Brainstorm Goals and Objectives:** Develop a vision for all the things that the lake association could do to help the lake, its residents, and the users. Ask an outsider — perhaps from Cooperative Extension or the local conservation district — to facilitate these meetings.

3. **Organize and Hold a Public Meeting:** This is the time to share the steering committee's ideas with other lake residents and gauge their interest and support. Find a central meeting location in a convenient area and schedule the meeting at a time when most people can attend, usually in the evening or on a weekend. In addition to presenting the steering committee's proposals for the association, you may want to ask someone knowledgable about lakes (and/or lake associations) to speak; this person might be with a government agency, a nearby college, a consulting firm, or even another lake association. Keep the program to a reasonable length (under two hours) and provide ample time for discussion (at least 45 minutes). Coffee and soft drinks (and maybe some cookies?) also help!
   Outcomes of the meeting:
   - Knowing whether there is support to form an association,
   - Common goal(s), and
   - A few achievable actions to be implemented in the coming year.

4. **Develop Articles of Incorporation and Bylaws, and Apply for Nonprofit Tax-exempt Status:** If there is adequate public support, then continue to meet with the steering committee to compile the organization's goals and objectives and complete and submit the necessary forms. Look for an attorney in your lake community — such skills can be valuable for this task.

5. **Elect a Board:** Once the papers have been filed and accepted and the organization is official, elect an official board following the procedure mandated by the bylaws. Many groups use their steering committee as their official board until their first annual meeting.

6. **Publish a Newsletter and Send Out a Membership Request:** Even a one-pager is sufficient to announce the development of and plans for the new organization. Get it out as soon as possible — and remember to ask the residents to become members.

7. **Develop a Strategic Plan:** Either the board or a special subcommittee should be responsible for developing this plan of action.

8. **Implement a Project:** Implement one of the action plans identified in the Strategic Plan. The action may be as simple as participating in the state's volunteer monitoring program or distributing some educational brochures. The key is to identify and accomplish an action. No matter how small, this step will encourage the group to accomplish more.

9. **Build on Success:** Organize a second public meeting to celebrate the organization of the group, discuss accomplishments and needs, and, if you haven't already done so, elect the board. This may be your annual meeting; if so, you may want to decide which actions in your Strategic Plan to focus on in the next year.

Source: Phillips, The Lake Pocket Book, 2000.

Historically, lake districts have been used primarily for lake restoration projects. Lake boards make funding a project equitable among lake residents. Because the process creates a new governmental entity it also dictates public notice of meetings, public hearings, and other avenues for public participation.

On large lakes with multiple governmental jurisdictions, lake boards provide a valuable tool to manage a large lake improvement project.

Lake boards are often organized around a lake problem, but lake residents may differ on the best way to solve it. To limit controversy, try to ensure that both sides are represented on the lake board. The board may also lack full community support because many people refuse to support any additional government board or agency. Just remember — as with any lake management tool, lake districts can be extremely effective if used appropriately.

Where to Go for Help

Many resources are available to help lake associations get started and stay active (see the reference section). However, help may be next door. Perhaps another lake near you has a lake association. Most groups are more than willing to share their successes with others.

Other important community resources include Cooperative Extension offices, local conservation and environmental groups, conservation districts, and universities.

Internet

The information super highway is now one of the most efficient and effective ways to access information. Unlike a passing fad, the internet is like the radio of the '30s and the television of the '50s.

Merging onto the information highway requires a computer and a modem. If you have neither, you can probably use a computer at the library or university. But if you just need some help getting started, look for an instructional book or contact a local internet service provider.

Internet information is accessed through search engines in two ways: (1) by opening browser software and going directly to a **website address** (usually starts with www.) or (2) by searching for specific information using a search engine. Yahoo.com, MSN.com, and Netscape.com are good places to start since they all connect to other search engines. Be aware the search engines operate differently and they do not always find everything that is out there.

Since the information is always changing on the internet some of the government home pages provide stable locations with sound information on lake management; plus, they also have links (direct connections) to other related sites. Browser programs, such as Netscape and Internet Explorer, let you mark the location of your favorite pages with a "bookmark" that allows you to go directly back to a particular web page once you have visited it.

Some lake web sites worth visiting include:

- www.epa.gov/owow/lakes/ — EPA's Office of Wetlands, Oceans, and Watersheds (it has many interesting links)
- www.nalms.org — North American Lake Management Society
- www.terrene.org — Terrene Institute
- www.worldlakes.org — LakeNet Program

Internet Lingo:

- **Search Engine:** software that scans the internet for sites that contain key words or phrases that you provide.
- **Browser:** interfacing software that allows you to navigate through the internet by pointing, clicking, and typing.
- **Plug-ins:** additional software (often included with the browser) that allow you to hear sound and see animation and video.

The web isn't just a source of information. It is also a way to get information out to other people. Many internet service providers give their customers a complimentary address for a home page (excluding the cost to develop the home page). For help with developing and maintaining a home page, contact your internet service provider or a consultant — or your high school or college.

- ***Home pages*** can be an effective way to get information out to the association's membership or build interest in a project. The most effective web pages are colorful with graphics, contain useful information that is frequently updated, and have interesting links to related sites. Home pages require maintenance, however. Usually, you can tell how useful a web site is by the number of people who visit it and how often it's updated.

- ***E-mail*** is one of the internet's most useful tools. It allows people to stay in touch easily. For example, a lake association board could use it to maintain communication about important lake projects and issues. It is also very cost effective. An e-mail usually costs less to send than a long distance phone call.

- ***List servers*** ("list servs") are e-mail subscriptions (usually free) to online sources of information usually contributed by and among subscribers. For example, the Wisconsin Department of Natural Resources has a list server on lake topics. Many list servers are managed by government, nonprofit, or universities as an inexpensive way to disseminate information to many people about news or current issues related to specific topics.

Land-use Planning and Stewardship

Many land uses contribute pollutants to lakes, including shoreline development, agricultural activities, new construction, and stormwater runoff (see Chapters 2, 4, and 6). Protecting a lake from these activities requires an equally diverse approach, using tools such as local regulations, education, best management practices, and citizen action.

States and communities across the nation are realizing the impact of land-use decisions on lakes. In Michigan, a special committee appointed by the Governor has identified inadequate land-use planning as the greatest threat to the environment, including water quality. Communities in New York's Catskill Mountains are implementing watershed management plans to protect the Delaware River Reservoirs, the source of New York City's drinking water.

Many regulatory land management tools ranging from zoning to conservation easements can be used effectively in lake management:

Regulatory Approaches

Environmental protection is rooted in regulations. The Clean Water Act in 1972 set standards for industrial and wastewater treatment plant discharges. States soon followed by passing their own regulations to reduce pollution to lakes and rivers. Although federal and state regulations are important for protecting lakes, local regulations can build upon them by requiring stronger standards or different components to accommodate unique local needs.

Zoning ordinances, special regulations, and innovative and creative land management tools continue to be used by states and local communities to control activities that can damage valuable water resources. These regulatory procedures can be combined in any number of ways to fit a particular lake or specific set of lake uses.

Zoning Ordinances

Protecting or improving our lakes requires managing more than just the immediate shoreline area (see previous chapters on watershed management and polluted runoff). More than 1.5 million acres of land are newly developed in the United States each year (Center for Watershed Protection, 1998a,b). These changes in land use can greatly influence the long-term health of our lakes.

More than 1.5 million acres of land are newly developed in the United States each year.

Originally developed to minimize conflicts between incompatible land uses such as industrial and residential areas, zoning ordinances today do much more than prevent conflict. They establish the pattern of development, protect the environment and public health, and determine the character of communities. Since protecting the lake requires looking at what happens on land, zoning is a natural lake management tool.

Zoning's effectiveness depends on many factors, particularly the restrictions in the language, the enforcement, and public support. Many people believe the law protects sensitive areas, only to find otherwise when development is proposed. Although zoning has its critics, it can be used very effectively for managing land uses in a way that is compatible with lake management goals. Some zoning approaches that support lake and watershed management protection:

▼ ***Site Plan Review*** standards require that the proposed construction project meet additional requirements beyond the basic zoning ordinance. A site plan consists of all the drawings, descriptions, and other information pertaining to the proposed development, plus specifics on how the project will affect adjacent properties and a pollution prevention plan. Site plan review standards are often applied to commercial and industrial uses, land uses requiring more than a specified number of parking spaces, structures greater than a specified size, and development in sensitive environmental areas — but not usually to single-family homes. Examples include requirements for:

- Keeping the property in as natural a state as possible;
- Controlling soil erosion;
- Stormwater management; and
- Street and access standards.

▼ ***Planned Unit Developments (PUDs)*** are commonly used in zoning ordinances to trigger a site plan review, achieve stricter standards, provide flexibility, and streamline the approval process for larger developments. Many include density bonuses in return for conservation lands set aside for permanent protection.

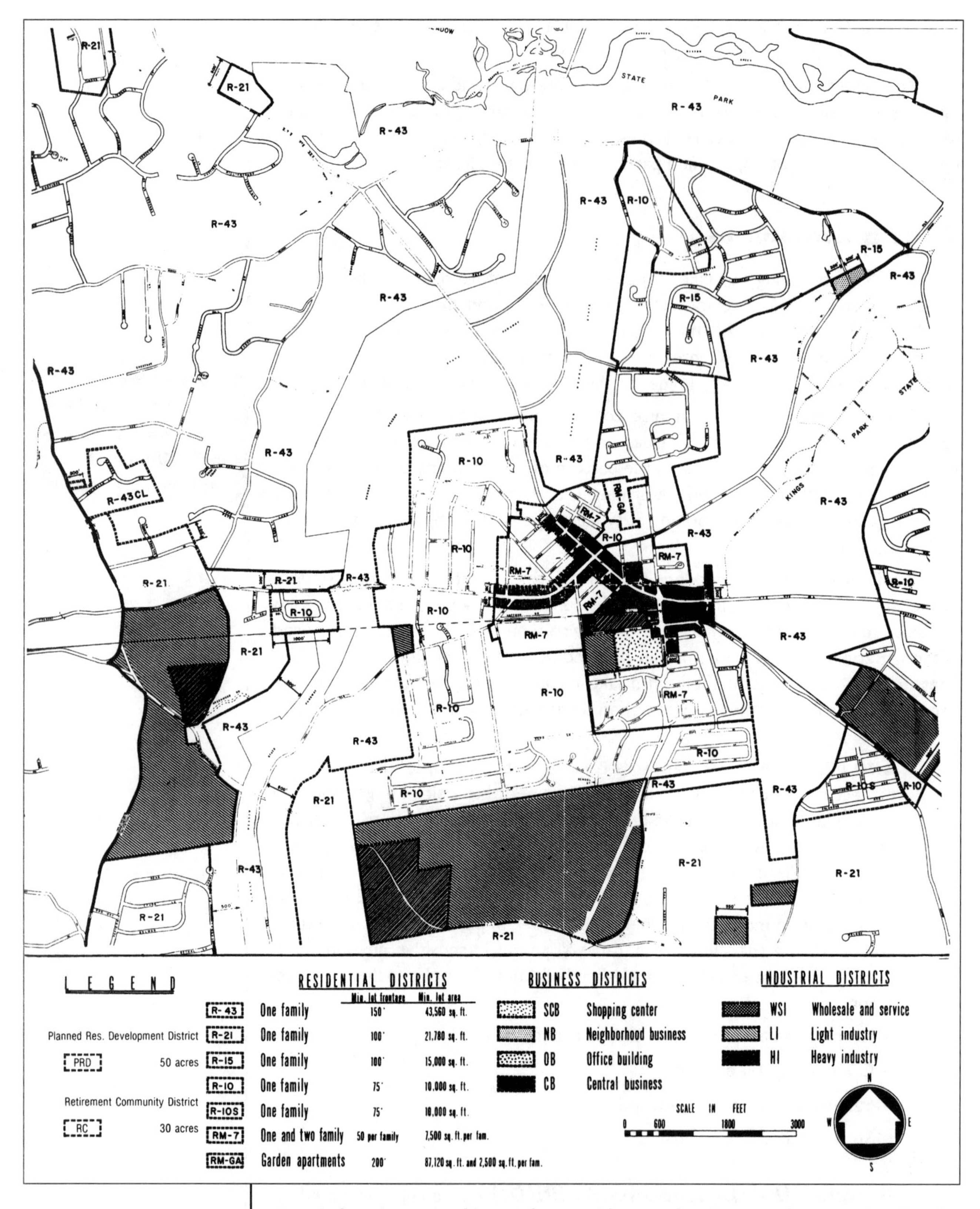

A typical zoning map, this one from Smithtown Planning Board, N.Y., printed in the Long Island Regional Planning Board's Nonpoint Source Management Handbook (1984).

- ***Overlay Zones:*** Although it appears to be very simple, this tool is actually very powerful. Overlay zones require stricter standards than existing zoning because of special natural or cultural features. This type of zoning is often used to require different regulations in sensitive areas such as lake and stream corridors, where they can help reduce polluted runoff. Such overlay zones often require building setbacks, greenbelt maintenance, and limits on certain activities.

- ***Open Space Ordinances:*** Many municipalities are adopting open space development and conservation design standards to protect natural areas and create more livable neighborhoods. This type of ordinance generally requires that a certain percentage of the property be set aside as open space to be used for recreation by all the property owners. Many ordinances also include a density bonus: the higher the percentage of open space set aside, the greater the number of homes allowed. An easement must be placed on the open space to keep it undeveloped forever.

  Many communities are discovering that the lots in open space subdivisions sell more quickly and at a higher price than those in typical subdivisions. These developments can help protect lakes by preserving valuable wetlands and reducing the amount of impervious surface that generates polluted runoff.

Density bonus: The higher the percentage of open space set aside, the greater the number of homes allowed.

Traditional subdivision design for a 10-acre site zoned for half-acre sites and bordered by a county road on the north, a lake on the south.

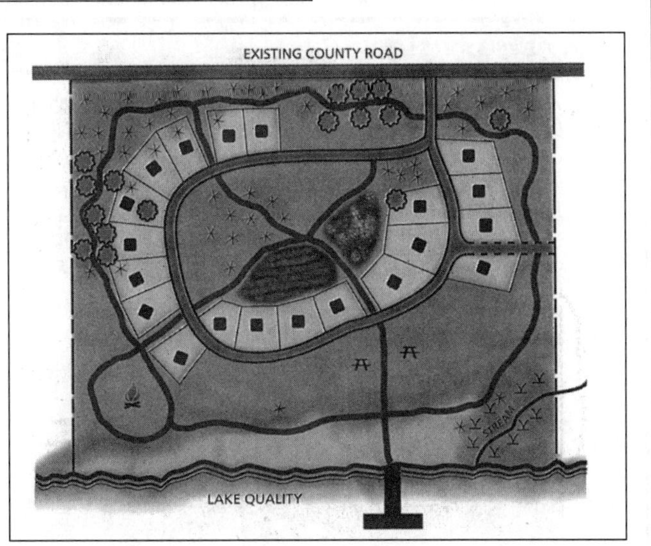

A conservation design that respects the natural resource base while providing home sites in a natural setting allots approximately 5 acres of the total 10 to common conservation areas to be used by all residents. Source: Planning for Success, Tip of the Mitt Watershed Council (1999).

Model Development Principles

Developed by The Center for Watershed Protection, these principles go beyond reducing the amount of impervious surface to suggesting how to manage areas for cars, buildings, and conservation so as to reduce runoff. Many of these principles accomplish other goals, including treating stormwater and protecting sensitive habitats.

RESIDENTIAL STREETS AND PARKING LOTS (Habitat for Cars)

1. Design residential streets for the minimum required width needed to support travel lanes, on-street parking, maintenance, and emergency and service vehicle access. Travel volume will determine the widths.

2. Reduce the total length of new residential streets by examining alternative street layouts to determine the most efficient and appropriate placement of homes.

3. Wherever possible, residential street right-of-way widths should reflect the minimum required to accommodate the travel-way, the sidewalk, and vegetated open channels for stormwater. Utilities and storm drains should be located within the pavement of the right-of-way wherever feasible.

4. Minimize the number of residential street cul-de-sacs and incorporate landscaped areas to reduce the amount of pavement. The radius of cul-de-sacs should be the minimum required to accommodate emergency and maintenance vehicles. Alternative turnarounds should be considered, such as "T" designs that use less pavement.

5. Where density, topography, soils, and slopes permit, vegetated channels should be used in the street right-of way to convey and treat stormwater runoff.

6. The required parking ratio (number of spaces required per business, etc.) governing a particular land use or activity should be enforced as both a maximum and minimum to curb excess parking space construction. Existing parking ratios should be reviewed for conformance, taking into account local and national experience to see if lower ratios are warranted and feasible.

7. Parking codes should be revised to lower parking requirements (number of parking spaces a business must provide) where mass transit or enforceable shared parking arrangements exist.

8. Reduce the overall amount of hard surfaces in parking lots by providing compact car spaces, minimizing stall dimensions, and using pervious surfaces in extra, spillover parking areas where possible.

9. Provide meaningful incentives to encourage structured and shared parking to make it more economically viable.

10. Wherever possible, provide stormwater treatment for parking lot runoff using bioretention areas, filter strips, and/or other practices that can be integrated into the landscape.

LOT DEVELOPMENT

11. Advocate open space design development incorporating smaller lot sizes to minimize total impervious area, reduce total construction costs, conserve natural areas, provide community recreational space, and promote watershed protection.

12. Relax side yard setbacks and allow narrower frontages to reduce total road length in the community and overall site imperviousness. Relax setback requirements to minimize driveway lengths to reduce overall lot imperviousness and require greater setbacks from the lakeshore.

13. Promote more flexible design standards for residential subdivision sidewalks. Where practical, consider locating sidewalks on only one side of the street and providing common walkways linking pedestrian areas.

14. Reduce overall lot imperviousness by promoting alternative driveway surfaces and shared driveways that connect two or more homes together.

15. Clearly specify how community open space will be managed and designate a sustainable legal entity, land trust, or government agency responsible for managing both natural and recreational open space.
16. Direct rooftop runoff to pervious areas such as yards, open channels, or vegetated areas, and avoid routing rooftop runoff to the roadway and the stormwater conveyance system.
CONSERVATION OF NATURAL AREAS
17. Create a variable width, naturally vegetated buffer system along all water bodies. This buffer should also encompass critical environmental features such as the 100-year floodplain, steep slopes, and wetlands.
18. Preserve or restore the riparian stream buffer with native vegetation. The buffer system should be maintained through the plan review, construction, and post-development stages.
19. Clearing and grading of forests and native vegetation at a site should be limited to the minimum amount needed to build, allow access, and provide fire protection. A fixed portion of any community open space should be managed as protected green space, preferably in a consolidated manner.
20. Conserve trees and other vegetation at each site by planting additional vegetation, clustering tree areas, and promoting the use of native plants.
21. Provide incentives and flexibility in the form of density bonuses and property tax reduction to encourage conservation of stream buffers, forests, meadows, and other areas of environmental value. In addition, off-site mitigation consistent with locally adopted watershed plans should be encouraged.
22. New stormwater outfalls should not discharge unmanaged stormwater into lakes or streams, wetlands, sole source aquifers, or other sensitive resources.

Special Ordinances

States and communities can also adopt regulations designed to manage a specific activity or resource. Adopting, implementing, and enforcing state and local regulations require a commitment to the costs and time to administer them. Some of these ordinances that have the most potential benefit for lake management:

- ***Soil Erosion and Stormwater:*** Soil erosion and stormwater are two leading water pollutants in the United States. Local ordinances can reduce the impacts of erosion and stormwater, especially from new construction. Soil erosion ordinances often require detailed plans that show how erosion will be prevented for any earth disturbance that is near surface water or is of a certain size (e.g., one acre) — plus provisions to ensure the plan is implemented appropriately. Stormwater regulations may require preserving drainage patterns, keeping stormwater onsite, and storing stormwater in retention and detention basins (or otherwise treating it) before it leaves a property.

- ***Impervious Surface:*** New development replaces forests, meadows, and wetlands with rooftops, roads, driveways, parking lots, and other hard (impervious) surfaces that rain can't penetrate. As the amount of impervious surface increases so does the runoff to a lake and its tributaries. Studies have shown that fish habitat is harmed when impervious surfaces account for more than 10 percent of a watershed (Center for Watershed

One of the most valuable things a community can do for lakes is to reduce the amount of impervious surface and identify and protect critical habitats.

Protection, 1998a,b). One of the most valuable things a community can do for lakes is to reduce the amount of impervious surface and identify and protect critical habitats (see the preceding Model Development Principles).

Sanitary Ordinances

Most lakeshore areas in this country rely on on-site septic systems for wastewater treatment. Most communities have standards and regulations that determine proper siting and design of septic systems, primarily to protect public health from diseases that can be spread by wastewater. Some also focus on protecting the environment. These standards may include:

- Setbacks from water bodies and wells;
- Minimum allowable depth to groundwater;
- Soil suitability criteria; and
- Size requirements based on estimated usage by current and future owners.

Some communities also have a septic system evaluation program to encourage the upgrade of older septic systems. When a home is sold, the septic system must be evaluated to ensure that it meets current code requirements. If it doesn't then the system must be upgraded. This type of program can be very valuable in eliminating failing septic systems around lakes.

The criteria for evaluating septic systems must be developed carefully. The National Small Flows Clearinghouse is a valuable source of information on design, construction, and management of septic systems. Operated by the University of West Virginia for the U.S. EPA, the Clearinghouse helps small communities find practical, affordable solutions to their wastewater problems.

Wetland Regulations

Preserving wetlands along the shoreline and in the watershed can help protect the quality of lake water. Wetlands filter runoff, provide wildlife habitat, prevent shoreline erosion, and help control flooding. Federal laws protect wetlands' valuable functions by regulating activities that degrade or destroy wetlands. Some state and local governments have also adopted additional regulations to protect wetlands.

Although local wetland ordinances add another layer of regulation for citizens, they can benefit property owners and communities in many ways. Often, local ordinances will require a community-wide wetland inventory that will identify wetlands that require permits, thus saving time and money in the long run.

Almost more important, local involvement can help streamline the wetland permit process, which at best, is difficult.

Key elements of a local ordinance may require:

- Additional setbacks for construction, and
- Protection of smaller wetlands that may not be regulated by other state or federal laws.

Lakeshore wetlands protect the health of the entire lake ecosystem in many ways, including nurturing aquatic life and preventing pollutants from reaching the lake. Photo by Jim Nelson.

Growth Management Tools

When combined with a sound zoning ordinance and a clear master plan for the community, growth management tools can improve the way land is used in the community and thus protect the lake and its watershed. These tools will not stop growth; they just guide the rate, location, type, timing, quality, and character of development to comply with the regulations and master plan.

But growth management tools can only influence new growth and development. A community that has almost reached its limits with urban sprawl will benefit less from growth management tools than areas that are just beginning to experience a population surge.

The following growth management tools, which can be used individually as well, can be a valuable part of a lake and watershed protection effort:

- Purchase and transfer of development rights;
- Urban growth boundaries; and
- Coordinated infrastructure management.

Purchase and Transfer of Development Rights

Every piece of property comes with a certain amount of "rights" such as the right to harvest timber, build structures, farm, etc. Purchase-of-development-rights programs actually buy the "right" to develop the property from the property owner in perpetuity.

A program that supports the purchase of development rights gives farmers an alternative to subdividing their land and selling it to gain retirement income, which results in loss of both agricultural acreage and open space. They can sell the development rights to provide for their retirement without having to sell their land; they can also continue to farm it, harvest timber, etc. Selling development rights does not yield as much profit as selling the land outright, but it maintains the rural character of a community.

Transfer of development rights is very similar, although it allows the development density to be transferred to a property that is more suitable for development.

Funding these programs is expensive. Some states purchase agricultural lands and some communities are beginning to pass mill levies to support such programs.

Purchase and transfer of development rights can be used as lake management tools to preserve important areas around a lake from pollution resulting from development, but they work best when combined with other management techniques. Since agriculture, for example, can also pollute lakes, these programs should be combined with installing best management practices on farms.

Urban Growth Boundaries

Boundaries are carefully drawn around a city or village to prohibit urban development beyond the line. As a result, development and economic growth are focused on an area in a way that avoids urban sprawl. **Service district boundaries** are similar, using public services such as sewer and water to concentrate development within a certain area. Another way to concentrate urban development near infrastructure is the **village center concept,** which uses mixed zoning to provide a variety of land uses such as residential, commercial, and industrial within a small area. The village center concept builds a community where a person could conceivably live, work, shop, visit doctors, etc., all within walking distance.

The State of Oregon required its communities to adopt urban growth boundaries by the mid-1980s. Drawing the lines was very controversial. However, after more than a decade of having urban growth boundaries in place, positive results can be measured. A 1991 study of the Portland metropolitan area found that it had expanded only 2 percent in area over the last 17 years, while significantly growing in population and development. Those statistics cannot be matched by any other urban area in the United States.

Urban growth boundaries can be a very beneficial land management tool for lakes near urban areas. They can help concentrate infrastructure and new development where it is best suited.

Coordinated Infrastructure Management

The decision to approve a new shopping center or large resort development usually rests with the municipality where the property is located. Even though a project may also affect a neighboring township and its lake through increased traffic, road damage, and stormwater runoff (among other things), its approval is seldom required. In addition, the neighboring township will not likely receive any financial benefits in taxes from the development.

Since lake and watershed boundaries generally do not coincide with political boundaries, the effects of development on lakes in neighboring jurisdictions usually are not considered. One way to address this issue is for the municipalities to adopt **impact coordination rules**. These require local governments to consider the impacts of projects on neighboring communities by soliciting comments from that municipality. Impact coordination rules can vary greatly, from having to notify a neighboring community about a project to mandating their approval.

Carrying Capacity Studies

As our population has grown — and more people have had more money to spend — they have increasingly turned to lakes for recreation and relaxation. Lakes are straining to keep up with the demands for boating facilities and homes along their shorelines. Boaters now have to schedule their time on the water, and homes are being wedged in using keyhole (funnel) designs.

Motorboats, including personal watercraft, affect the lake in several ways, including contributing to pollution. Lake users also conflict, which is to be expected when you have sailboats, anglers, speedboats, water skiers, swimmers, personal watercraft, canoeists, windsurfers (and more!) sharing the same waters. Some of the more common issues:

- Shoreline erosion from speedboats and water skiing (and loss of natural habitat as shorelines are artificially altered to prevent erosion).
- Damage to bottom-dwelling plants and animals near the shore (littoral zone).
- Oil and gas discharge from powerboats.
- Noise.

Good boating practices protect the health of this lake.

A carrying capacity study for recreational boating can determine the number of boats that can be used safely in a lake. Generally, these studies assess lake characteristics and levels of boating activity by residents and the public, and help clarify the conflicts. Some studies also assess potential damage to water quality from boating (Engel, 1989).

Carrying capacity studies often produce lake-specific regulations that dictate when and where motorboats can operate and designate zones in the lake that are appropriate for specific uses. For example, pleasure motor boating and water skiing might be banned between 6 p.m. and 10 a.m., to minimize conflicts with anglers. Another approach is to set aside certain areas, such as sheltered coves of the lake, for particular uses such as swimming or fishing, with power boating and water skiing restricted to more open water areas (Wyckoff, 1995).

Although most states require minimum isolation distance between watercraft and restrict watercraft speed, stricter regulations could be specified: for example, a powerboat should be at least 200 feet away from an anchored fishing boat or restricted to slow no-wake speed. Motor sizes are commonly restricted (no motors, only electric motors, or only motors less than 10 hp) on small lakes or lakes in wilderness settings. State environmental agencies often cooperate with local governments to establish stricter boating regulations.

A carrying capacity study can also provide data to limit the number of boats lakefront property owners can maintain. Or limit boat usage by restricting the number of boats per lake residence and controlling the amount of parking at public access sites. These would require ordinances and may be easier to establish on a lake that is just being developed.

Boating and recreational use issues can be emotional. If you decide to conduct a carrying capacity study, be sure it takes a scientific approach to assessing impacts. And remember, establishing special regulations for a lake may require working cooperatively with the state environmental department and state and local marine patrol.

Voluntary Activities

Regulations alone cannot protect a lake; riparian property owners significantly affect water quality and must do their part. Although encouraging them to change their behavior and voluntarily help protect a lake may seem like a daunting task, lake communities all over the country have validated Margaret Mead's words: "Never doubt that a small group of thoughtful, committed citizens can change the world. Indeed, it's the only thing that ever has."

Good shoreline property management practices can go a long way toward maintaining and protecting lake water quality — from how you care for your lawn to how you operate your boat.

Personal Property Management Practices

The privilege of lakefront living comes with an enormous responsibility, because anything you do along the lakefront can have an immediate effect on a lake's ecosystem. Good shoreline property management practices can go a long way toward maintaining and protecting lake water quality — from how you care for your lawn to how you operate your boat. Use the following checklist as a guide, and supplement it with the publications and programs cited in the references and text throughout this book.

- ***Lawn Care:*** To many people a yard without a lawn is not a yard. But a large, intensively managed lawn along a lake shore may not be the best thing for the lake. In fact, studies have indicated that lawn fertilization is one of the largest sources of pollution.

Lawn Care Tips:

- Maintain a strip of natural vegetation along the shoreline, or plant one, preferably with native species.
- Leave an unmowed, unfertilized area at least 25 feet wide along the water's edge.
- Locate your compost pile away from the shoreline.
- Test the soil to determine the most appropriate fertilizer needed.
- Use a low maintenance, slow growing grass seed that is recommended for your soil conditions and climate.
- Keep the grass as high as possible (3 or more inches) to shade out weeds and improve rooting so less water is required.
- Use a self-mulching lawn mower, which will reduce or eliminate the need for fertilizer.
- Avoid using fertilizer. If fertilizer is absolutely necessary, use a slow release product low in nitrogen and phosphorus free.
- After raking leaves, add them to the compost pile. Avoid dumping leaves on or near the shoreline — they can attract leeches.
- Avoid or minimize the use of pesticides whenever possible.
- Landscape with native species that are suited for the specific site: its soil, light, and moisture conditions.
- Consider a lawn of pine needles or native ground covers instead of turf grass.
- Do not try to grow grass on a wooded lot.

A lakeshore greenbelt such as this one along Lake Lawrence, Wis., filters potential pollutants as well as being aesthetically pleasing. Photo by Wisconsin Department of Natural Resources.

▼ ***Septic Systems:*** When designed, constructed, and maintained properly, septic systems can provide excellent wastewater treatment for homes in rural areas. But when they have problems, they can leach nutrients and bacteria to the lake. Proper maintenance can prevent many problems and help ensure that the system operates effectively throughout its expected lifetime.

Septic System Maintenance Tips:

- Have the septic tank sludge level inspected and pumped out as needed based on occupancy (commonly every three to five years depending on seasonal or permanent use).
- Consider adding a septic effluent filter to help keep solids out of the drainfield. However, a filter requires that the septic tank be pumped more frequently.
- Conserve water by installing water conservation devices.
- Reduce water use by turning off the faucet while brushing teeth and washing only full loads of laundry.
- Carefully use household hazardous materials such as drain cleaner, paints, varnish, and motor oil.
- Dispose of leftover household chemicals at a household hazardous waste collection site. Do not dump down drain.
- Recycle used motor oil and antifreeze.
- Never build, pave, or cultivate land over a drainfield; don't drive vehicles over a drainfield.
- Don't apply fertilizer around a drainfield because the nutrients saturate the soil and cause it to stop removing nutrients from the wastewater.

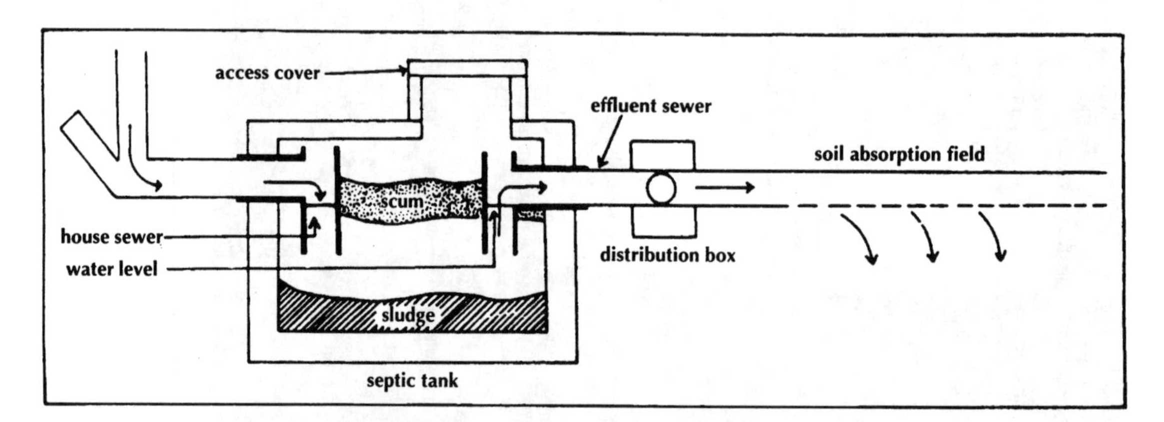

A schematic of a typical septic system. From Tip of the Mitt Watershed Council.

▼ ***Shoreline Erosion:*** Waves, currents, and ice move soil particles toward, away from, and along the shoreline. People augment this natural erosion by removing vegetation, dredging, filling, or building along the shoreline. As discussed earlier in this manual, the resulting sediment can destroy habitat and affect fish and other aquatic organisms, cloud the water, and stimulate (because it may contain nutrients) undesirable plant and algal growth. Shoreline erosion can also destroy valuable waterfront property, including buildings. You may need to install either structural or biotechnical (vegetative) erosion controls.

Shoreline Erosion Prevention Tips:

- Preserve the natural rocks and vegetation along the shoreline.
- Prevent runoff from roofs and driveways from flowing over land to the shoreline by directing it into a swale or collection area or away from the lake.
- Maintain nearshore "berms" pushed up by ice action in northern areas.
- Preserve shoreline wetlands.
- During construction, implement proper soil erosion control measures to prevent sediment from washing into waterways.
- Avoid using seawalls, because they can cause erosion on neighboring properties and destroy valuable habitat for aquatic life.
- Limit the amount of foot traffic and other recreational activities in erosion-prone areas.

▼ ***Powerboating:*** Motor boating and personal watercraft are some of the most popular recreational activities on lakes. They can be fun and exciting ways to experience lakes — but they can also negatively affect them. Proper maintenance and responsible use of power boats can greatly reduce their impact.

Watercraft Pollution Prevention Tips:

- Maintain the boat motor, since well-tuned motors contribute the least amount of pollution.
- Refuel on land to reduce any chance of spilling oil or gas into the water.
- Check and clean the engine well away from shorelines. Oil can harm micro-organisms and other aquatic life that feed on them.
- Be careful when filling the tank; do not overfill. Catch accidental spills with an absorbent pad and dispose of it properly.
- Obey no-wake zones and avoid shallow areas to prevent disturbing vegetation, wildlife, and bottom sediments.
- When it is necessary to ride in shallow water, keep watercraft at idle (headway) speed. This will help reduce the stirring-up of bottom sediments.
- Avoid vegetated areas, including docking around reeds and grasses. The boat motor may suck them in, causing engine or pump problems. But the plants themselves may be pulled out or damaged by stirred-up sediments or less light.
- Avoid marshy areas and aquatic plant beds. They are important habitat for fish, birds, turtles, snake, frogs, and other aquatic life.
- Protect wildlife by maintaining a buffer of 300 feet or more between the boat and animal. Learn about protected species found in lakes and what you should do to avoid disturbing them.

Let Fallen Trees Lie: Resist the urge to immediately clean up a fallen tree: such trees — and overhanging plants — are mini-food chains. Small fish gather beneath them to feed on insects and use their cover to hide from larger predator fish.

Researchers from Ontario have found that shoreline sites with more fallen trees produce more preyfish than shorelines where these trees have been removed. They also found that along undeveloped shorelines, fish feed at levels seven times higher.

Shorelines Equal Habitat: Using the green frog (*Rana clamitans*) to check the health of aquatic life in northern Wisconsin lakes, researchers have found that as the number of homes increases, the number of green frogs declines. Undeveloped lakes averaged one frog per 126 feet of lakeshore compared to one frog per every 220 feet for developed lakes and 470 feet on densely developed lakes. Protecting a certain amount of shoreline from development may help protect aquatic life, including the green frog.

Miscellaneous Shoreline Management Tips:

- Place campfire rings as far back from the lake as possible. When fires are cool, collect ashes and add to compost pile or dispose properly.
- Maintain aquatic vegetation in the shallow areas of the shoreline. These provide valuable wildlife habitat and protect the shoreline from erosion.
- Don't feed the waterfowl. This will increase the amount of waterfowl feces that contribute nutrients to your lake; waterfowl also help spread swimmer's itch.
- Pick up pet waste and dispose of it properly to prevent it from making its way to a waterway in a rainstorm.
- Use an indoor shower rather than the lake for bathing.
- Divert rain gutters to unpaved areas where water can soak into the ground before reaching the lake.
- Participate in your lake association. For example, the lake association could develop a "welcome wagon" service that gives new owners a packet of information on lakeshore living and gifts and coupons from area businesses. Brochures, newsletters, press releases, and other media tools are other excellent ways to reach property owners.

Prevent the introduction of exotic species (e.g., zebra mussels, round goby, Eurasian watermilfoil, etc.), by draining the bilge motor, live well, and transom while on land; inspecting the boat, trailer, and anchors; removing any visible plants or animals before leaving the lake; and emptying bait buckets on land.

Land Protection Through Purchase, Donation, or Easement

Permanently protecting ecologically valuable land in the watershed is another lake and watershed management tool that preserves sensitive habitats that are critical for lake protection; it also can be combined with other lake management efforts.

First, you must inventory sensitive areas around the lake and in the watershed. The inventory might identify and characterize sensitive areas such as wetlands, steep slopes, wildlife habitat, ecological corridors, threatened and endangered species, etc.

Next, prioritize the sensitive areas and identify the property owners. Then, you will want to work with the private landowners to discuss land management and protection options that would protect the property from uses that might harm water quality. Land protection must be voluntary, ranging from placing future restrictions on use of the property to selling the land.

Protecting sensitive areas from development can help reduce runoff and pollution entering lakes. Purchasing lands for conservation value can also be an effective, though expensive, watershed management tool. Many citizens are interested in preserving the conservation values of their land for their grandchildren and future generations.

▼ ***Conservation easements*** (also known as conservation restrictions) are legal agreements between a landowner and a qualified government agency or nongovernmental organization (most commonly a land trust) that permanently limit a property's uses; they remain with it if it is sold. A conservation easement does not transfer title to the property or open it to the public. The landowner continues to own the property, and may live on it, sell it, or pass it on to heirs. The donation of a conservation easement can lower estate taxes and provide income tax or property tax deductions. Conservation easements are extremely flexible and can be written to meet specific needs of the landowner.

▼ ***Outright donation of land*** to a land trust or government agency is another way to protect sensitive habitats. This option has several benefits for the landowner:

- It is a simple transaction;
- It can provide significant income tax deductions; and
- It can reduce estate taxes.

▼ ***Land trusts,*** one of the fastest growing sectors in the conservation arena, specialize in land protection. They manage property that has been donated to them, and hold and enforce conservation easements on lands they purchase. Generally, land trusts will only accept or purchase lands that have conservation value.

Lake Monitoring

Monitoring programs have been covered in previous chapters, but they are discussed here again to emphasize their importance and to introduce the role of volunteers. Water quality monitoring is important for characterizing these dynamic systems we know as lakes, understanding how they work, and documenting changes and trends.

One of the most popular reasons for monitoring water quality is to detect or document problems. It is easier and much more cost effective to treat lake problems as they develop rather than when they have become a crisis. Water quality monitoring is also the best approach for determining whether protection and restoration approaches are effective.

But who should do the monitoring — professionals or volunteers? The answer is both. Professional monitoring is an essential part of a lake management plan. And volunteer monitoring can provide supplemental information that will strengthen a professional monitoring program.

Volunteer monitoring also strengthens your entire program because as more people become involved in monitoring, they learn more about the lake, its problems and opportunities, and ongoing efforts to protect it. (An excellent reference on volunteer lake monitoring is *Volunteer Lake Monitoring: A Methods Manual* [Simpson, 1991]).

Volunteers prepare to monitor for newly hatched zebra mussel larvae on Black Lake, Mich.

Volunteer monitoring used to be limited to measuring Secchi disk transparency and the occasional chlorophyll *a* test to check algae. Not anymore. The opportunities for volunteer monitoring are almost limitless. As government budgets diminish, the need and demand for volunteer monitoring programs will continue to increase. Expanded chemistry parameters, such as dissolved oxygen, pH (alkalinity), and conductivity, are becoming more popular, as are programs to check for the presence and abundance of certain exotic species such as zebra mussels or Eurasian watermilfoil.

Check with state agencies to learn about their volunteer monitoring programs. If state programs are not available, contact local universities, conservation districts, and other local resources for assistance with developing a volunteer monitoring program.

Establishing a Volunteer Monitoring Program

1. Setting Objectives

The first step in developing a volunteer lake monitoring program is to establish objectives. Is the objective to provide credible information on water quality conditions to state and local agencies? To educate the public about water quality issues? Or to build a constituency of involved citizens? All of these objectives (plus many others) can be achieved by a well-organized and carefully implemented program, but you must first determine your priorities.

2. Working with Potential Data Users

Potential users of the data besides your lake organization may include state environmental department employees such as water quality specialists, biologists, and fisheries managers; local planners and environmental organizations; agricultural agencies; universities; and certain federal agencies.

Before you set up your program, communicate with these potential data users about your program goals, what data they might need, and what methods of data collection, analysis, management, and reporting they recommend or would accept. Federal and state governments may not consider the data valid if certain methods weren't used to collect it.

- ***Data Management.*** This step is often forgotten until after a few years of data have accumulated. But the best time to determine data management is before any data have been collected. What is data management? Data management involves storing data in a manner that makes information easily accessible and accurately retained: a good data management system might store the data in a computer data base or well-organized file folders.

  When determining how to store and manage your data it is helpful to consider how you plan to use the information. Computer databases (such as Lotus, Excel, FoxPro) offer many advantages, including efficient storage, data-sharing opportunities, and the ability to print graphs. When choosing any data management system consider who will enter the data, maintain the database, and distribute the data. Although a computer database may seem the most efficient process, if there is no one to enter the data then a simpler management system may be better.

Human error may show up more in the data management portion of a volunteer monitoring program than in the actual monitoring. It is important that if data are entered into a computer or other filing system that it be checked to insure that the data stored are complete, calculations are correct, proper units are reported, and the results are reasonable (Lease, 1995).

▼ ***Reporting and Using the Data.*** Using the data may seem like the easiest part of the water quality monitoring program, but it can be an overwhelming task and is often left behind in the hustle. Data should be used on a regular basis (quarterly or annually) or volunteers may question the relevance of their efforts. Of course the data should be used in a manner that is consistent with the original goal of the program. Data use can involve everything from volunteers presenting findings at fairs or community meetings to state agency use of the data in its biennial water quality report to EPA.

When presenting data remember that it shouldn't be too technical or too simple for the intended audience. Graphics, including charts and drawings, can help dramatically. If you use charts or tables, accompany them with descriptive text, written in plain English (not water quality monitoring jargon) if the audience is not familiar with such terminology. Describe the program's purpose to give it meaning.

Data should be presented with a purpose: to show trends, seasonal variations, or indicate problems that relate to the purpose of the monitoring program. For example, a graph might be used to show trophic status index values from year to year, dissolved oxygen/temperature profile for different months, or bacteria counts at public beaches in different locations.

Data presentations should also be timely and relevant to the lake condition. For example, sending out a press release in January reporting on high bacteria levels detected at beaches in August would be too late. However, sending out an annual summary of the data that included the high bacteria levels in August and emphasized the need to find solutions prior to the next summer would be effective.

3. Selecting Parameters and Methods

Determining the proper parameters and sampling methods to meet the goals of the program can be tricky. Professional lake managers can help you with this.

The objectives of the monitoring program determine the lake conditions and parameters to be monitored. For example:

- If lake residents are concerned about water quality affecting a lake's coldwater fishery, then you need to monitor dissolved oxygen and temperature; or,
- If the beaches have been closed because of bacteria, then a special study measuring bacteria levels along the shore and inlets (or tributaries) would be helpful.
- If very little water quality data have been collected on your lake, then you need to do baseline monitoring that would include many chemical and biological parameters.

When selecting the parameters and methods consider how the data will be used. That requires understanding the meaning of the test results, e.g., know the background levels for the parameters monitored, how weather may influence the data, water quality protection criteria, seasonal trends, data ranges, etc.

A carefully planned monitoring program can provide useful information to accomplish its objective. But interpret data cautiously. It cannot always prove what you would like it to. A set of data collected at one time provides a snapshot of the lake's condition at that one time. You may want to contact local professionals to help interpret the data you've collected. For more information, visit EPA's volunteer monitoring web site at www.epa.gov/owow/monitoring/vol.html.

4. Quality Assurance/Quality Control

Given proper training and supervision, volunteers can conduct monitoring and collect samples that yield high quality data. To ensure its quality will be acceptable to other agencies, you should adopt quality assurance/quality control (QA/QC) measures.

Quality assurance/quality control are often used interchangeably but they have different meanings. **Quality assurance** reviews all aspects of the monitoring — planning, implementation, and completion — to ensure high quality data are collected. **Quality control** is performed during data collection to ensure accuracy, precision, and unbiased monitoring.

Five major areas must be addressed when developing a quality assurance/quality control program:

- Accuracy,
- Precision,
- Representativeness,
- Completeness, and
- Comparability.

See Quality Assurance/Quality Control box for descriptions of these areas. A quality assurance/quality control program will reduce potential problems with data collection and can improve the experience for volunteers.

For more information on developing a Quality Assurance Project Plan (QAPP) — essentially a document that outlines procedures to ensure that data meet project requirements — see EPA's document, *A Volunteer Monitor's Guide to Quality Assurance Project Plans* (EPA 841-B-96-003) at www.epa.gov.owow/monitoring/volunteer/qappcovr.htm. You can also order this document by calling (800) 490-9198.

Quality Assurance/Quality Control
Accuracy: The degree of agreement between the sampling result and the true value of the parameter being measured. Accuracy is most affected by the equipment and the measurement procedure: calibrate equipment carefully according to equipment or standard method requirements and test against a known standard. If conducting pH monitoring with an electronic probe, use standard solutions of known pH values to examine and calibrate its accuracy.
Precision: The ability of the monitor to reproduce the data result on the same sample (regardless of accuracy). Human error in sampling techniques plays an important role in estimating precision. For example, a replicate sample, which entails collecting two or more samples at the same site, same time, using the same methods, and analyzed with the same technique can be used to check a monitor's precision.
Representativeness: The degree to which the collected data accurately and precisely represents the lake condition being measured. It is most affected by sample site location. For example, if the monitoring objective is to characterize the algal condition in a lake, then the sample would be collected in the deepest, open water area of the lake, rather than taking a sample along the shore near a stream mouth.
Completeness: A measure of the amount of valid data obtained versus the amount expected to be obtained as specified in the original sampling design objectives. Completeness is usually expressed as a percentage based on the number of sampling dates expected and the number of actual samples taken. For example, if 25 sampling dates were planned and only 20 samples were selected, due to bad weather and equipment failures, the completeness would be 80%.
Comparability: This is often very important for citizen monitoring programs because it represents how well data from one lake compare to data from another. For example, state and regional agencies and local monitors should work together to establish standard sampling methods and procedures for volunteer monitoring programs.
Source: Simpson, 1991.

Comparability: This is often very important for citizen monitoring programs because it represents how well data from one lake compare to data from another. For example, state and regional agencies and local monitors should work together to establish standard sampling methods and procedures for volunteer monitoring programs.

5. Training Volunteers

Volunteers must be trained to perform water quality monitoring tests and follow QA/QC procedures. Training will require time, materials, and probably professional assistance — items that can add costs to a monitoring program. But training is another monitoring investment that has high returns for a program.

First, **develop a volunteer monitor description** that details the duties and tasks of the monitor. Use this to inform interested volunteers of their responsibilities.

Second, **organize and hold the training.** Training can be done in a group setting or one-on-one. Both have their benefits.

- Group training builds camaraderie among the monitors. It is more efficient, requiring less time than one-on-one training. It encourages group problem-solving and ensures each volunteer receives consistent information.
- One-on-one training can be more detailed and intensive and often provides better results.

Combining the methods can be most effective but will require a greater investment in time and money. Bringing in professionals to help with group training can be advantageous, especially if they will be using the data collected for a project.

A thorough training session would include:

- A review of the monitor's responsibilities;
- Presentation on the relevance and value of each parameter;
- Review of quality assurance/quality control procedures;
- How to do the sampling and use equipment;
- How to record the data;
- Hands-on opportunities to collect samples and encouragement to participate;
- Discussion of safety issues;
- Written instructions to take home, review, and use in the field; and
- Refreshments and time for socializing.

Allow ample time for questions and answers. And be sure to give volunteers information on who to contact if they have problems or need additional assistance. Give the volunteers an evaluation form; this will help trainers plan future sessions.

Training is an ongoing part of volunteer monitoring programs. Annual training sessions bring in new monitors and provide refreshers for veterans.

A training session takes volunteer monitors out on a boat into their lake.

6. Evaluating the Program

Taking the time to assess whether a program met its objectives is an important component of all programs. An evaluation can look at many different levels of a project or simply at the effectiveness of the overall program. And it can be done by the volunteers or lake residents.

As with other components of the monitoring program, designing the evaluation will take some effort. Not surprisingly, the most common ways of conducting evaluations are written evaluations and interviews.

Some simple and effective evaluation questions might include:

- Did the program accomplish its goals;
- What worked well;
- What didn't work well;
- Did the project budget cover all expenses;
- Were the proper parameters monitored;
- Did the data help facilitate the desired change and/or meet the program's objectives;
- Were the data distributed to the appropriate agencies;
- Did the data meet quality assurance standards; and
- What could be done to improve the program?

More specific questions that address the monitoring program's goals should also be included.

Putting It All Together

Integrating protection and maintenance into a project may seem like an overwhelming task, but taken one step at a time it is doable. Once you've identified the project's goals and objectives, selected the most appropriate actions, and begun work, then the most important task becomes assessing progress throughout the lake management process.

Just as lake management activities are ongoing, so is the evaluation. An education program on lakes might require assessing the students' and teachers' comments about the program, and checking on the reactions of the volunteers or staff who conduct the program as well. Lake monitoring is another way to evaluate the effectiveness of management strategies. But remember, focus on the positive results and opportunities the evaluation finds, not the success or failure of the program.

Maintenance and protection of lakes is ongoing, not short term. As time passes and goals, objectives, and actions are accomplished, new opportunities to improve and protect lakes will continually arise.

Respect for the resource and sound programs are the basis of a strong lake association.

A Lake Association Profile

Every lake has a unique character. For Walloon Lake in Northern Michigan it may be the call of the loons that nest in its quiet coves, the blue, clear color of the water, or the lore of it being Ernest Hemingway's childhood vacation spot.

The Walloon Lake Association and its sister organization the Walloon Lake Trust and Conservancy have worked to maintain its unique character since 1910 — and since 1960, have been extremely proactive in creating a variety of programs to preserve and improve the environment of Walloon Lake.

The Walloon Lake Association:

- Publishes six newsletters and a membership directory annually;
- Disseminates information to lake residents through its "neighborhood leadership committee"; and
- Sponsors numerous field trips and programs to encourage members and their children and grandchildren to learn more about the lake and its watershed.

The Association's political sense is also on target: they appoint representatives to attend local government meetings of the five townships surrounding the lake — in fact, several representatives have become official members of township boards and planning commissions. As a result, local regulations such as building setbacks have been passed to help protect the lake.

Their water quality monitoring program includes:

- The usual Secchi disk, chlorophyll *a*, and spring phosphorus;
- An expanded program that monitors dissolved oxygen, temperature, conductivity, pH, and turbidity using a multi-parameter water quality testing probe;
- An annual boat survey to keep tabs on the number of lake resident-owned boats; and
- Special studies on shoreline algal communities, exotic species, and beavers on tributaries.

Their sister organization, the Walloon Lake Trust and Conservancy, has put watershed protection into action by protecting 1,100-plus acres through gifts of land, conservation easements, and purchases (with help from a regional land conservancy).

The secret to their success? More than 90% of the lake residents belong to the Association — which maintains a full-time executive director and an office in a nearby town.

Walloon Lake Association's strength is built on the respect and care its members have for the lake. But it is the education, water quality monitoring, governmental affairs, land protection, and membership programs that will encourage future generations to love the lake and do all that's necessary to protect it.

Respect for the resource and sound programs are the basis of a strong lake association.

Lake Maintenance and Protection: An Ongoing Opportunity

"We did not inherit the earth from our ancestors, but are borrowing it from our children." This quote attributed to Chief Seattle is particularly relevant for lake management. Lakes are an important part of the American landscape and our cultural and natural history. Restoring, protecting, and maintaining these resources is an ongoing process that requires vigilance and dedication of lake property owners, lake users, and governments.

But, oh, the rewards. Through sound lake management, future generations will be able to hear the cry of a loon, see the reflection of a full moon, experience the thrill of fishing, or feel the refreshment from a summer swim in lakes across North America.

Henry David Thoreau wrote in Walden, "a lake is a landscape's most beautiful, expressive feature; it is Earth's eye, on looking into which the beholder measures the depth of his own nature." May this powerful spiritual connection that humans have with lakes be expressed with commitment and personal responsibility to protect and enhance these national water treasures.

References

Arendt. R. 1996. Conservation Design for Subdivisions. Island Press, Washington, DC.

______. 1998. Growing Greener. National Land Trust, Media, PA.

Arendt, R., E.A. Brabec, H. L. Dodson, C. Reid, and R.D. Yaro. 1994. Rural by Design. Planners Press, Chicago.

Baker, J.P., H. Olem, C.S. Creager, M.D. Marcus, and B.R. Parkhurst. 1993. Fish and Fisheries Management in Lakes and Reservoirs. EPA 841-R-93-002. Terrene Institute and U.S. Environmental Protection Agency, Washington DC.

Caduto, M. 1985. Pond and Brook. Prentice-Hall, Englewood Cliffs, NJ.

Center for Watershed Protection. 1998a. Better Site Design: A Handbook for Changing Development Rules in Your Community. Ellicott City, MD.

———. 1998b. Rapid Watershed Planning Handbook: A Comprehensive Guide for Managing Urbanizing Watersheds. Ellicott City, MD.

Cwikiel, W. 1996. Living with Michigan's Wetlands: A Landowners Guide. Tip of the Mitt Watershed Council, Conway, MI.

Dresen, M.D. and R.M. Korth. 1994. Life on the Edge . . . Owning Waterfront Property. Wis. Dep.Natural Resour., Madison.

Engel, S. 1989. Lake use planning in local efforts to manage lakes. Pages 101-5 *in* Proc. Natl. Conf. Enhancing States Lake Management Programs, May 1988. Northeast Ill. Plann. Commiss. Chicago.

Ewing, R., M. DeAnna, C. Heflin, D. Porter. 1996. Best Development Practices. Planners Press, Chicago.

Fuller, D. 1997. Understanding, Controlling, and Living With Shoreline Erosion. Tip of the Mitt Watershed Council, Conway, MI.

Hendler, B. 1977. Caring for the Land. American Society of Planning Officials, Chicago.

Henderson, C. 1981. Landscaping for Wildlife. Minnesota Dep. Natural Resour., St. Paul.

Jeffries, M. and D. Mills. 1990. Freshwater Ecology: Principles and Applications. Belhaven Press, New York.

Lease, F. 1995. Designing a data management system. *In* The Volunteer Monitor 7(1).

Long Island Regional Planning Board. 1984. Nonpoint Source Management Handbook. Long Island, NY.

Marsh, W. 1998. Landscape Planning Environmental Applications. 3rd ed. John Wiley & Sons, Inc., New York.

McComas, S. 1993. Lake Smarts. Terrene Institute, Alexandria, VA.

McHarg, I. 1992. Design With Nature. John Wiley & Sons, Inc., Garden City, NJ.

Mitchell, M. and W. Stapp. 1994. Field Manual for Water Quality Monitoring. 8th ed. Thomson-Shore, Inc., Dexter, MI.

New York State Department of Environmental Conservation and Federation of Lake Associations. 1990. Diet for a Small Lake: A New Yorker's Guide to Lake Management. Albany, NY.

Novotny, V. and G. Chesters. 1981. Handbook of Nonpoint Source Pollution: Sources and Management. Van Nostrand Reinhold Environmental Engineering, New York.

Novotny, V. and H. Olem. 1994. Water Quality: Prevention, Identification, and Management of Diffuse Pollution. Van Nostrand Reinhold, New York.

Phillips, N., M. Kelly, J. Taggart and R. Reeder. 2000. The Lake Pocket Book. Terrene Institute, Alexandria, VA.

Simpson, J.T. 1991. Volunteer Lake Monitoring: A Methods Manual. EPA 440/4-91-002. Washington, DC. Available on www.epa.gov/owow/monitoring/volunteer/lake/index.html.

Small, S. 1992. Preserving Family Lands: Essential Tax Strategies for the Landowner. Landowner Planning Center, Boston.

Tip of the Mitt Watershed Council. 1999. Planning for Success. Conway, MI.

The Volunteer Monitor. 1993–present. River Network, Portland, OR. See www.epa.gov/owow/volunteer/vm_index.html

U.S. Environmental Protection Agency. 1996. A Volunteer Monitor's Guide to Quality Assurance Project Plans. EPA 841-B-96-003. Www.epa.gov/owow/monitoring/volunteer/qappcovr.htm. Washington, DC.

Wetzel, R. 1975. Limnology. W.B. Saunders Company, Philadelphia.

Wyckoff, M. 1995. Regulating Keyhole Development: Carrying Capacity Analysis and Ordinances Providing Lake Access Regulations. Planning Zoning Center, Inc., Lansing, MI.

APPENDIX 9-A

Example of Lake Association Bylaws

BYLAWS OF MUDD LAKE PRESERVATION ASSOCIATION
A Michigan Nonprofit Corporation

I. NAME AND PURPOSE

1. **Name:** The name of this nonprofit corporation shall be Mudd Lake Preservation Association.
2. **Purpose:** The corporation is organized to operate exclusively for charitable, scientific, and educational purposes within the meaning of Section 501(c)(3) of the Internal Revenue Code, and more specifically:
   a. To preserve and improve Mudd Lake and its watershed for quality use by future generations.
   b. To preserve natural and scenic areas, and recreational resources.
   c. To carry on activities permitted by exempt organizations under Section 501(c)(3) of the Internal Revenue Code, as amended.

II. BOARD OF TRUSTEES

1. **Responsibilities:** The Board of Trustees established practices for this nonprofit corporation and to elect the Board of Trustees and the Officers. The Board of Trustees shall function as a board of directors under Michigan Law.
2. **Number:** The Board of Trustees shall determine how many Board of Trustee members are to be elected.
3. **Term:** Each Board of Trustees members shall serve for a term of two years. A person may serve for more than one term.
4. **Election:** The Board of Trustees members may be elected at an annual meeting of the Board of Trustees. In the absence of election, the existing Board of Trustees shall continue. The incoming Board of Trustees shall be elected by the preceding Board of Trustees. A member of the Board of Trustees may vote for himself or herself.
5. **Quorum:** If there are seven or fewer Board of Trustees members, fifty percent (50%) of the members of the Board of Trustees shall constitute a quorum. If there are over seven Board of Trustees members, then one third (1/3) of the Board of Trustees shall constitute a quorum (Michigan Codified Laws, M.C.L. 450.2523).

6. **Voting:** For all matters coming before the Board of Trustees, a majority vote of those present at a meeting at which a quorum is present shall govern. Procedures will be by Roberts Rules of Order.
7. **Vacancy:** In the event of a vacancy on the Board of Trustees, the President shall have the right to appoint another Board of Trustees member.
8. **Qualifications:** To be a member of the Board of Trustees, a person must be recognized as a member of the Mudd Lake Preservation Association.
9. **Disqualification:** A Board of Trustees member having three consecutive unexcused Board of Trustees meeting absences shall be terminated from serving on the Board of Trustees.
10. **Ad Hoc Committee(s):** The President may appoint ad hoc advisory committees for any purpose.
11. **Termination:** Two-thirds (2/3) of the Board of Trustees (not merely a quorum) shall have authority to terminate a person's position on the Board of Trustees. **Good cause for termination is not required.**

III. OFFICERS

1. **Officers:** The officers shall consist of a President, one or more Vice Presidents and Secretaries, and a Treasurer.
2. **Term:** The officers shall serve for a term of one year. Officers may be elected to serve for more than one term.
3. **Elections:** The officers shall be elected from and by the Board of Trustees following the election of Board of Trustees members. Elections may be held at the Annual Meeting. If an election is not conducted, the existing officers shall remain in office.
4. **Duties:** The duties of the officers shall be such as are implied by their respective titles. The President shall preside over all meetings and may attend all committee meetings. The Vice-President shall preside in the absence of the President. The Secretary shall keep the roll of sponsors and members, the minutes of all meetings, and shall maintain committee reports. The Secretary shall also tend to all correspondence designated by the Board. The Treasurer shall collect the dues, all other monies, pay the bills, and oversee filing of all appropriate government reports and forms. The Treasurer shall maintain an itemized account of all receipts and disbursements.

IV. MEMBERSHIP

1. **Qualifications:** Membership shall be open to all persons having an interest in and who support the mission of this nonprofit corporation.
2. **Membership Dues:** Dues, if any, shall be established by the Board of Trustees.
3. **Applications:** Applications for membership may be on a form prescribed by the Board of Trustees. An application for membership is not required.

4. **Qualifications:** Persons, families, or entities contributing dues shall be recognized as members, subject to the discretion of the Board of Trustees.
5. **Board of Trustees Members:** Only individual members may be elected to the Board of Trustees.
6. **Directorship Basis:** As this nonprofit corporation is formed on a directorship basis, the members will have no formal vote in corporate affairs.

V. MEETINGS

1. **Regular Meetings:** Regular meetings shall be held at such times and places as may be stated by the Board of Trustees.
2. **Annual Meetings:** Annual meetings shall be held at a time and location to be determined by the Board of Trustees.
3. **Special Meetings:** Special meetings may be called by the President, on 48 hours notice, by phone, fax, or by first class mail. The notice shall specify the purpose of the special meeting. Actual receipt of the fax or mail is not required. Service shall be deemed effective when the fax is sent or the letter is deposited with the U.S. Postal Service.
4. **Consent Actions:** The Board of Trustees may act by a consent action signed by a majority of the members of the Board of Trustees. Written consents shall be filed with the Board of Trustees minutes (Michigan Codified Law, M.C.L. 450.2525).
5. **Presence:** A person may be deemed present at a Board of Trustees meeting when participating by phone, fax, or other means which would necessarily require the personal presence of all Board of Trustees members.

VI. AMENDMENTS

1. **Procedure:** The Bylaws may be amended by a majority of the Board of Trustees present at a meeting at which a quorum is present.
2. **Effective Time:** Amendments to the Bylaws shall be effective immediately upon the vote of the Board.

Appendix A

Glossary

This Glossary defines commonly used terms and important processes and concepts relating to lakes and lake management. To provide further detail, items have been cross-referenced.

Acid neutralizing capacity (ANC): the equivalent capacity of a solution to neutralize strong acids. The components of ANC include weak bases (carbonate species, dissociated organic acids, alumino-hydroxides, borates, and silicates) and strong bases (primarily, OH). In the National Surface Water Survey, as well as in most other recent studies of acid-base chemistry of surface waters, ANC was measured by the Gran titration procedure.

Acid rain: rainfall that contains acidic chemicals, such as nitric acid from automobile emissions and sulfuric acid that have escaped into the air from burning fossil fuels.

Acidic deposition: transfer of acids and acidifying compounds from the atmosphere to terrestrial and aquatic environments via rain, snow, sleet, hail, cloud droplets, particles, and gas exchange.

Adsorption: the adhesion of one substance to the surface of another; clays, for example, can adsorb phosphorus and organic molecules.

Aerobic: describes life or processes that require the presence of molecular oxygen.

Algae: small aquatic plants that occur as single cells, colonies, or filaments. They contain chlorophyll but lack special water-carrying tissues. Through the process of photosynthesis, algae produce most of the food and oxygen in water environments.

Algal: of or related to algae.

Allochthonous: materials (e.g., organic matter and sediment) that enter a lake from atmosphere or drainage basin. See Autochthonous.

Anaerobic: describes processes that occur in the absence of molecular oxygen.

Anoxia: a condition of no oxygen in the water. Often occurs near the bottom of fertile, stratified lakes in the summer and under ice in late winter.

Aphotic zone: that area of the lake too dark to support photosynthesis.

Aquatic life: organisms that live and grow in, or frequent, water.

Aquifer: an underground, water-bearing bed of permeable rock, sand, or gravel. Aquifers contain large amounts of groundwater that feed into wells and springs.

Autochthonous: materials produced within a lake; e.g., autochthonous organic matter from plankton versus allochthonous organic matter from terrestrial vegetation.

Bacteria: a large group of microscopic organisms of many different shapes, generally without chlorophyll. Some bacteria are helpful (as in a fermentation process), but certain species can cause diseases such as swimmer's itch, pneumonia, or typhoid fever, among others.

Bareground banks: river or stream banks that have no vegetation (no plant covering) to hold the soil against erosive action.

Bathymetric map: a map showing the bottom contours and depth of a lake; can be used to calculate lake volume.

Benthic: refers to life or things found on the bottom of a lake. Examples: benthic animals, benthic sediments.

Benthos: macroscopic (seen without aid of a microscope) organisms living in and on the bottom sediments of lakes and streams. Originally, the term meant the lake bottom, but it is now applied almost uniformly to the animals associated with the substrate.

Berm: a narrow shelf, ledge, or barricade, typically at the top or bottom of a slope; a mound or wall of earth; for example, small dams or ridges.

Best management practices (BMPs): systems, activities, and structures that human beings can construct or practice to prevent nonpoint source pollution.

Biochemical oxygen demand (BOD): the rate of oxygen consumption by organisms during the decomposition (see Respiration) of organic matter, expressed as grams oxygen per cubic meter of water per hour.

Biodiversity: a multiplicity of different, mutually dependent living things characteristic of a particular region or habitat.

Biomass: the weight of biological matter. Standing crop is the amount of biomass (e.g., fish or algae) in a body of water at a given time. Often measured in terms of grams per square meter of surface.

Biota: all plant and animal species occurring in a specified area.

Cadmium: bluish-white toxic metal or metallic element used especially in protective plating and in bearing metals.

Chemical oxygen demand (COD): nonbiological uptake of molecular oxygen by organic and inorganic compounds in water.

Chlorophyll *a*: A type of chlorophyll present in all types of algae, sometimes in direct proportion to the biomass of algae.

Chlorophyll: a green pigment in algae and other green plants that is essential for the conversion of sunlight, carbon dioxide, and water to sugar (see Photosynthesis). Sugar is then converted to starch, proteins, fats, and other organic molecules.

Clarifier tanks: holding tanks associated with wastewater and sewage treatment centers. Wastewater in these tanks is treated to remove harmful substances before being released into a watershed.

Clean Water Act: the federal Clean Water Act of 1972 (formerly referred to as the Federal Water Pollution Control Act): requires the development of comprehensive programs for preventing, reducing, or eliminating the pollution and improving the condition of the nation's navigable, surface, and groundwaters.

Cluster development: placement of housing and other buildings of a development in groups to provide larger areas of open space.

Coliform: a bacteria carried in human and animal wastes.

Combined sewer overflow (CSO): what happens when too much stormwater flows into drainage systems that also contain wastewater effluent. The combined flow from the stormwater and wastewater may result in releases of untreated wastewater directly into lakes, rivers, or streams. This is most common in older systems since most modern drainage systems separate stormwater and wastewater flows. (CSOs refer to overflows before the water reaches the treatment plant.)

Comparability: this is often very important for citizen monitoring programs because it represents how well data from one lake compare to data from another. For example, state and regional agencies and local monitors should work together to establish standard sampling methods and procedures for volunteer monitoring programs.

Compliance officer: one who plans, manages, or oversees a company's submission to laws, regulations, and practices; a person delegated to ensure a company's conformity with the law.

Compliance: the act of fulfilling an official requirement; submission to operative laws, regulations, practices, terms, or conditions.

Compost: a mixture of soil and decayed organic matter (food, vegetative, and animal wastes) used for fertilizing and conditioning land.

Conservation easement: (also known as conservation restrictions) are legal agreements between a landowner and a qualified government agency or nongovernmental organization (most commonly a land trust) that permanently limit a property's uses; they remain with it if it is sold. A conservation easement does not transfer title to the property or open it to the public. The landowner continues to own the property, and may live on it, sell it, or pass it on to heirs.

Conservation tillage: a practice or method of plowing in which crop residue is left on the field as protective mulch or cover instead of being plowed under.

Consumers: animals that cannot produce their own food through photosynthesis and must consume plants or animals for energy. See Producers.

Decomposition: the transformation of organic molecules (e.g., sugar) to inorganic molecules (e.g., carbon dioxide and water) through biological and non-biological processes.

Delphi: a technique that solicits potential solutions to a problem situation from a group of experts and then asks the experts to rank the full list of alternatives.

Denitrification: the process by which nitrate in water or sediments is converted to nitrogen gas, which is then lost to the atmosphere.

Density flows: a flow of water of one density (determined by temperature or salinity) over or under water of another density (e.g., flow of cold river water under warm reservoir surface water).

Detritus: organic material composed of dead plants or animals, or parts thereof (e.g., leaves, grass clippings) that settle to the bottom of a lake. Bacteria and fungi slowly decompose detritus, thus recycling it back into the lake's ecosystem.

Drainage basin: land area from which water flows into a stream or lake. See Watershed.

Drainage lakes: lakes having a defined surface inlet and outlet.

Ecology: a branch of science concerned with the interrelationship of organisms with their environment.

Ecoregion: Comprised of relatively homogenous ecological systems delineated by geology, soils, climate, vegetation, and landform, and involving interrelationships among organisms and their environment.

Ecosystem: a system of interrelated organisms and their physical-chemical environment. In this manual, the ecosystem is usually defined to include the lake and its watershed.

Effluent: liquid wastes from sewage treatment, septic systems, or industrial sources that are released to a surface water.

Environment: the complex of one's surroundings; the climatic, soil-related, and life-related factors that act on organisms or ecological communities and ultimately determine their form and survival.

Environmental movement: an organized or grass roots, public, or private movement or group acting to preserve the quality and continuity of life through the conservation of natural resources and the prevention and/or reduction of pollution.

Environmental Protection Agency (EPA): a division or office of government, either federal or state, responsible for safeguarding and managing a region's natural resources and quality of life. The U.S. EPA is an agency of the federal government; the names of state EPAs vary.

Epilimnion: uppermost, warmest, well-mixed layer of a lake during summertime thermal stratification. The epilimnion extends from the surface to the thermocline. See Stratification.

Epiphytes: small plants or animals that grow attached to larger plants.

Erosion: the gradual wearing down of land by water, wind, or melting snow. Soil losses, for example, from streambanks and forests, hilly ground, lawns, and farm fields.

Eutrophic: from Greek for "well-nourished": describes a lake of high photosynthetic activity and low transparency. See Trophic State.

Eutrophication: the process of physical, chemical, and biological changes associated with nutrient, organic matter, and silt enrichment and sedimentation of a lake or reservoir that cause a water body to age. If the process is accelerated by human influences, it is termed cultural eutrophication.

Eutrophication cultural: human activities, such as discharge of sewage and stormwater, and nonpoint source pollutants, can dramatically hasten the process of eutrophication.

Fall overturn: the autumn mixing, top to bottom, of lake water caused by cooling and wind-derived energy.

Fecal coliform test: most common test for the presence of fecal material from warm-blooded animals. Fecal coliforms are measured because of convenience; they are not necessarily harmful but indicate the potential presence of other disease-causing organisms.

Floodplain: land adjacent to lakes or rivers that is covered as water levels rise and overflow the normal water channels.

Flushing rate: the rate at which water enters and leaves a lake relative to lake volume, usually expressed as time needed to replace the lake volume with inflowing water.

Flux: the rate at which a measurable amount of a material flows past a designated point in a given amount of time.

Food chain: the general progression of feeding levels from primary producers, to herbivores, to planktivores, to the larger predators.

Food web: the complex of feeding interactions existing among the lake's organisms.

Forage fish: Fish, including a variety of panfish and minnows, that are prey for game fish.

Forest Service (FS): an agency of the federal government located within the Department of Agriculture that manages and protects our forests, wooded areas, and timber resources.

Groundwater: the supply of fresh water found beneath the earth's surface (usually in aquifers); often used to supply drinking water to wells and springs; may be connected to lakes.

Habitat: the physical environment or typical place within which a plant or animal naturally or normally lives and grows.

Hydrographic map: a map showing the location of areas or objects within a lake.

Hydrologic cycle: the circular flow or cycling of water from the atmosphere to the earth (precipitation) and back to the atmosphere (evaporation and plant transpiration). Runoff, surface water, groundwater, and water infiltrated in soils are all part of the hydrologic cycle.

Hypolimnion: lower, cooler layer of a lake during summertime thermal stratification. See Stratification.

Influent: a tributary stream.

Internal nutrient cycling: transformation of nutrients such as nitrogen or phosphorus from biological to inorganic forms through decomposition, occurring within the lake itself.

Isothermal: the same temperature throughout the lake.

Lake: a considerable inland body of standing water, either naturally formed or built by humans.

Lake district: a special purpose unit of government with authority to manage a lake(s) and with financial powers to raise funds through mill levy, user charge, special assessment, bonding, and borrowing. May or may not have police power to inspect septic systems, regulate surface water use, or zone land.

Lake management: the practice of keeping lake quality in a state such that attainable uses can be achieved.

Lake protection: the act of preventing degradation or deterioration of attainable lake uses.

Lake restoration: the act of bringing a lake back to its attainable uses.

Lentic: relating to standing water (versus **lotic,** running water).

Limnetic zone: also called Epilimnion. See Stratification.

Limnology: the scientific study of the physical, chemical, geological, and biological factors that affect aquatic productivity and water quality in freshwater ecosystems — lakes, reservoirs, rivers, and streams.

Limnologist: one who practices limnology.

Littoral zone: the shallow zone along the shore of a lake; that portion of a water body extending from the shoreline lakeward to the greatest depth occupied by rooted plants. Plants growing here support a rich biological community.

Loading: the total amount of material (sediment, nutrients, oxygen-demanding material) brought into the lake by inflowing streams, runoff, direct discharge through pipes, groundwater, the air, and other sources over a specific period of time (often annually).

Macroinvertebrates: aquatic insects, worms, clams, snails, and other animals visible without aid of a microscope, that may be associated with or live on substrates such as sediments and macrophytes. They supply a major portion of fish diets and consume detritus and algae.

Macrophytes: plants large enough to be seen without magnification. Some forms, such as duckweed and coontail (*Ceratophyllum*), are free-floating forms without roots in the sediment.

Mandatory property owners association: organization of property owners in a subdivision or development with membership and annual fee required by covenants on the property deed. Association will often enforce deed restrictions on members' property and may have common facilities such as bathhouse, clubhouse, golf course, etc.

Marginal zone: area where land and water meet at the perimeter of a lake. Includes plant species, insects, and animals that thrive in this narrow, specialized ecological system.

Meaningful indicators: link objectives to management objectives; are meaningful to stakeholders; are measurable, or ranked subjectively; and can be predicted.

Mercury: a heavy silver-white poisonous metallic element sometimes found as a contaminant in rainfall.

Mesotrophic: the medium range of eutrophication. See Trophic State.

Metalimnion: layer of rapid temperature and density change in a thermally stratified lake; lies between epilimnion and hypolimnion. Resistance to mixing is high in the region. See Stratification.

Minimum tillage: a practice of plowing or turning the soil only enough to plant new crops, while leaving plant residue on the surface as compost.

Morphometry: relating to a lake's physical structure (e.g., depth, shoreline length).

Mulch: a protective covering (as of sawdust, compost, or paper) spread or left on the ground. Mulch prevents evaporation, maintains even soil temperature, prevents erosion, controls weeds, and enriches the soil.

National Pollutant Discharge Elimination System (NPDES): federal operating permits issued by EPA to industrial and municipal facilities to help these facilities comply with the Clean Water Act.

Natural Resources Conservation Service (NRCS): a federal agency responsible for safeguarding and managing soil and water resources. NRCS operates within the Department of Agriculture and maintains local offices throughout the country.

Nekton: large aquatic and marine organisms whose mobility is not determined by water movement—for example, fish and amphibians.

NEPA: The National Environmental Policy Act of 1969 that established a national policy for the environment, created the Council on Environmental Quality, and directed that every recommendation or report on proposals for legislation and other major federal actions significantly affecting the quality of the human environment include a detailed statement on the environmental impact of the proposed action. For the complete act, see Council on Environmental Quality, 1991.

NOAA: National Oceanic and Atmospheric Administration

Nominal group process: a process of soliciting concerns/issues/ideas from members of a group and ranking the resulting list to ascertain group priorities. Designed to neutralize dominant personalities.

Noncompliance: a condition of not submitting to applicable laws, regulations, terms, or conditions.

Nonpoint source (NPS): pollution that cannot be traced to a specific origin or starting point, but seems to flow from many different sources. NPS pollutants are generally carried off the land by stormwater (or melting snow) runoff. The commonly used categories for nonpoint sources are agriculture, forestry, urban, mining, construction, dams and channels, land disposal, and saltwater intrusion.

Nutrient: an element or chemical essential to life, such as carbon, oxygen, nitrogen, and phosphorus.

Nutrient budget: quantitative assessment of nutrients (e.g., nitrogen or phosphorus) moving into, being retained in, and moving out of an ecosystem; commonly constructed for phosphorus because of its tendency to control lake trophic state.

Nutrient cycling: the flow of nutrients from one component of an ecosystem to another, as when macrophytes die and release nutrients that become available to algae (organic to inorganic phase and return).

Nutrients: substances or ingredients that nourish or promote growth and repair the natural destruction of organic life.

Oligotrophic: "poorly nourished," from the Greek. Describes a lake of low plant productivity and high transparency. See Trophic State.

Ooze: lake bottom accumulation of inorganic sediments and the partially decomposed remains of algae, weeds, fish, and aquatic insects. Sometimes called muck. See Sediment.

Ordinary high water mark: physical demarcation line, indicating the highest point that water level reaches and maintains for some time. Line is visible on rocks, or shoreline, and by the location of certain types of vegetation.

Organic: of, relating to, or derived from living things; relating to, produced with or based on the use of plant and animal fertilizers rather than chemically formulated fertilizers or pesticides.

Organic matter: molecules manufactured by plants and animals and containing linked carbon atoms and elements such as hydrogen, oxygen, nitrogen, sulfur, and phosphorus.

Paleolimnology: the study of lake sediments and the relics preserved in them.

Pathogen: a microorganism capable of producing disease. They are of great concern to human health relative to drinking water and swimming beaches.

Pelagic zone: the open area of a lake, from the edge of the littoral zone to the center of the lake.

Perched: a condition where the lake water is isolated from the groundwater table by impermeable material such as clay.

Permeable: a surface or material that has pores or openings that allow liquids to penetrate or pass through.

Pesticide: an agent used to destroy insects and other pests.

pH: a measure of the concentration of hydrogen ions of a substance, which ranges from very acid (pH = 1) to very alkaline (pH = 14). pH 7 is neutral and most lake waters range between 6 and 9. pH values less than 6 are considered acidic, and most life forms cannot survive at pH of 4.0 or lower.

Photic zone: the lighted region of a lake where photosynthesis takes place. Extends down to a depth where plant growth and respiration are balanced by the amount of light available.

Photosynthesis: a chemical reaction that occurs only in plants. Plants use a green pigment called chlorophyll to convert water and carbon dioxide into cellular material and oxygen in the presence of light. Hence, photosynthesis occurs only during daylight hours.

Phytoplankton: microscopic algae and microbes that float freely in open water of lakes and oceans. In some lakes, they provide the primary base of the food chain for all animals. They also produce oxygen by a process called photosynthesis.

Phytoremediation: plants are used to clean up certain types of heavily contaminated soils by absorbing the contaminants from the soil.

Plankton: small, mostly microscopic plants and animals that are too small to outswim most currents, so the movement of water tends to move them from place to place. Plankton consists of phytoplankton (planktonic plants) and zooplankton (planktonic animals).

Plankton rain: the almost constant settling of plankton, live and dead, through the water to the bottom sediments.

Plantivores: fish and invertebrate that collectively prey on zooplankton.

Point source (PS): pollution discharged into water bodies from specific, identifiable pipes or points, such as an industrial facility or municipal sewage treatment plant.

Pollutants: solid, liquid, or gaseous substances that contaminate the local or general environment.

Pollution: the condition of being polluted. A generic word for any type of contamination of water, land, or air.

Precipitation: a water deposit on earth in the form of hail, rain, sleet, and snow.

Primary productivity: the rate at which algae and macrophytes fix or convert light, water, and carbon dioxide to sugar in plant cells. Commonly measured as milligrams of carbon per square meter per hour.

Producers: green plants that manufacture their own food through photosynthesis.

Profundal zone: mass of lake water and sediment occurring on the lake bottom below the depth of light penetration. Also called Hypolimnion. See Stratification.

Reservoir: lake created by artificially damming a stream or river where water is collected and kept in quantity for a variety of uses, including flood control, water supply, recreation, and hydroelectric power.

Residence time: commonly called the hydraulic residence time — the amount of time required to completely replace the lake's current volume of water with an equal volume of "new" water.

Respiration: process by which organic matter is oxidized by organisms, including plants, animals, and bacteria. The process releases energy, carbon dioxide, and water.

Runoff: that portion of precipitation that flows over the land carrying with it such substances as soil, oil, trash, and other materials until it ultimately reaches streams, rivers, lakes, or other water bodies.

Secchi depth: a measure of transparency of water (the ability of light to penetrate water) obtained by lowering a black and white, or all white, disk (Secchi disk, 20 cm in diameter) into water until it is no longer visible. Measured in units of meters or feet.

Secchi disk: a white or black and white disk used to measure transparency of water. See Secchi depth.

Sediment: bottom material in a lake that has been deposited after the formation of a lake basin. It originates from remains of aquatic organisms, chemical precipitation of dissolved minerals, and erosion of surrounding lands. See Ooze.

Sediment oxygen demand (SOD): after a long time, plant and algal cells that die can sink to low places in a lake where they begin to decompose, eventually accumulating as a thick layer of soft, highly organic sediments. This decomposition by bacteria uses oxygen from the overlying water, and thus, can drain a lake's dissolved oxygen. If accidentally resuspended (as in a storm or by power boats), these sediments can kill fish and other animals. When bottom dissolved oxygen falls too low, millions of small invertebrates (animals without backbones) living in and on the bottom may also be reduced or even eliminated.

Seepage lakes: lakes having either an inlet or outlet (but not both) and generally obtaining their water from groundwater and rain or snow.

Septic tank: a holding tank for collecting residential wastewaters. Used as an alternative to municipal sewer systems in some areas. Wastewater collected in septic tanks disperses into the soil through a septic drainfield.

Sewage treatment plant: a facility (usually municipal) that treats sewer waste to remove harmful substances before discharge.

Soil retention capacity: the ability of a given soil type to adsorb substances such as phosphorus, thus retarding their movement to the water.

Spawning: the production and deposit of eggs by fish within their aquatic habitat.

Standing crop: the amount of biomass (e.g., fish or algae) in a body of water at a given time.

Stratification: process in which several horizontal water layers of different density may form in some lakes. During stratification, the bottom mass (hypolimnion or profundal zone) is cool, high in nutrients, low in light, low in productivity, and low in dissolved oxygen. The top mass (epilimnion or limnetic zone) is warm, higher in dissolved oxygen, light, and production, but lower (normally) in nutrients. The sharp boundary between the two masses is called a thermocline. The metalimnion exists in this area.

Swimmer's itch: a rash caused by penetration into the skin of the immature stage (cercaria) of a flatworm (not easily controlled due to complex life cycle). A shower or alcohol rubdown should minimize penetration.

Thermal stratification: lake stratification caused by temperature-created differences in water density.

Thermocline: a horizontal plane across a lake at the depth of the most rapid vertical change in temperature and density in a stratified lake. See Metalimnion and Stratification.

Tillage: the operation of plowing or cultivating land.

Topographic map: a map showing the elevation of the landscape at fixed contour intervals, usually 2, 5, 10, or 20 feet. This information can be used to delineate a watershed.

Toxic: poisonous substances harmful to living things. Of, relating to, or caused by poison.

Trophic state: the degree of eutrophication of a lake. Transparency, chlorophyll *a* levels, phosphorus concentrations, amount of macrophytes, and quantity of dissolved oxygen in the hypolimnion can be used to assess trophic state.

Trophic state index: a number used to categorize lakes as oligo-, meso-, or eutrophic, on a scale generally from 1 to 100: the higher the number, the more eutrophic. It can be calculated a variety of ways, using chlorophyll (a measure of algae abundance), Secchi depth (an indirect measure of algae abundance by measuring water clarity), or nutrients. Lakes with TSI of 60 or more are considered eutrophic.

Turbid: thick or cloudy with sediment.

Turbidity: cloudiness; characterized by obscurity.

Upset: a waste/sewage treatment plant malfunction. In an upset, untreated or incompletely treated wastewater enters the watershed.

USDA: U.S. Department of Agriculture

USFS: U.S. Forest Service, a USDA agency

USGS: U.S. Geological Survey

Vegetative/vegetation filter strips: plantings used to trap water (and the substances it carries) to prevent it from running off the land; a BMP that helps prevent nonpoint source pollution.

Voluntary lake property owners association: organization of property owners in an area around a lake that members join at their option.

Wastewater treatment plant: sometimes synonymous with sewage treatment plant, but often an industrial treatment facility that processes the water to remove toxic and hazardous wastes.

Water body: a land basin filled with water. Any river, lake, stream, or ocean that receives runoff waters from a watershed.

Water column: water in the lake between the interface with the atmosphere at the surface and the interface with the sediment layer at the bottom. Idea derives from vertical series of measurements (oxygen, temperature, phosphorus) used to characterize lake water.

Water hardness: originally defined as the capacity of water to preciptate soap, water hardness is now defined as the sum of the calcium and magnesium concentrations, both expressed as calcium carbonate, in mg/L.

Water table: the upper surface of groundwater; below this point, the soil is saturated with water.

Watershed: a drainage area or basin in which all land and water areas drain or flow toward a central collector such as a stream, river, or lake at a lower elevation.

Wetlands: lands or areas, such as tidal flats or swamps, that are often or periodically saturated with water. Wetlands contain much soil moisture and plants that grow well in that condition.

Zooplankton: microscopic animals that float freely in lake water, graze on detritus particles, bacteria, and algae, and may be consumed by fish.

Appendix B

Metric Units

COMMON UNITS OF MEASURE IN LAKE MANAGEMENT

Limnology, the primary science upon which lake management is based, uses metric units in professional publications. Although most units in this Manual are expressed in British/U.S. form, the reader is strongly encouraged to become more comfortable with common metric units — they are far easier to manipulate, and any further encounter with the literature and books on lake management will entail using the metric system of measurement.

The following table compares the two systems; to convert English units to metric, use the conversion factors supplied in this table.

METRIC TO ENGLISH CONVERSIONS

METRIC UNIT	SYMBOL	ENGLISH UNIT	CONVERSION FACTOR*
LENGTH			
Millimeter	mm = 0.001 m	inch	0.03937
Centimeter	cm = 0.01 m	inch	0.3937
Meter	m = 1.0 m	yard	1.094
Kilometer	km = 1000 m	mile	0.6214
WEIGHT			
Microgram	μg = 0.000001 g	(no reasonable equivalent)	
Milligram	mg = 0.001 g	grain	0.015432
Gram	g = 1.0 g	ounce (avoir)	0.03527
Kilogram	kg = 1000 g	pound	2.205
VOLUME			
Milliliter	mL = 0.001 L	ounce	29.57
Liter	L = 1.0 L	quart	1.057
Kiloliter (cubic meter)	kL = 1000 L (m^3)	cu. yard	1.308

* To convert metric to English units, multiply by factor.

OTHER USEFUL CONVERSIONS

1 gallon = 3.785 liters
1 milligram/liter = 1 part per million
1 hectare = 2.47 acres
1 acre-foot = 32,590 gallons
1 cubic meter = 264 gallons